Applied Biotechnology and Bioinformatics

Scrivener Publishing
100 Cummings Center, Suite 541J
Beverly, MA 01915-6106

Publishers at Scrivener
Martin Scrivener (martin@scrivenerpublishing.com)
Phillip Carmical (pcarmical@scrivenerpublishing.com)

Applied Biotechnology and Bioinformatics

Agriculture, Pharmaceutical Research and Environment

Edited by

Hrudayanath Thatoi
Center for Industrial Biotechnology Research, Siksha 'O' Anusandhan University, Bhubaneswar, Odisha, India

Sonali Mohapatra
Dept. of Biological Systems Engineering, University of Wisconsin, Madison, USA

Swagat Kumar Das
Dept. of Biotechnology, Odisha University of Technology and Research, Bhubaneswar, Odisha, India

and

Sukanta Kumar Pradhan
Dept. of Bioinformatics, Odisha University of Agriculture and Technology, Bhubaneswar, Odisha, India

Scrivener Publishing

WILEY

This edition first published 2025 by John Wiley & Sons, Inc., 111 River Street, Hoboken, NJ 07030, USA and Scrivener Publishing LLC, 100 Cummings Center, Suite 541J, Beverly, MA 01915, USA

For more information about Scrivener publications please visit www.scrivenerpublishing.com.

Wiley Global Headquarters
111 River Street, Hoboken, NJ 07030, USA

For details of our global editorial offices, customer services, and more information about Wiley products visit us at www.wiley.com.

Library of Congress Cataloging-in-Publication Data

ISBN 978-1-119-89640-1

Front cover images courtesy of Wikimedia Commons
Cover design by Russell Richardson

Set in size of 11pt and Minion Pro by Manila Typesetting Company, Makati, Philippines

Printed in the USA

10 9 8 7 6 5 4 3 2 1

Contents

Preface

The field of modern biotechnology is a multidisciplinary, groundbreaking, and advantageous area of biology that includes several cutting-edge methods. The discipline of biotechnology is expanding at a rapid, previously unseen pace due to developments in forensics, molecular modeling, clinical healthcare, pharmaceuticals, agriculture, environmental bioremediation, renewable energy, and many other areas. In biological science, bioinformatics is an advanced field of research that has grown to an unprecedented height.

Presently, bioinformatics, a full-fledged multidisciplinary field that combines advances in computer and information technology, has made great progress in its applications to the field of biotechnology and biological sciences. Numerous applications of bioinformatics—primarily in the areas of gene and protein identification, structural and functional prediction, drug development and design, folding of genes and proteins and their complexity, vaccine design, and organism identification—have contributed to the advancement of biotechnology. Bioinformatics play a critical role in analyzing different types of data created by high-throughput research methods—such as genomic, transcriptomic, and proteomic datasets—that will be useful in addressing various problems related to disease management, clean environment, alternative energy sources, agricultural productivity, and more.

Biotechnology is essential to crop improvement in agriculture because it allows genes to transfer across plants to increase traits such as disease resistance and yield. Biotechnology plays a broad role in healthcare, including genetic testing, gene therapy, pharmacogenomics, and drug development. Bioremediation and biodegradation, using microbial technologies to clean up environmental contamination, are critical in the current scenario. In this context, bioinformatics employs tools to analyze and mechanize different degradation pathways that help in microbial applications.

This book is comprised of eighteen unique chapters, each written by renowned researchers in the fields of microbiology, bioinformatics, agriculture, food, pharmaceuticals, and bioremediation. It provides an in-depth discussion on emerging topics while concentrating on the most recent research about the applications of biotechnology and bioinformatics in agriculture, nutraceuticals, pharmaceuticals, and the environment. General readers, scholars, biotechnology researchers, and bioinformatics professionals working in the areas of applied biotechnology, bioengineering, biomedical sciences, microbiology, and environmental sciences will come away from this book with a wider understanding of recent innovations, tools, techniques, and applications.

The editors express their gratitude to the esteemed writers for providing chapters in this book that showcase their superb work and extensive expertise in their respective fields of research. Finally, the editors thank Martin Scrivener and Scrivener Publishing for their assistance and publication of this book.

Hrudayanath Thatoi
Sonali Mohapatra
Swagat Kumar Das
Sukanta Kumar Pradhan

Part I
AGRICULTURE

1

Next-Generation Sequencing in Vegetable Crops

Meenu Kumari[1]*, Tanya Barpanda[2], Meghana Devireddy[3], Ankit Kumar Sinha[3], R. S. Pan[1] and A. K. Singh[1]

[1]ICAR-Research Complex for Eastern Region, RS, Ranchi, India
[2]Orissa University of Agriculture & Technology, Odisha, India
[3]ICAR-Indian Agricultural Research Institute, New Delhi, India

Abstract

The last few decades have witnessed revolutionary advances in all biological disciplines in DNA sequencing technologies at a fraction of the cost with respect to traditional sequencing. There are two methods of crop breeding, i.e., conventional approach through hybridization followed by selection and marker-assisted selection (MAS). Limitations of conventional approach like long periods of selection to fix a trait in the breeding population, environmental effect, and low efficiency for complex and less heritable traits lead breeders to choose MAB. It necessitates the use of different molecular markers based on the availability of information on linkage of traits with markers. However, MAB efficiency was convenient to explore traits that are governed by few numbers of quantitative trait loci (QTLs), whereas for complex traits like yield, quality, biotic stress, and abiotic stress, which are governed by a large number of minor QTLs, Marker-Assisted Selection is not effective. The next-generation sequencing (NGS) approach opened an era of data science where millions of bases are being sequenced in one round and extremely reduced the time and cost of sequencing. In this chapter, we describe the status, recent development, and application of NGS in vegetable crops for their utilization in practical improvement approaches.

Keywords: MAB, QTL, GBS, transcriptomics, GBS, NGS

**Corresponding author*: meenu10iari@gmail.com

Hrudayanath Thatoi, Sonali Mohapatra, Swagat Kumar Das and Sukanta Kumar Pradhan (eds.) Applied Biotechnology and Bioinformatics: Agriculture, Pharmaceutical Research and Environment, (3–36)

1.1 Introduction

The next-generation sequencing (NGS) approach has exemplary development in the field of life sciences and shifted sequencing studies from "model organism" to "every organism" with the power of high-throughput NGS technology. NGS techniques became commercially available in 2005, and since then, there has tremendous development at an astonishing rate in this field like the evolutionary process to address important and unexplored questions of plant systems. These approaches can be broadly divided into three main categories: sequencing by synthesis, sequencing by ligation, and single-molecule sequencing. However, emerging technologies made it possible without library preparation, like the use of quantum dot (qdot)-derived fluorescence resonance energy transfer (FRET) to detect fluorescently labeled nucleotide incorporation. Another sequencing approach is nanopore sequencing, where chemical or electronic properties of bases (DNA or RNA) can be analyzed directly while passing through nanopores. The NGS evolution trend with the advancement of technologies has been depicted in Figure 1.1. This has exemplary development in the

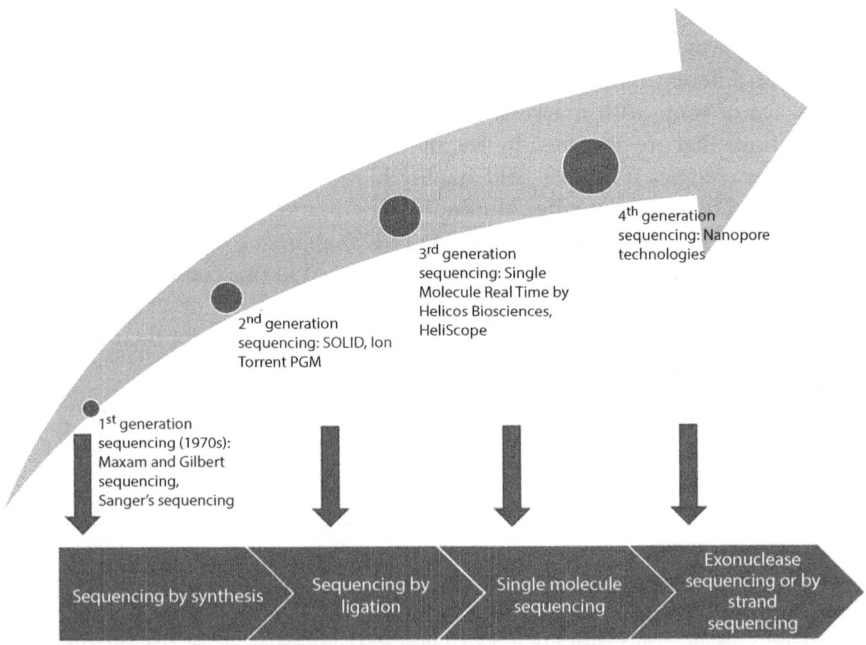

Figure 1.1 Timeline of next-generation sequencing technology.

field of life sciences and shifted sequencing studies from "model organism" to "pan organism" with the power of high-throughput NGS technology.

With the advancement of high-throughput sequencing platforms, data have been generated for millions of plant species and, therefore, understanding of plant system at the nucleotide level has been enriched in horticultural crops like fruits, vegetables, spices, plantations, etc.

1.2 Next-Generation Sequencing Approach in Genomics

1.2.1 Solanaceous

Among vegetables, the tomato was a pioneer in identifying the genetic basis of quantitative traits and in the map-based cloning of genes and quantitative trait loci (QTLs) [1]. Through the creation of extensive molecular marker libraries [2], genetic and physical maps [3], and mapping populations [4], tomato has served as a model plant for improvement and inheritance studies since the early 1990s. The Tomato Genome Consortium started the project called "SOL-100" associated with NGS-based sequencing of 100 different species of Solanaceae family and relating their sequences to the reference genome (solgenomic.net).

Completed genome	Draft genome	Projects based on resequencing
Solanum lycopersicum *Solanum tuberosum* *Capsicum annuum* *Solanum melongena* *Solanum lycopersicoides* *Solanum pennellii* *Solanum pimpinellifolium* *Solanum lycopersicum* var. *cerasiforme* *Iochroma cyaneum*	*Nicotiana attenuata* *Nicotiana benthamiana* *Nicotiana tabacum* *Petunia axillaris* *Petunia inflata* *Solanum chilense* *Coffea humblotiana*	*Solanum lycopersicum* inbreds 150 Tomato Genome Resequencing Project BGI Tomato 360 genomes Varitome Project

The tomato cultivar "Heinz 1706" is estimated to have a genome size of approximately 900 Mb with a simple structure composed of two main components: pericentromeric heterochromatin with repetitive sequences, occupying 75% of the whole genome, and distal euchromatin comprising the remaining 25% (220 Mb). It was sequenced through the BAC-by-BAC approach, resulting in the sequencing of 117 Mb of euchromatic regions precisely [5]. The main reason behind choosing this particular cultivar was because of its well-characterized *Hind*III BAC library available at that time [6]. In 2008, 30,800 BAC clones were selected, pooled, and short gun sequencing was done using the Sanger method of sequencing to accelerate the sequencing progress (The Tomato Genetic Consortium.2012). The sequences were congregated into contigs that elucidated 540 Mb of the genome. In 2009, the emerging NGS platforms paved the way for the sequencing consortium to plan for whole genome sequencing, which was earlier confined to only sequencing of euchromatin regions. Different variants of NGS technologies like 454/Roche GS FLX, Illumina Genome Analyser, and SOLiD sequencing were used. Using Sanger data, a *de novo* sequencing of "Heinz 1706" was assembled. Assemblies were generated by independent programs like Newbler and CABOG, and merged later on. Read-mapping and base error correction resulted in more accurate data where one base calling error per 29.4 kb and one indel error per 6.4 kb were obtained [7, 8]. Two BAC-based physical maps were used to connect the resulting high-quality scaffolds, and a high-density genetic map was used to anchor them [9] as well as with introgression line mapping and genome-wide BAC-FISH (Fluorescence *In Situ* Hybridization). The final assembly of tomato genome consisted of 760 Mb involving of 91 scaffolds. Most of the gaps identified after aligning the scaffolds with 12 chromosomes were confined to pericentromeric regions [7]. The consortium also sequenced the LA1589 accession of wild *S. pimpinellifolium* to explore the diversity it possesses through variation with the reference genome of "Heinz 1706". It was performed by Illumina technology using the whole-genome short gun sequencing approach. An assembly of 739 Mb was congregated, which, when compared with reference genome, showed the nucleotide divergence of only 0.6%, indicating the high level of similarity among two species (The Tomato Genome Consortium 2012). The first resequencing in tomato was attempted on 4 *S. lycopersicum* and 4 *S. l. cerasiformae* genotypes through the Illumina GAIIx platform, which generated a total of 4 million unique SNPs, 1,686 putative copy-number variations (CNV), and almost 1,28,000 InDels [10]. These variations can be utilized for QTL and gene mapping. Long-read technologies made further improvements in sequencing high-quality reference genomes including *S. pimpinellifolium* accessions like LA2093, LA1589 [11], LA1670 [12], *S. l. var. cerasiformae* acc. LA1673 [12], *S. l cv.* Moneyberg [13], *S. l cv. Heinz* 1706 [14], and *S. lycopersicoides* acc. LA2951 [15]. A total of 100 tomato accessions were sequenced using nanopore technology to detect variations and *de novo* assemblies were released for 14 reference genomes [16]. A total of 2,38,490 structural variants (mostly insertions and deletions) were discovered. The functional analysis allowed the linking of these variants to three major traits important for domestication and improvement targets: smoky flavor (not preferable among consumers), *sb1* (for branching patterns), and *fw3.2* (major QTL for fruit mass).

Capsicum annuum cv. CM334 was sequenced with high genome coverage and is considered as a reference for genome sequence of pepper [17]. The genome size was estimated to be 3.48 Gb, which is very high when compared to its fellow crops of same family. An increase in the genome size of pepper is supposed to be because of long terminal repeats of retrotransposons. In addition, CM334 genome has 76.4% transposable elements. This study also provided deeper insights into pepper pungency causing capsaicinoid biosynthesis pathway. For a better understanding of evolution and domestication, another cultivar of *C. annuum* var. *glabriusculum* (Zunla-1 and wild Chiltepin) was sequenced using Illumina technology through the whole-genome short gun approach [18]. They found 1104 target genes responsible in the capsaicinoid biosynthesis pathway, which indicates miRNAs for regulation. Resequencing data of 20 genomes were compared and found that domestication happened through artificial selection. Phylogenetic analysis was done by comparing the genomes of tomato, potato, and Arabidopsis genomes with pepper, which led to the discovery of pepper-specific duplications in 13 gene families. A heterozygous F1 from a wide cross was used for sequencing and was assembled to assess the ability to derive both haplotypes [19]. A pungency gene at the PUN1 locus was derived with large insertion/deletions that facilitated marker-assisted selection for pepper improvement.

The potato being a highly heterozygous polyploid, sequencing is much difficult when compared to diploids because for any given gene in a genotype, four different alleles per locus will be depicted. Thus, genotyping systems should differentiate among alleles and must be quantifying the copy number of allele. In order to overcome this problem, double monoploids were used for easy sequencing. Such a unique doubled monoploid was used for sequencing high-quality draft genome of 844 Mb using a combination of methods like Sanger, Roche/454, and Illumina and *de novo* assembly was done and is considered as a potato reference genome [20]. This had provided understanding into the evolution of genome of eudicots. Additionally, 3.67 million SNPs and 275 gene-specific (presence/absence) alterations were identified, concluding that homozygous alleles in the double monoploids is the reason for reduced vigor. A total of 15,235 genes were found in their full expression in developing tubers, giving valuable insights in the study of evolution of tuber development. Evolutionary innovation of tuberization has been confined to only the Petota section of Solanum. Although tomato is a neighboring species, it did not acquire this trait. These insights ignite the urge to study the evolutionary pathway further, to draw a deeper understanding about this genus.

Eggplant sequencing stands out among Solanaceae because it is phylogenetically unique being indigenous to Old World, whereas other crops of this family like tomato, chili pepper, and potato originated in South America. First draft genome of eggplant cv. Nakate-Shinkuro was built using a HiSeq 2000 Illumina sequencer, resulting in 33,873 scaffolds that depicted 74% of the whole genome [21]. Also, the reference genome of inbred line "67/3" was sequenced through Illumina and *de novo* assembled [22]. An inbred line HQ-1315 was assembled using an amalgamation of Illumina, Nanopore, and 10x genomic sequencing technologies to produce high-quality reference genome of ~1.17 Gb size that was assembled using Hi-C technology [23]. High genome acreage is associated with long terminal repeat (LTR) retrotransposons comprising 70.09% of the whole genome.

1.2.2 Malvaceae

Okra is one of the major vegetable crops belonging to the Malvaceae family. Plastids and mitochondrial genome of okra were sequenced through a combination of Illumina and Nanopore NGS [24]. This study found that plastid genome is a bit more conserved whereas mitochondrial genome has subgenomic configurations. They also observed immense transfer of sequences between the organelles, for instance, the presence of plastid genes (psaA, rps7, and psbJ) in mitochondrial genome.

1.2.3 Brassicaceae

Brassica genome sequencing consortium was started in 2003, initially confined to sequencing of *B. rapa* diploid A genome through BAC approach and Sanger technique. With the advent of NGS techniques, many genome sequencing projects emerged. The International *B. rapa* Genome Sequencing Project Consortium released *B. rapa* (Chinese Cabbage) genome annotated with 41,174 protein coding genes [25]. This study provided new perception into the expansion of the Brassica lineage. Second sequencing attempts were initiated in *B. oleracea* diploid C genome using the combination of Roche/454 and Illumina. Later on, reference genomes for all three Brassica progenitors and allopolyploids species like *B. napus* and *B. juncea* were released [25–28, 30].

The first genome sequencing of radish was witnessed in 2014. The "Aokubi" inbred line having an S-h haplotype was sequenced through Illumina NGS [31]. This study also states that radish and Chinese cabbage share common ancestral genes. Second genome sequencing involved a combination of Illumina and Roche 454 methods [32]. Later, chromosome scale draft genome of radish was produced by Illumina, Roche, and PacBio NGS technologies [33]. In addition to cultivated radish, a wild relative, *R. raphanistrum*, was also sequenced through Illumina approach [34]. Still, efforts have to be driven to understand the relationships with the wild relatives, mining for QTL's and developing useful markers for crop improvement.

1.2.4 Cucurbitaceae

The cucumber (Chinese Long inbred line 9930) was sequenced using both sanger and next-generation Illumina sequencing [35]. A semi-wild cultivar "GY14" (University of Wisconsin) and a wild cultivar "PI 183967" were also sequenced by Qi *et al.* [36]. A Spanish public–private initiative called MELANOMICS was started in 2009, which aimed at developing draft genome of melon using the NGS whole genome short gun. Melon's chloroplast and mitochondrial genome was assembled, and surprisingly, mitochondria of melon contain one of the largest genomes ever reported in plants [37]. The key reasons for the difference in the genome size of cucumber (367 Mb) and melon (454 Mb) could be possibly due to transposable element amplification in melon as long terminal repeats [38] and lack of WGD (whole genome duplication) in the melon lineage. This variation resulted in the phenotypic as well as quality differences between melons and cucumber, such as genes related to stress and flavor [39]. Ancestral proto chromosome of cucurbits was observed to be 15 based on the syntenic relationships among the cucurbit genomes [40]. The watermelon cultivar "97103" was sequenced using Illumina and high-quality draft genome assembled (*de novo*) to a size of 353.5 Mb [41]. The watermelon genome has also undergone 27 fissions and 28 fusions depicting that it has more shuffling than bottle gourd [42]. This study validates the 11-chromosome structure of watermelon.

1.2.5 Apiaceae

Carrot belongs to the Umbelliferae or Apiaceae family and the genome of orange carrot double haploid line "DC 27" was sequenced using the NGS whole genome sequencing short gun method [43]. A *de novo* assembly of it is performed using Illumina & Roche 454. In this study, for the first time in carrot, Chalcone-flavone isomerase (CHI), flavonoid 3'-monooxygenase (F3'H), and UDP-galactose gene sequences were identified. In 2016, chromosome scale whole-genome sequencing of DH 1 was prepared using Illumina at Beijing Genome Institute (BGI), 2000. Sequencing of intragenic region of mitochondrial genome provided the evidences for the transfer of DNA fragments between plastid and mitochondria [44].

1.2.6 Moringaceae

A draft genome for *Moringa oleifera* was sequenced, representing an assembly of >90% of genome size [45]. Comparative analysis of it with woody species revealed the evolutionary relationship and also helped in the identification of species-specific genes. Subsequently, a broad range of genomics studies that have been done in different vegetable crops with estimated genome size and applied sequencing technology are listed in Table 1.1.

Table 1.1 Genomics studies on different vegetable crops using sequencing technology.

Family	Accession/Species	Estimated genome size	Sequencing technology	Reference
Solanaceae	Heinz 1706 (*S. lycopersicum*)	900 Mb	Sanger + Roche 454	Tomato Genome Consortium, 2012
	LA 1589 (*S. pimpinellifolium*)	923 Mb	Illumina	
	LA0480 (*S. pimpinellifolium*)	900 Mb	Illumina	[46]
	LA2093 (*S. pimpinellifolium*)	923 Mb	Illumina + PacBio	[11]
	LA0716 (*S. pennellii*)	1.2 Gb	Illumina	[47]
	LYC1722 (*S. pennellii*)	1.2 Gb	Nanopore	[48]

(*Continued*)

Table 1.1 Genomics studies on different vegetable crops using sequencing technology. (*Continued*)

Family	Accession/Species	Estimated genome size	Sequencing technology	Reference
	LA31111 (*S. chilense*)	1.2 Gb	Illumina	[49]
	CM334 (*C. annuum*)	3.48 Gb		[17]
	Zunla-1 (*C. annuum*)	3.26 Gb		[18]
	Chiltepin (*C. a.* var. *glabriusculum*)	3.07 Gb		[18]
	PI159236 (*C. chinense*)	3.14 Gb 3.2 Gb		[17, 50]
	PBC81 (*C. baccatum*)	3.9 Gb		[50]
	C. annuum	3.26 Gb	10× Genomics	[19]
	cv. Nakate-Shinkuro (*S. melongena*)	1.13 Gb	Illumina + Roche 454	[21]
	cv. 67/3 (*S. melongena*)	1.04-1.21 Gb	Illumina	[22]
	Guiqie 1 (*S. melongena*)	1.21 Gb	Illumina + Hi-C	[51]
	HQ-1315 (*S. melongena*)	1.21 Gb	Illumina + Nnopore + Hi-C + 10 × Genomics	[23]
Brassicaceae	*B. rapa* spp. *pekinensis*	485 Mg	Illumina +PacBio+ Hi-C	[25, 52, 53]
	B. napus	1.13 Gb	Sanger +Illumina + 454	[26]

(*Continued*)

Table 1.1 Genomics studies on different vegetable crops using sequencing technology. (*Continued*)

Family	Accession/Species	Estimated genome size	Sequencing technology	Reference
	B. juncea var. *tumida*	922 Mb	PacBio+ Illumina	[29]
	B. oleracea var. *capitata*	630 Mb	Sanger +454 +Illumina	[27]
	Raphanus sativus	528.6 Mb	Sanger + Illumina	[31]
Cucurbitaceae	cv 9930 (*C. s.* var. *sativus*)	367 Mb	Sanger + Illumina + PacBio + 10X Genomics + Hi-C	[35, 51, 54]
	cv. Gy 14	367 Mb	454	[55]
	cv. B10	367 Mb	Sanger + 454	[56]
	PI 183967 (*C. sativus* var. *hardwickii*)	367 Mb	Sanger + Illumina	[36]
	Cucumis melo	450 Mb	Sanger + 454	[39, 57]
	Citrullus lanatus	425 Mb	Illumina + PacBio + Hi-C	[41, 42, 58]
	Cucurbita maxima	386.8 Mb	Illumina	[30]
	Cucurbita moschata	372 Mb		[30]
	Cucurbita pepo spp. *pepo*	283 Mb		[59]
	Cucurbita agryrosperma spp. *agryrosperma*	238 Mb	Illumina + PacBio	[60]
	Lagenaria siceraria	334 Mb	Illumina	[61]

(*Continued*)

Table 1.1 Genomics studies on different vegetable crops using sequencing technology. (*Continued*)

Family	Accession/Species	Estimated genome size	Sequencing technology	Reference
	Momordica charantia	339 Mb	Illumina , PacBio	[62, 63]
	Benincasa hispida	1.02 Gb	Illumina + PacBio	[64]
Fabaceae	*Vigna unguiculata*	560 Mb	PacBio	[65]
	Phaseolus vulgaris	587 Mb	Sanger + 454 + Illumina	[66]
Amaranthaceae	*Spinacia oleracea*	1 Gb	Illumina	[67]
	Beta vulgaris spp. *vulgaris*	714-758 Mb	Sanger + 454 + Illumina	[68]
Asteraceae	cv. Salinas (*Lactuca sativa*)	2.7 Gb	Illumina	[69]
Apiaceae	*Daucus carota* spp. *sativus*	473 Mb	Sanger + Illumina	[44]
	Coriandrum sativum	2.13 Gb	PacBio + Illumina + 10X genomics + Hi-C	[70]
Asparagaceae	*Asparagus officinalis*	1.3 Gb	PacBio + Illumina	[71]
Moringaceae	*Moringa oleifera*	315 Mb	Illumina	[45]

1.3 NGS Approach in Single-Nucleotide Polymorphic Markers Development

Genotyping by sequencing (GBS) was used in tomato cultivars consisting of four types i.e., large fruited, cherry fruited, grape fruited, and rootstocks, to generate SNPs [72]. A total of 10,615 SNPs were generated and five subsets were made, out of which one subset comprising 224 markers showed polymorphism in 91 cultivars tested and also able to distinguish 139 F1 cultivars from that of the reference genome. The results suggest the usefulness of these markers in DNA barcoding for identification of varieties (Table 1.2). GBS was also used in *C. annuum* and 109,610 SNPs were generated. These were used in QTL mapping and GWAS for capsaicinoid content of peppers [73]. Five candidate genes were identified successfully, from 69 QTL regions, which are involved in capsaicinoid biosynthesis [74]. The sequencing of sweet potato varieties led to the generation of various SNPs and InDels, which are related to starch biosynthesis [75].

Table 1.2 List of markers used for studying different cultivars of vegetable crops.

Family	Species	Marker type	Number of markers	Method	References
Solanaceae	*S. lycopersicum*	SNP	10,615	GBS	[72]
	S. lycopersicum	SNP	3614	WGS	[76]
	S. lycopersicum× *S. pennellii* *S. lycopersicum* × *S. pimpinellifolium*	SNP	141,083	GBS	[77]
	S. lycopersicum	SNP	4,812,432	Modified SAM	[17]
	S. pimpinellifolium	SNP	4,680,647	Modified SAM	[17]
	S. lycopersicum	SNP	8,784	GBS	[78]
	S. melongena *S. incanum*	SSR	11,262 11,829	*De novo* transcriptome	[79]
	S. melongena	SNP	10,000	RAD-seq	[80]
	C. annuum	SNP	1.76 million	Resequencing	[81]
	C. annuum	InDel	14,498	Whole genome resequencing	[82]
	C. annuum	SNP	109,610	GBS	[74]
	S. tuberosum	SNP	575,340	*De novo* transcriptome	[83]
	S. tuberosum	SNP InDel	27 million 3 million	Resequencing	[84]

(Continued)

Table 1.2 List of markers used for studying different cultivars of vegetable crops. (*Continued*)

Family	Species	Marker type	Number of markers	Method	References
Convolvulaceae	*Ipomoea batatus*	SNP InDel	622 85	GBS	[75]
Brassicaceae	*B. napus*	SSR	21,523	GWAS	[85]
	B. oleraceae	SNP InDel	496,463 37,493	WGS	[86]
	B. napus	SNP	37,721	GBS	[87]
	R. sativus	SNP	52,559	Rad-seq	[88]
Cucurbitaceae	*Cucurbita* spp.	SNP	37,869	GBS	[89]
	C. melo	SNP	375	GBS	[90]
	C. sativus	SSR	2171	GWAS	[91]
	C. lanatus	SNP	203,894 -279,412	WGRS	[92]
	M. charantia	InDel	389,487	WGS	[93]
	L. siceraria	SSR	45,066	Rad-seq	[94]
Fabaceae	*P. vulgaris*	SNP InDel	43,698 1267	WGS	[95]
	V. unguiculata	SNP	1031	GBS	[96]
Amaranthaceae	*Amaranthus* spp.	SNP	27,658	WGS	[97]
	S. oleracea	SSR	3852	WGS	[98]
Apiaceae	*D. carota*	SNP	3636	WGS	[99]

1.4 Next-Generation Sequencing Approach in Trait-Specific Breeding

In red tomato skin, yellow-colored naringenin chalcone (NGC) of flavonoids determines the exterior color of the fruit [100], and this flavonoid also builds up naturally in the cuticle of red fruit skin as it ripens and causes the peel's yellow color [101]. On the other hand, a transparent epidermis devoid of the yellow pigment NGC causes pink tomatoes. DNA markers obtained from the SlMYB12 gene of the Y locus on chromosome 1 would be helpful for marker-assisted selection (MAS) of tomato fruit color since NGC biosynthesis is controlled by this gene. The SlMYB12 gene, which has 4.9 kb, was transcribed from the line "FCR" (red-fruited YY) and the line "FCP" (pink-fruited YY) in order to create a gene-based marker. These SlMYB12 alleles' sequence alignment showed no sequence differences between the "FCR" and "FCP" alleles. The DNA sequence of SlMYB12 was physically centered in between CAPS-456 and CAPS-38123, indicating that fruit peel color in cultivated tomato is controlled by SlMYB12 [102]. Most recently, the *S. lycopersicum* × *S. pimpinellifolium* RIL population was utilized to create a "ultra-high density" tomato genetic map. This population had 141,083 SNP markers divided into 2,869 genomic bins. Additionally, this map was employed for the fine mapping of genes and QTLs relating to tomato fruit weight and lycopene concentration [77]. A new high-density genetic bin map for tomato was created using Genotyping-by-Sequencing (GBS) and a distinct population of tomato plants (*S. lycopersicum* × *S. pimpinellifolium*). The map includes 1,195 genetic bins and 8,470 SNPs and was used to accurately locate the late blight resistance gene Ph-5 in tomatoes through fine mapping [24].

To identify single-nucleotide polymorphism (SNP) markers, a high-throughput genotyping by sequencing (GBS) in the 188 F_5 population descended from the parents AR1 (PM resistant) and TF68 (PM susceptible) of *Capsicum annum* to provide powdery mildew resistance [103]. These SNP markers were then used to construct a genetic linkage map and perform QTL analysis. Each chromosome's 1,308 SNP markers were used to create 12 linkage groups, with a total map length of 2506.8 cM. Moreover, two QTLs for Powdery mildew resistance, Pm-2.1 and Pm-5.1, were discovered on chromosome 2 and chromosome 5, respectively. Development of novel pepper cultivars with improved resistance to bacterial wilt disease will benefit from the identification of SNP markers linked to the resistance to the disease. Using the technique of whole-genome resequencing, Ahn *et al.* [104] discovered SNPs across the entire genome. Two pepper cultivars, Saengryeg 211 (sensitive) and 82PR66 (resistant), having different bacterial wilt resistance properties have genomes sequenced and compared to the reference sequence, *C. annuum* cv. CM334. The density of SNPs varied among the chromosomes, with the Saengryeg 211 and 82PR66 SNPs on chromosomes 10 and 11, respectively, having the highest density of SNPs. Intra- and inter-specific linkage maps for QTL mapping related to anthocyanin pigment in Brinjal (*S. melongena*) were studied by Barchi *et al.* [22]. SNPs were produced using high-throughput sequencing (Illumina) and restriction site-associated DNA (RAD). In order to study major breeding attributes in cucumber, such as fruit development [105, 106], parthenocarpy [107], the formation of trichomes that serve as a plant's defense against biotic and abiotic stresses [108], primary plant regulatory processes in reaction to N deficit and transcriptome responses in *C. sativus* leaves [109], and the mechanism of melatonin-induced lateral root formation, the NGS technology and specific RNA-sequence methods are mentioned in Table 1.3.

In watermelon breeding, dwarfism is a valuable trait due to its contribution to higher yields and reduced labor in cultivation and harvesting. However, the genetic regulation for this trait is not well explained. To investigate this, researchers used NGS to analyze watermelon samples and identified a candidate dwarfism gene, Cla010726. They conducted a whole-genome re-sequencing from dwarf and vine pools DNA bulks (F2 population) and detected a genomic region containing the candidate gene through a genome-wide analysis of SNPs [112]. To develop a high-density

Table 1.3 Different NGS techniques used for targeted breeding of vegetable crops.

Sr. no.	Targeted breeding trait	NGS technique	References
1.	Fruit crops' parthenocarpy, a crucial characteristic affecting yield and quality	RNA-seq HiSeq Illumina 2000	[107]
2.	Trichomes, which resemble epidermal hairs, serve as a plant's defence mechanism for biotic and abiotic conditions	RNA-seq HiSeq Illumina 2000	[108]
3.	Early regulatory mechanism of plants in response to N starvation, transcriptome response of leaves under N deficiency	RNA-seq HiSeq Illumina 2000	[109]
4.	Molecular mechanisms of plant sex determination	RNA-seq Illumina	[110]
5.	Examine the factors that underlie melatonin-induced lateral root development in salt-stressed plants	RNA-seq Illumina	[111]

linkage map for bitter gourd using genotyping-by-sequencing (GBS) technology and to perform QTL analysis for six major yield-contributing traits, studies were conducted by Rao *et al.* [113]. The study used a mapping population generated from the cross DBGy-201 × Pusa Do Mausami and identified 19 QTLs for the six quantitative traits. The QTLs derived from each parent had either a positive or a negative additive effect on trait scores. The phenotypic variation explained by the QTLs ranged from 0.09% to 32.65%, with a total of six major QTLs detected.

1.5 Next-Generation Sequencing Approach in Metagenomics

Metagenomics involves understanding of microbial and virus populations in targeted samples through a nucleic acid sequencing approach [114]. Very limited studies have been conducted in context to metagenomics of vegetable crops. Tomato production is threatened by Begomoviruses occurrence worldwide and therefore studies were made to understand the population dynamics of begomoviruses affecting tomato. Next-generation sequencing was employed to assess the diversity of single-stranded DNA viruses in tomatoes with or without the Ty-1 gene, which provides tolerance to begomoviruses. Leaf samples with begomovirus-like symptoms were enriched for circular DNA and sequenced using Illumina technology. Fifteen distinct viruses/subviral agents, mainly in mixed infections, suggest the implications of utilizing virus-specific resistance in tomato breeding. Additional viruses, including two distinct Begomovirus species, a new alpha-satellite, and a new gemycircularvirus, were found in tomatoes lacking the Ty-1 gene. The sole species of begomovirus that caused serious symptoms in Ty-1 crops was a unique begomovirus, which was only identified in the Ty-1 pool. This study sheds light on the potential adaptation of begomoviruses to Ty-1 and its effects on a subset of single-stranded DNA viral/subviral agents [115]. To investigate the microbial communities present in Chinese cabbage, NGS using 16S rRNA gene amplicon sequencing was employed. The results indicated that the bacterial communities on Chinese cabbage were primarily composed of Proteobacteria and Bacteroidetes, with *Chryseobacterium*, *Aurantimonadaceae*, *Sphingomonas*, and *Pseudomonas* being the most abundant genera. The study also detected diverse potential pathogens, such as *Pantoea*, *Erwinia*, *Klebsiella*, *Yersinia*, *Bacillus*, *Staphylococcus*, *Salmonella*, and *Clostridium*. Although further studies are required to determine the association between these potential pathogens and foodborne illness, the study suggests that metagenomic approaches can be used to detect pathogenic bacteria on fresh vegetables [116]. Gawande *et al.* [117] used the Illumina MiSeq platform to analyze the gut microbial community structure of adult Onion Thrips tabaci collected from 10 different agro-climatically diverse locations in India. The analysis revealed 1662 OTUs belonging to 21 bacterial phyla, with Proteobacteria being the predominant phylum. The Colorado potato beetle (CPB) is a serious insect pest that can develop resistance to insecticides. Using biological insecticides based on viruses may be a promising approach to control this pest, but there is limited information on viruses that infect leaf feeding beetles. The metagenomic analysis of 297 CPB genomic and transcriptomic samples was performed. The study discovered 3137 virus-positive contigs linked to various viruses from six types, 17 orders, and 32 families, corresponding to over 97 virus species. The indicated sequences were homologous to insect viruses, plant viruses, and endogenous retroviral element genetic sequences. The study's findings imply that additional investigation may be useful in locating novel viruses to increase the range of biopesticides available to combat CPB [118].

1.6 Next-Generation Sequencing Approach in Transcriptomics

The transcriptome is the functional derivative of the genome and may be defined as the total amount of transcripts existing in a cell at a given developmental stage as well as a particular physiological condition. It comprises all the different RNA molecules present in a cell at a particular time. Therefore, the study of the transcriptome has the potential to reveal the functional role of the genome. In addition to this, transcriptome sequencing can yield information about the genes of an organism at a significantly lower cost than genome sequencing, as here, only the transcribed regions have to be sequenced and studied. Today, many research projects focus on the study of the transcriptome rather than the genome as only around 1–2% of genes in a genome are coding. Some of the recent transcriptomic studies that highlight the multidimensional applications of NGS in the transcriptomics of vegetable crops are described here in detail.

a) Discovery of Novel Genes

Transcriptome analysis has most significantly helped in the identification of several key genes that play an important role in living systems. Several such genes have been identified in vegetable crops. For example, the genes *CsACS*, *CsAsr*1, and *CsIAA*2, identified by transcriptome analysis, were found to be responsible for sex determination in cucumber [110]. Another set of genes, namely, *CsCUC*3 and *CsSTM*, also in cucumber, were found to regulate fruit spine development [108]. This has also been extended into some of the tissue-specific genes. The revelation of several genes specifically active in root and leaves in carrots, involved in anthocyanin biosynthesis, will help accelerate breeding attempts to improve the nutritive value of carrots [119].

b) Understanding the Molecular Mechanisms Underlying Biological Functions

Understanding ripening can help us breed fruits with a better shelf life, which retain their nutritional quality for long periods. Mutant LeMADS-RIN, a MADS-box gene, has been found to inhibit ripening in tomato. A complete understanding of this mutation has the potential to explain the genetic regulatory pathways that result in the ripening of fruits [120] compared to the fruit transcriptomes of a mutant and non-mutant genotype to get better insights into the role of ethylene in ripening. Mutant cauliflowers having green curds due to ectopic chloroplasts were subjected to genome-wide profiling of gene expression by RNA-seq. As a result, several genes associated with the development and differentiation of chloroplasts were discovered. The results hinted the significance of regulatory genes in light signaling pathways [121]. In less-studied vegetable crops such as sweet potatoes, characterization of the transcriptome helps us unravel the molecular aspects of cellular processes such as the development and differentiation of roots and leaves, differential gene expression in different tissues, and potential responses of the plant to biotic and abiotic stress [122]. All the genomic and transcriptomic data available for *Daucus carota* sp. *Sativus* were compiled into Carrot DB (Carrot Data Base). This database contains tools for Genome Mapping and Basic Local Alignment Search Tool (BLAST). This is a potent opportunity that has infinite possibilities. It may be used for the discovery of sequences of novel or putative genes or scaffolds, for the development of SSR markers, to predict the sequences of different proteins found in carrot, etc. [43].

c) Development of Markers

Trick *et al.* [123] developed a method of SNP discovery in the polyploid *B. napus* cultivars, using transcriptome analysis and the Illumina platform. Single gene sequences of *Brassica* were aligned with sequence reads with the help of MAQ software. Consequently, 23,330–42,593 putative SNPs with different read depths were detected. Out of these, ~90% of the SNPs were hemi-SNPs, which means that they were homozygous in one line but heterozygous in the other [124]. These hemi-SNPs can serve as powerful tools for genetic mapping. Blanca *et al.* [125] carried out their studies in the unexplored *C. pepo* plant. They sequenced many genes, found several allelic variations, and developed SSR and SNP markers as a result of their study. Their findings will accelerate the breeding prospects of squash. Such marker discovery was also carried out by Nicolai *et al.* [126] in capsicum.

d) Understanding Abiotic Stress

Any abiotic stress such as high temperature, soil salinity, excess humidity, drought, and metal deficiency or toxicity has an adverse effect on plants, and plants fight these stresses by altering their genetic responses. As a result, on exposure to any abiotic stress, the gene expression pattern in plants changes drastically. A transcriptomic analysis of these changes has huge research potentials. Lee and Choi [127] carried out a comparative transcriptome analysis in non-stressed and cold stressed pepper and identified many genes, hormones, and processes involved in cold stress tolerance. Drought tolerance has been observed in potato plants transformed with the yeast trehalose-6-phosphate synthase (*ScTPS*1) gene. The transcriptome profiles of the transgenics and wild types can be compared for differential expression patterns of genes in the two types of plants under drought stress conditions [128]. The molecular response of radish to the toxic heavy metal lead was studied by Wang *et al.* [129] using RNAseq technology. The upregulated differentially expressed genes (DEGs) under lead stress were mainly found to be involved in glutathione metabolism-related processes and cell wall defense. On the other hand, the downregulated DEGs were predominantly involved in carbohydrate metabolism-related pathways.

e) Identifying Viruses Infecting Vegetables

When any plant is attacked by viruses or viroids, several specific RNA molecules are synthesized. These RNAs are 21 to 24 nucleotides long and are called short interfering RNAs (siRNA). These RNAs survey the cytoplasm for viral genetic sequence, which may be homologous to the inducer and upon recognition immediately pairs with it to destroy it. This phenomenon is called RNAi or RNA silencing. NGS of siRNAs has the potential for the identification of viruses or viroids that infect plants. This can be a very sensitive tool for the identification as it will work even in very minute concentrations, in case of infections that do not show any symptoms.

Most importantly, this will help to discover viruses and viroids previously unknown to mankind.

1.7 Next-Generation Sequencing Approach in Exome and Captured Sequencing

Polyploidization has played a major role during evolution in shaping complex genomes, especially in vegetable crops. Therefore, vegetables, like *Brassicas*, generally have large genome sizes with a large number of repeats. Such large genome sizes are very difficult, inconvenient, and costly to sequence and analyze by whole-genome sequencing of a large number of individuals even after the advent of NGS. One way to solve this problem, without compromising on the amount of information gathered, is called "Sequence Capture" or "Targeted Sequencing". In this technique, there is a need to sequence only 1–2% of high-value sequences. These may be sequences rich in functional part of the genome or having low repetitive regions. This means that they include either specific genes of interest or even targets within these genes. When the captured sequence represents the entire protein coding region of the genome, it is called the exome.

Since a major portion of the genome is composed of repetitive and non-coding regions, this technique substantially reduces the bulk of the DNA to be sequenced while providing almost the same amount of information. This allows us to deeply focus on the important regions of the genome that are ultimately responsible for the production of genetic variation.

Captured sequencing may be carried out by one of the following methods:

1. Hybridization based sequence capture
2. PCR-based amplification
3. Selective circularization [130]

Out of these, the hybridization-based approach is most commonly used as it allows a large amount of DNA to be analyzed by virtue of it being simple, fast, inexpensive and requires a minimum input of DNA [131]. Hybridization-based exome capture may be done with the help of arrays or in solution. The arrays contain probes from the sequence library, which act as baits for the enrichment of the target sequence on the arrays that can be then sequenced by NGS techniques.

1.8 Applications of Exome and Captured Sequencing in Crop Research

Exome and captured sequencing can contribute to crop research in the following ways:

a) Identification of mutations in exomes and quick generation of novel and targeted polymorphisms
b) Exploration of biodiversity and gene mining
c) Development of SNP markers
d) Construction of genetic maps
e) Study the population structure, evolutionary history, and phylogenetic relationships
f) QTL mapping and gene identification
g) Genomic selection

At present, very limited studies using exome and captured sequencing have been conducted in vegetable crops.

In *Manihot esculenta* (Cassava), the transcriptome of 16 accessions was sequenced using the Illumina HiSeq platform. As a result, 675,559 EST-derived SNP markers were identified [132]. A subset of these markers were genotyped in 100 F_1 progeny, which were developed by the cross of parents contrasting in starch viscosity, by capture-based targeted enrichment sequencing. As a result of this study, a major quantitative trait locus (QTL) regulating starch pasting properties was discovered. Moreover, a novel QTL associated with starch pasting time was identified. This information can further be used in research and breeding activities.

The nucleotide binding-site leucine-rich repeat (NBS-LRR) protein encodes for majority of disease resistance genes. NGS allows the identification of pathogen resistance gene families in this region. RenSeq technology was used in *Solanum* spp. (potato and tomato) and subsequently many NBS-LRRs were identified [133]. In such a way, new disease resistance genes can be identified.

Uitdewilligen *et al.* [134] used an in-solution hybridization method called "SureSelect" for DNA sequencing. This reduced the complexity of the tetraploid cultivars of potatoes and allowed them to focus on 807 target genes of 83 tetraploid cultivars and one reference (DM 1–3 511). The resultant libraries were paired-end sequenced using Illumina HiSeq2000. As a result, 12.4 Gigabases of DNA sequence was obtained, which mapped to 2.1 Mb of the potato reference genome. The obtained data were correlated with traits like maturity and flesh color, and contrasting alleles were identified. Hybridization-based target enrichment system has also been applied to tomatoes to find out sequence variations in carotenoid biosynthesis genes [135]. Sharma *et al.* (2021) [136] employed a twofold approach by combining conventional QTL analysis to bulk segregant analysis (BSA) using a unique whole-exome capture (WEC) platform for potato. They analyzed a diploid potato population derived from a cross between two highly heterozygous parents for tuber sprout growth. Their findings revealed that the traits of tuber sprouting and growth are very complexly genetic regulated. These findings can have wide implications for both potato and other root, bulb, and tuber crops such as onion, sugar beet, and yam, where long-term storage is essential.

Targeted sequencing has now allowed the study of pathogen populations at the time of colonization by the use of DNA sequence capture that facilitates sequencing of pathogen DNA directly from field samples (Pathogen Enrichment Sequencing: PenSeq). Using this technology, the race diversity in *Albugo candida* that infects a wide range of vegetable crops has been explored [137]. The study also helped in studying phylogenetic relationships. This method finds application in the study of many microbes or pathogens, especially those that cannot be cultured.

1.9 Conclusion and Future Prospects

Crop improvement is a continuous process and classical breeding made significant achievements but at a slow pace in targeting the complex and low heritable traits like yield, disease resistance, and other quantitative traits. In the last few decades, next-generation sequencing has made revolutionary remarks in the understanding of metabolic processes of plants at low cost with high accuracy due to genome-wide high-density SNP markers. The GBS methods are robust across the crop species and genomic selection is made feasible for polygenic traits, which seems to be a promising routine process for vegetable crop species. Furthermore, NGS technologies have been used for the discovery of SSR markers and have shown potential for cultivar identification and genetic diversity system in a cost-effective way. Therefore, NGS technologies have been a great tool for the characterization, conservation, and improvement of biotic and abiotic stresses and horticultural traits.

References

1. Rothan, C., Diouf, I., Causse, M., Trait discovery and editing in tomato. *Plant J.*, 97, 1, 73–90, 2019.
2. Tanksley, S.D., Ganal, M.W., Prince, J.P., De Vicente, M.C., Bonierbale, M.W., Broun, P., Martin, G., High density molecular linkage maps of the tomato and potato genomes. *Genetics*, 132, 4, 1141–1160, 1992.
3. Haanstra, J.P.W., Wye, C., Verbakel, H., Meijer-Dekens, F., Van den Berg, P., Odinot, P. *et al.*, An integrated high-density RFLP-AFLP map of tomato based on two *Lycopersicon esculentum*L. *pennellii* F_2 populations. *Theor. Appl. Genet.*, 99, 1, 254–71, 1999.
4. Bernacchi, D., Beck-Bunn, T., Eshed, Y., Lopez, J., Petiard, V., Uhlig, J. *et al.*, Advanced backcross QTL analysis in tomato. I. Identification of QTLs for traits of agronomic importance from *Lycopersicon hirsutum*. *Theor. Appl. Genet.*, 97, 3, 381–97, 1998.
5. Mueller, L.A., Tanksley, S.D., Giovannoni, J.J., van Eck, J., Stack, S., Choi, D., Kim, B.D., Chen, M., Cheng, Z., Li, C., Ling, H., Xue, Y., Seymour, G., Bishop, G., Bryan, G., Sharma, R., Khurana, J., Tyagi, A., Chattopadhyay, D., Singh, N.K., Stiekema, W., Lindhout, P., Jesse, T., Lankhorst, R.K., Bouzayen, M., Shibata, D., Tabata, S., Granell, A., Botella, M.A., Giuliano, G., Frusciante, L., Causse, M., Zamir, D., The tomato sequencing project, the first cornerstone of the International Solanaceae Project (SOL). *Comp. Funct. Genomics*, 6, 153–158, 2005.
6. Budiman, M.A., Mao, L., Wood, T.C., Wing, R.A., A deep-coverage tomato BAC library and prospects toward development of an STC framework for genome sequencing. *Genome Res.*, 10, 129–136, 2000.
7. Sato, S. and Tabata, S., Tomato Genome Sequence. in: Functional Genomics and Biotechnology in Solanaceae and Cucurbitaceae Crops, in: *Biotechnology in Agriculture and Forestry*, H. Ezura, T. Ariizumi, J. Garcia-Mas, J. Rose, (Eds.), p. 70, Springer, Berlin, Heidelberg, 2016.
8. Kumar, R. and Khurana, A., Functional genomics of tomato: Opportunities and challenges in post-genome NGS era. *J. Biosci.*, 39, 917–929, 2014.
9. Shirasawa, K., Asamizu, E., Fukuoka, H., Ohyama, A., Sato, S., Nakamura, Y., Tabata, S., Sasamoto, S., Wada, T., Kishida, Y., Tsuruoka, H., Fujishiro, T., Yamada, M., Isobe, S., An interspecific linkage map of SSR and intronic polymorphism markers in tomato. *Theor. Appl. Genet.*, 121, 731–739, 2010.
10. Causse, M., Desplat, N., Pascual, L. *et al.*, Whole genome resequencing in tomato reveals variation associated with introgression and breeding events. *BMC Genomics*, 14, 791, 2013.

11. Wang, X., Gao, L., Jiao, C., Stravoravdis, S., Hosmani, P.S., Saha, S., Zhang, J., Mainiero, S., Strickler, S.R., Catala, C. *et al.*, Genome of *Solanum pimpinellifolium* provides insights into structural variants during tomato breeding. *Nat. Commun.*, 11, 5817, 2020.
12. Takei, H., Shirasawa, K., Kuwabara, K., Toyoda, A., Matsuzawa, Y., Iioka, S., *et al. De novo* genome assembly of two tomato ancestors, *Solanum pimpinellifolium* and *Solanum lycopersicum* var. cerasiforme, by long-read sequencing. *DNA Res.*, 28, 1, dsaa029, 2021.
13. van Rengs, W.M., Schmidt, M.H.W., Effgen, S., Le, D.B., Wang, Y., Zaidan, M.W.A.M. *et al.*, A chromosome scale tomato genome built from complementary PacBio and Nanopore sequences alone reveals extensive linkage drag during breeding. *Plant J.*, 110, 2, 572–88, 2022.
14. Su, X., Wang, B., Geng, X. *et al.*, A high-continuity and annotated tomato reference genome. *BMC Genomics*, 22, 898, 2021.
15. Powell, A.F., Feder, A., Li, J., Schmidt, M.H.W., Courtney, L., Alseekh, S. *et al.*, A *Solanum lycopersicoides* reference genome facilitates insights into tomato specialized metabolism and immunity. *Plant J.*, 110, 6, 1791–1810, 2022.
16. Alonge, M., Wang, X., Benoit, M., Soyk, S., Pereira, L., Zhang, L. *et al.*, Major impacts of widespread structural variation on gene expression and crop improvement in tomato. *Cell*, 182, 1, 145–61, 2020.
17. Kim, J.E., Oh, S.K., Lee, J.H., Lee, B.M., Jo, S.H., Genome-wide SNP calling using next generation sequencing data in tomato. *Mol. Cells*, 37, 1, 36–42, 2014.
18. Qin, C., Yu, C., Shen, Y. *et al.*, Whole-genome sequencing of cultivated and wild peppers provides insights into Capsicum domestication and specialization. *Proc. Natl. Acad. Sci.*, 111, 14, 5135–5140, 2014.
19. Hulse-Kemp, A.M., Maheshwari, S., Stoffel, K., Hill, T.A., Jaffe, D., Williams, S.R., Weisenfeld, N., Ramakrishnan, S., Kumar, V., Shah, P. *et al.*, Reference quality assembly of the 3.5-Gb genome of *Capsicum annuum* from a single linked-read library. *Hortic. Res.*, 5, 1–13, 2018.
20. Potato Genome Sequencing Consortium. Genome sequence and analysis of the tuber crop potato. *Nature*, 475, 189–195, 73552011.
21. Hirakawa, H., Shirasawa, K., Miyatake, K., Nunome, T., Negoro, S., Ohyama, A., Yamaguchi, H., Sato, S., Isobe, S., Tabata, S., Fukuoka, H., Draft genome sequence of eggplant (*Solanum melongena* L.): the representative solanum species indigenous to the old world. *DNA Res.*, 21, 6, 649–60, 2014.
22. Barchi, L., Delledonne, M., Lanteri, S., Dal Molin, A., Minio, A., Ferrarini, A., Rotino, G.L., A high quality eggplant genome sequence: a new tool for the analysis of the Solanaceae family evolution and for the molecular deciphering of complex traits, in: *Plant breeding: the art of bringing science to life*, Kölliker, R. and Boller, B. (Eds.), p. 122, 20th Eucarpia General Congress, Zurich, 2016.

23. Wei, Q., Wang, J., Wang, W., Hu, T., Hu, H., Bao, C., A high-quality chromosome-level genome assembly reveals genetics for important traits in eggplant. *Hortic. Res.*, 7, 153, 2020.
24. Jia, M., Genetic characterization and mapping of late blight resistance genes in the wild tomato accession PI 270443. The Pennsylvania State University, 2019.
25. Wang, X., Wang, H., Wang, J., Sun, R., Wu, J., Liu, S., Bai, Y., Mun, J.H., Bancroft, I., Cheng, F., Huang, S., Li, X., Hua, W. *et al.*, The genome of the mesopolyploid crop species *Brassica rapa. Nat. Genet.*, 43, 10, 1035–1039, 2011.
26. Chalhoub, B., Denoeud, F., Liu, S., Parkin, I.A., Tang, H. *et al.*, Early allopolyploid evolution in the post-Neolithic *Brassica napus* oilseed genome. *Science*, 345, 6199, 950–953, 2014.
27. Liu, S., Liu, Y., Yang, X. *et al.*, The *Brassica oleracea* genome reveals the asymmetrical evolution of polyploid genomes. *Nat. Commun.*, 5, 3930, 2014.
28. Parkin, I.A., Koh, C., Tang, H. *et al.*, Transcriptome and methylome profiling reveals relics of genome dominance in the mesopolyploid *Brassica oleracea. Genome Biol.*, 15, R77, 2014.
29. Yang, J., Liu, D., Wang, X. *et al.*, The genome sequence of allopolyploid *Brassica juncea* and analysis of differential homeolog gene expression influencing selection. *Nat. Genet.*, 48, 1225–1232, 2016.
30. Sun, F., Fan, G., Hu, Q., Zhou, Y., Guan, M. *et al.*, The high-quality genome of *Brassica napus* cultivar 'ZS11' reveals the introgression history in semi-winter morphotype. *Plant J.*, 92, 3, 452–468, 2017.
31. Kitashiba, H., Li, F., Hirakawa, H., Kawanabe, T., Zou, Z., *et al.* Draft sequences of the radish (*Raphanus sativus* L.) genome. *DNA Res.*, 21, 481–490, 2014.
32. Mitsui, Y., Shimomura, M., Komatsu, K., Namiki, N., Shibata-Hatta, M. *et al.*, The radish genome and comprehensive gene expression profile of tuberous root formation and development. *Sci. Rep.*, 5, 10835, 2015.
33. Jeong, Y.M., Kim, N., Ahn, B.O., Oh, M., Chung, W.H., Chung, H., Jeong, S., Lim, K.B., Hwang, Y.J., Kim, G.B., Baek, S., Choi, S.B., Hyung, D.J., Lee, S.W., Sohn, S.H., Kwon, S.J., Jin, M., Seol, Y.J., Chae, W.B., Choi, K.J., Park, B.S., Yu, H.J., Mun, J.H., Elucidating the triplicated ancestral genome structure of radish based on chromosome-level comparison with the Brassica genomes. *Theor. Appl. Genet.*, 129, 357–1372, 2016.
34. Moghe, G.D., Hufnagel, D.E., Tang, H., Xiao, Y., Dworkin, I., Town, C.D., Conner, J.K., Shiu, S.H., Consequences of whole-genome triplication as revealed by comparative genomic analyses of the wild radish *Raphanus raphanistrum* and three other Brassicaceae species. *Plant Cell*, 26, 1925–1937, 2014.
35. Huang, S., Li, R., Zhang, Z., Li, L., Gu, X. *et al.*, The genome of the cucumber, *Cucumis sativus* L. *Nat. Genet.*, 41, 12, 1275–81, 2009.

36. Qi, J., Liu, X., Shen, D. *et al.*, A genomic variation map provides insights into the genetic basis of cucumber domestication and diversity. *Nat. Genet.*, 45, 1510–1515, 2013.
37. Rodriguez-Moreno, L., Gonzalez, V., Benjak, A., Marti, M., Puigdomenech, P., Aranda, M., Garcia-Mas, J., Determination of the melon chloroplast and mitochondrial genome sequences reveals that the largest reported mitochondrial genome in plants contains a significant amount of DNA having a nuclear origin. *BMC Genomics*, 12, 424, 2011.
38. Sebastian, P., Schaefer, H., Telford, I., Renner, S., Cucumber (*Cucumis sativus*) and melon (*C. melo*) have numerous wild relatives in Asia and Australia, and the sister species of melon is from Australia. *Proc. Natl. Acad. Sci. U.S.A.*, 107, 14269–14273, 2010.
39. Garcia-Mas, J., Benjak, A., Sanseverino, W., Bourgeois, M., Mir, G., González, V.M., Hénaff, E., Câmara, F., Cozzuto, L. *et al.*, The genome of melon (*Cucumis melo* L.). *Proc. Natl. Acad. Sci. U.S.A.*, 109, 29, 11872–11872, 2012.
40. Xie, D., Xu, Y., Wang, J. *et al.*, The wax gourd genomes offer insights into the genetic diversity and ancestral cucurbit karyotype. *Nat. Commun.*, 10, 5158, 2019.
41. Guo, S., Zhang, J., Sun, H., Salse, J., Lucas, W.J., Zhang, H., Zheng, Y., Mao, L., Ren, Y., Wang, Z. *et al.*, The draft genome of watermelon (*Citrullus lanatus*) and resequencing of 20 diverse accessions. *Nat. Genet.*, 45, 51–58, 2013.
42. Wu, X., Wu, X., Wang, Y., Wang, B., Lu, Z., Xu, P., Li, G., Molecular Genetic Mapping of Two Complementary Genes Underpinning Fruit Bitterness in the Bottle Gourd (*Lagenaria siceraria* [Mol.] Standl.). *Front. Plant Sci.*, 10, 1493, 2019.
43. Xu, Z. S., Tan, H. W., Wang, F., Hou, X. L., and Xiong, A. S., CarrotDB: a genomic and transcriptomic database for carrot. *Database*, 2014.
44. Iorizzo, M., Senalik, D., Szklarczyk, M., Grzebelus, D., Spooner, D., Simon, P., *De novo* assembly of the carrot mitochondrial genome using next generation sequencing of whole genomic DNA provides first evidence of DNA transfer into an angiosperm plastid genome. *BMC Plant Biol.*, 12, 61, 2012.
45. Tian, Y., Zeng, Y., Zhang, J., Yang, C.G., Yan, L., Wang, X.J., Shi, C.Y., Xie, J., Dai, T.Y., Peng, L. *et al.*, High quality reference genome of drumstick tree (*Moringa oleifera* Lam.), a potential perennial crop. *Sci. China Life Sci.*, 58, 627–638, 2015.
46. Razali, R., Bougouffa, S., Morton, M.J.L., Lightfoot, D.J., Alam, I., Essack, M., Arold, S.T., Kamau, A.A., Schmöckel, S.M., Pailles, Y. *et al.*, The genome sequence of the wild tomato *Solanum pimpinellifolium* provides insights into salinity tolerance. *Front. Plant Sci.*, 9, 1402, 2018.
47. Bolger, A., Scossa, F., Bolger, M.E., Lanz, C., Maumus, F., Tohge, T., Quesneville, H., Alseekh, S., Sørensen, I., Lichtenstein, G. *et al.*, The genome of the stress-tolerant wild tomato species *Solanum pennellii*. *Nat. Genet.*, 46, 1034–1038, 2014.

48. Schmidt, M.H.W., Vogel, A., Denton, A.K., Istace, B., Wormit, A., van de Geest, H., Bolger, M.E., Alseekh, S., Maß, J., Pfaff, C. *et al.*, *De novo* assembly of a new *Solanum pennellii* accession using nanopore sequencing. *Plant Cell*, 29, 2336–2348, 2017.
49. Stam, R., Nosenko, T., Hörger, A.C., Stephan, W., Seidel, M., Kuhn, J.M.M., Haberer, G., Tellier, A., The *de novo* reference genome and transcriptome assemblies of the wild tomato species *Solanum chilense* highlights birth and death of NLR genes between tomato species. *G3: Genes Genomes Genet.*, 9, 3933–3941, 2019.
50. Kim, S., Park, J., Yeom, S.I., Kim, Y.M., Seo, E., Kim, K.T., Kim, M.S., Lee, J.M., Cheong, K., Shin, H.S. *et al.*, New reference genome sequences of hot pepper reveal the massive evolution of plant disease-resistance genes by retroduplication. *Genome Biol.*, 18, 210, 2017.
51. Li, D., Qian, J., Li, W., Jiang, Y., Gan, G., Li, W., and Wang, Y., Genome sequence and analysis of the eggplant (*Solanum melongena L.*). *BioRxiv*, 824540, 2019, BioRxiv.
52. Cai, C., Wang, X., Liu, B., Wu, J., Liang, J., Cui, Y., Cheng, F., Wang, X., *Brassica rapa* genome 2.0: A reference upgrade through sequence re-assembly and gene re-annotation. *Mol. Plant*, 10, 649–651, 2017.
53. Zhang, L., Cai, X., Wu, J., Liu, M., Grob, S., Cheng, F., Liang, J., Cai, C., Liu, Z., Liu, B. *et al.*, Improved *Brassica rapa* reference genome by single-molecule sequencing and chromosome conformation capture technologies. *Hortic. Res.*, 5, 50, 2018.
54. Li, Z., Zhang, Z., Yan, P., Huang, S., Fei, Z., Lin, K., RNA-Seq improves annotation of protein-coding genes in the cucumber genome. *BMC Genomics*, 12, 540, 2011.
55. Yang, L., Koo, D.H., Li, Y., Zhang, X., Luan, F., Havey, M.J., Jiang, J., Weng, Y., Chromosome rearrangements during domestication of cucumber as revealed by high-density genetic mapping and draft genome assembly. *Plant J.*, 71, 895–906, 2012.
56. Woycicki, R., Witkowicz, J., Gawronski, P., Dabrowska, J., Lomsadze, A., Pawelkowicz, M., Siedlecka, E., Yagi, K., Plader, W., Seroczynska, A. *et al.*, The genome sequence of the North-European cucumber (*Cucumis sativus* L.) unravels evolutionary adaptation mechanisms in plants. *PLoS One*, 6, e22728, 2011.
57. Ruggieri, V., Alexiou, K.G., Morata, J., Argyris, J., Pujol, M., Yano, R., Nonaka, S., Ezura, H., Latrasse, D., Boualem, A. *et al.*, An improved assembly and annotation of the melon (*Cucumis melo* L.) reference genome. *Sci. Rep.*, 8, 8088, 2018.
58. Guo, S., Zhao, S., Sun, H., Wang, X., Wu, S., Lin, T., Ren, Y., Gao, L., Deng, Y., Zhang, J. *et al.*, Resequencing of 414 cultivated and wild watermelon accessions identifies selection for fruit quality traits. *Nat. Genet.*, 51, 1616–1623, 2019.

59. Montero-Pau, J., Blanca, J., Bombarely, A., Ziarsolo, P., Esteras, C. *et al.*, *De novo* assembly of the zucchini genome reveals a whole-genome duplication associated with the origin of the *Cucurbita* genus. *Plant Biotechnol. J.*, 16, 1161–1171, 2018.
60. Barrera-Redondo, J., Ibarra-Laclette, E., Vázquez-Lobo, A., GutiérrezGuerrero, Y.T., Sánchez de la Vega, G., Piñero, D., Montes-Hernández, S., Lira-Saade, R., and Eguiarte, L.E., The genome of Cucurbita argyrosperma (silver-seed gourd) reveals faster rates of protein-coding gene and long noncoding RNA turnover and neofunctionalization within *Cucurbita*. *Mol. Plant*, 12, 506–520, 2019.
61. Wu, S., Shamimuzzaman, M., Sun, H. *et al.*, The bottle gourd genome provides insights into Cucurbitaceae evolution and facilitates mapping of a papaya ring-spot virus resistance locus. *Plant J.*, 92, 963–975, 2017.
62. Urasaki, N., Takagi, H., Natsume, S., Uemura, A., Taniai, N., Miyagi, N., Fukushima, M., Suzuki, S., Tarora, K., Tamaki, M. *et al.*, Draft genome sequence of bitter gourd (*Momordica charantia*), a vegetable and medicinal plant in tropical and subtropical regions. *DNA Res.*, 24, 51–58, 2017.
63. Matsumura, H., Hsiao, M.C., Toyoda, A., Taniai, N., Tarora, K., Urasaki, N., Anand, S.S., Dhillon, N.P.S., Schafleitner, R., Lee, C.R., Long-read bitter gourd genome and the genomic architecture of domestication. *bioRxiv*, 822411, 2019.
64. Xie, L., Wang, J., Liu, F., Zhou, H., Chen, Y., Integrated analysis of multiomics and fine-mapping reveals a candidate gene regulating pericarp color and flavonoids accumulation in wax gourd (*Benincasa hispida*). *Front. Plant Sci.*, 13, 1019787, 2022.
65. Lonardi, S., Muñoz-Amatriaín, M., Liang, Q., Shu, S., Wanamaker, S.I., Lo, S., Tanskanen, J., Schulman, A.H., Zhu, T., Luo, M.C. *et al.*, The genome of cowpea (*Vigna unguiculata* [L.] Walp.). *Plant J.*, 98, 767–782, 2019.
66. Slotte, T., Hazzouri, K.M., Ågren, J.A., Koenig, D., Maumus, F., Guo, Y.L., Steige, K., Platts, A.E., Escobar, J.S., Newman, L.K. *et al.*, The *Capsella rubella* genome and the genomic consequences of rapid mating system evolution. *Nat. Genet.*, 45, 831–835, 2013.
67. Xu, C., Jiao, C., Sun, H., Cai, X., Wang, X., Ge, C., Zheng, Y., Liu, W., Sun, X., Xu, Y. *et al.*, Draft genome of spinach and transcriptome diversity of 120 *Spinacia* accessions. *Nat. Commun.*, 8, 15275, 2017.
68. Dohm, J.C., Minoche, A.E., Holtgräwe, D., Capella-Gutiérrez, S., Zakrzewski, F., Tafer, H., Rupp, O., Sörensen, T.R., Stracke, R., Reinhardt, R. *et al.*, The genome of the recently domesticated crop plant sugar beet (*Beta vulgaris*). *Nature*, 505, 546–549, 2014.
69. Reyes-Chin-Wo, S., Wang, Z., Yang, X. *et al.*, Genome assembly with *in vitro* proximity ligation data and whole-genome triplication in lettuce. *Nat. Commun.*, 8, 14953, 2017.

70. Song, X., Wang, J., Li, N., Yu, J., Meng, F., Wei, C., Liu, C., Chen, W., Nie, F., Zhang, Z. *et al.*, Deciphering the high-quality genome sequence of coriander that causes controversial feelings. *Plant Biotechnol. J.*, 18, 1444–1456, 2019.
71. Harkess, A., Zhou, J., Xu, C., Bowers, J.E., Van der Hulst, R., Ayyampalayam, S., Mercati, F., Riccardi, P., McKain, M.R., Kakrana, A. *et al.*, The asparagus genome sheds light on the origin and evolution of a young Y chromosome. *Nat. Commun.*, 8, 1279, 2017.
72. Kim, M., Jung, J.K., Shim, E.J., Chung, S.M., Park, Y., Lee, G.P., Sim, S., Genome-wide SNP discovery and core marker sets for DNA barcoding and variety identification in commercial tomato cultivars. *Sci. Hortic.*, 276, 109734, 2021.
73. Park, M., Lee, J.H., Han, K., Jang, S., Han, J., Lim, J.H., Jung, J.W., Kang, B.C., A major QTL and candidate genes for capsaicinoid biosynthesis in the pericarp of Capsicum chinense revealed using QTL-seq and RNA-seq. *Theor. Appl. Genet.*, 132, 2, 515–529, 2019.
74. Han, K., Lee, H.Y., Ro, N.Y., Hur, O.S., Lee, J.H. *et al.*, QTL mapping and GWAS reveal candidate genes controlling capsaicinoid content in Capsicum. *Plant Biotechnol. J.*, 16, 9, 1546–1558, 2018.
75. Zhang, K., Luo, K., Li, S., Peng, D., Tang, D., Lu, H., Zhao, Y., Lv, C., Wang, J., Genetic variation and sequence diversity of starch biosynthesis and sucrose metabolism genes in sweet potato. *Agronomy*, 10, 5, 627, 2020.
76. Bhandari, P. and Lee, T.G., A Genetic Map and Linkage Panel for the Large-fruited Fresh-market Tomato. *J. Am. Soc. Hortic. Sci.*, 146, 2, 125–131, 2021.
77. Gonda, I., Ashrafi, H., Lyon, D.A., Strickler, S.R., Hulse-Kemp, A.M., Ma, Q., Foolad, M.R., Sequencing-based bin map construction of a tomato mapping population, facilitating high-resolution quantitative trait loci detection. *Plant Genome*, 12, 1, 180010, 2019.
78. Sim, S.C., Durstewitz, G., Plieske, J., Wieseke, R., Ganal, M.W., Van Deynze, A. *et al.*, Development of a large SNP genotyping array and generation of high-density genetic maps in tomato. *PLoS One*, 7, 7, e40563, 2012.
79. Tiwari, S.K., Mishra, P., Singh, S., Singh, V.K., Kashyap, S.P., Singh, M., Singh, B., Development and validation of microsatellite markers from *de novo* transcriptome assembly of eggplant (*Solanum melongena* L.) and its putative progenitor *S. incanum* L. cultivars. *bioRxiv*, 560805, 2019.
80. Baird, N.A., Etter, P.D., Atwood, T.S., Currey, M.C., Shiver, A.L., Lewis, Z.A., Selker, E.U., Cresko, W.A., Johnson, E.A., Rapid SNP Discovery and Genetic Mapping Using Sequenced RAD Markers. *PLoS One*, 3, e3376, 2008.
81. Lee, J., Park, S.J., Do, J.W., Han, J., Choi, D., Yoon, J.B., Development of a genetic map of chili pepper using single nucleotide polymorphism markers generated from next generation resequencing of parents. *Korean J. Hortic. Sci. Technol.*, 31, 473–482, 2013.
82. Li, W., Cheng, J., Wu, Z., Qin, C., Tan, S., Tang, X., Cui, J., Zhang, L., Hu, K., An InDel-based linkage map of hot pepper (Capsicum annuum). *Mol. Breed.*, 35, 1, 32, 2015.

83. Hamilton, J.P., Hansey, C.N., Whitty, B.R., Stoffel, K., Massa, A.N., Van Deynze, A., De Jong, W.S., Douches, D.S., Buell, C.R., Single nucleotide polymorphism discovery in elite North American potato germplasm. *BMC Genomics*, 12, 1, 1–12, 2011.
84. Zhang, F., Qu, L., Gu, Y. *et al.*, Resequencing and genome-wide association studies of autotetraploid potato. *Mol. Hortic.*, 2, 6, 2022.
85. Zhu, J., Zhang, J., Jiang, M., Wang, W., Jiang, J., Li, Y., Yang, L., Zhou, X., Development of genome-wide SSR markers in rapeseed by next generation sequencing. *Gene*, 798, 145798, 2021.
86. Sampath, P., Waminal, N.E., Lee, J., Koo, H.J., Choi, B., Park, J.Y., Ahn, K., Yang, T.J., Nuclear and chloroplast genome diversity revealed by low-coverage whole-genome shotgun sequence in 44 Brassica oleracea breeding lines. *Hortic. Plant J.*, 7, 6, 539–551, 2021.
87. Ryu, J., Lyu, J.I., Kim, D.G., Koo, K.M., Yang, B., Jo, Y.D., Kim, S.H., Kwon, S.J., Ha, B.K., Kang, S.Y., Kim, J.B., Ahn, J.W., Single Nucleotide Polymorphism (SNP) discovery and association study of flowering times, crude fat and fatty acid composition in rapeseed (Brassica napus L.) mutant lines using genotyping-by-sequencing (GBS). *Agronomy*, 11, 3, 508, 2021.
88. Kobayashi, H., Shirasawa, K., Fukino, N., Hirakawa, H., Akanuma, T., Kitashiba, H., Identification of genome-wide single-nucleotide polymorphisms among geographically diverse radish accessions. *DNA Res.*, 27, 1, dsaa001, 2020.
89. Nguyen, N.N., Kim, M., Jung, J.K., Shim, E.J., Chung, S.M., Park, Y., Lee, G.P., Sim, S.C., Genome-wide SNP discovery and core marker sets for assessment of genetic variations in cultivated pumpkin (Cucurbita spp.). *Hortic. Res.*, 7, 121, 2020.
90. Howlader, J. and Hong, Y., Natarajan, S. *et al.* Development of powdery mildew race 5-specific SNP markers in *Cucumis melo* L. using whole-genome resequencing. *Hortic. Environ. Biotechnol.*, 61, 347–357, 2020.
91. Wang, Z., Dong, S., Liu, Y., Beckles, D.M., Li, C. *et al.*, A Genome-wide association study identifies candidate genes for heat tolerance in adult cucumber (*Cucumis sativus* L.) plants. *Hortic. Plant J.*, 2023, ISSN 2468-0141.
92. Subburaj, S., Lee, K., Jeon, Y., Tu, L., Son, G., Choi, S., Lim, Y.P., McGregor, C., Lee, G.J., Whole genome resequencing of watermelons to identify single nucleotide polymorphisms related to flesh color and lycopene content. *PLoS One*, 14, 10, e0223441, 2019.
93. Cui, J., Peng, J., Cheng, J. *et al.*, Development and validation of genome-wide InDel markers with high levels of polymorphism in bitter gourd (Momordica charantia*). BMC Genomics*, 22, 190, 2021.
94. Bonthala, B., Abdin, M.Z., Arya, L. *et al.*, Genome-wide SSR markers in bottle gourd: development, characterization, utilization in assessment of genetic diversity of National Genebank of India and synteny with other related cucurbits. *J. Appl. Genet.*, 63, 237–263, 2022.

95. Zou, X., Shi, C., Austin, R.S. *et al.*, Genome-wide single nucleotide polymorphism and Insertion-Deletion discovery through next-generation sequencing of reduced representation libraries in common bean. *Mol. Breed.*, 33, 769–778, 2014.
96. Shi, A. and Buckley, B., Mou, B. *et al.* Association analysis of cowpea bacterial blight resistance in USDA cowpea germplasm. *Euphytica*, 208, 143–155, 2016.
97. Maughan, J., Yourstone, S., Jellen, R., Udall, J., SNP discovery via genomic reduction, barcoding, and 454-pyrosequencing in amaranth. *Plant Genome J.*, 2, 3, 2009.
98. Göl, Ş., Göktay, M., Allmer, J. *et al.*, Newly developed SSR markers reveal genetic diversity and geographical clustering in spinach (Spinacia oleracea). *Mol. Genet. Genomics*, 292, 847–855, 2017.
99. Iorizzo, M., Senalik, D., Ellison, S., Grzebelus, D., Cavagnaro, P., Spooner, D., Deynze, A., Simon, P., Development of a high-throughput SNP Resource to advance genomic, genetic and breeding research in carrot (*Daucus carota* L.). *Plant and Animal Genome XXI Conference*, 2013.
100. Ballester, A.R., Molthoff, J., de Vos, R., Hekkert, B.T.L., Orzaez, D., Ferna&xxxx0769;ndez-Moreno, J.P., Bovy, A., Biochemical and molecular analysis of pink tomatoes: deregulated expression of the gene encoding transcription factor SlMYB12 leads to pink tomato fruit color. *Plant Physiol.*, 152, 1, 71–84, 2010.
101. Hunt, G.M. and Baker, E.A., Phenolic constituents of tomato fruit cuticles. *Phytochem.*, 19, 1415–1419, 1980.
102. Kim, B., Kim, N., Kang, J., Choi, Y., Sim, S.C., Min, S.R., & Park, Y., Single Nucleotide polymorphisms linked to the SlMYB12 gene that controls fruit peel color in domesticated tomatoes (Solanum lycopersicum L.). *Hortic. Sci. Technol.*, 33, 4, 566–574, 2015.
103. Manivannan, A., Choi, S., Jun, T.H., Yang, E.Y., Kim, J.H., Lee, E.S., Ahn, Y.K., Genotyping by sequencing-based discovery of SNP markers and construction of linkage map from F5 population of pepper with contrasting powdery mildew resistance trait. *BioMed. Res. Int.*, 2021, 1, 6673010, 2021.
104. Ahn, Y.K. and Karna, S., Jun, T.H. *et al.*, Complete genome sequencing and analysis of Capsicum annuum varieties. *Mol. Breed.*, 36, 10, 2016.
105. Che, G. and Zhang, X., Molecular basis of cucumber fruit domestication. *Curr. Opin. Plant Biol.*, 47, 38–46, 2019.
106. Sui, X., Nie, J., Li, X., Scanlon, M.J., Zhang, C. *et al.*, Transcriptomic and functional analysis of cucumber (*Cucumis sativus* L.) fruit phloem during early development. *Plant J.*, 96, 5, 982–996, 2018.
107. Li, J., Wu, Z., Cui, L., Zhang, T., Guo, Q., Xu, J., Chen, J., Transcriptome comparison of global distinctive features between pollination and parthenocarpic fruit set reveals transcriptional phytohormone cross-talk in cucumber (*Cucumis sativus* L.). *Plant Cell Physiol.*, 55, 7, 1325–1342, 2014.

108. Chen, C., Liu, M., Jiang, L., Liu, X., Zhao, J., Yan, S., Yang, S., Ren, H., Liu, R., Zhang, X., Transcriptome profiling reveals roles of meristem regulators and polarity genes during fruit trichome development in cucumber (*Cucumis sativus* L.). *J. Exp. Bot.*, 65, 4943–4958, 2014.
109. Zhao, W., Yang, X., Yu, H., Jiang, W., Sun, N., Liu, X. *et al.*, RNA-Seq-based transcriptome profiling of early nitrogen deficiency response in cucumber seedlings provides new insight into the putative nitrogen regulatory network. *Plant Cell Physiol.*, 56, 3, 455–467, 2015.
110. Wu, T., Qin, Z., Zhou, X., Feng, Z., Du, Y., Transcriptome profile analysis of floral sex determination in cucumber. *J. Plant Physiol.*, 167, 905–913, 2010.
111. Zhang, N., Zhang, H.J., Zhao, B., Sun, Q.Q., Cao, Y.Y., Li, R. *et al.*, The RNA-seq approach to discriminate gene expression profiles in response to melatonin on cucumber lateral root formation. *J. Pineal Res.*, 56, 39–50, 2014.
112. Dong, W., Wu, D., Li, G., Wu, D., Wang, Z., Next-generation sequencing from bulked segregant analysis identifies a dwarfism gene in watermelon. *Sci. Rep.*, 8, 1, 2908, 2018.
113. Rao, P.G., Behera, T.K., Gaikwad, A.B., Munshi, A.D., Srivastava, A., Boopalakrishnan, G., Genetic analysis and QTL mapping of yield and fruit traits in bitter gourd (Momordica charantia L.). *Sci. Rep.*, 11, 1, 1–10, 2021.
114. Roossinck, M.J., Martin, D.P., Roumagnac, P., Plant Virus Metagenomics: Advances in Virus Discovery. *Phytopathology*, 105, 6, 716–27, 2015 Jun.
115. de Nazaré Almeida dos Reis, L., Fonseca, M.E.D.N., Ribeiro, S.G., Naito, F.Y.B., Boiteux, L.S., Pereira-Carvalho, R.D.C., Metagenomics of neotropical single-stranded DNA viruses in tomato cultivars with and without the Ty-1 gene. *Viruses*, 12, 8, 819, 2020.
116. Kim, D., Hong, S., Kim, Y.T., Ryu, S., Kim, H.B., Lee, J.H., Metagenomic approach to identifying foodborne pathogens on Chinese cabbage. *Korean Soc. Microbiol. Biotechnol.*, 28, 2, 227–235, 2018.
117. Gawande, S.J., Anandhan, S., Ingle, A., Roylawar, P., Khandagale, K., Gawai, T., Singh, M., Microbiome profiling of the onion thrips, Thrips tabaci Lindeman (Thysanoptera: Thripidae). *PLoS One*, 14, 9, e0223281, 2019.
118. Starchevskaya, M., Kamanova, E., Vyatkin, Y., Tregubchak, T., Bauer, T., Bodnev, S., Antonets, D., The metagenomic analysis of viral diversity in Colorado potato beetle public NGS data. *Viruses*, 15, 2, 395, 2023.
119. Bannoud, F., Ellison, S., Paolinelli, M., Horejsi, T., Senalik, D., Fanzone, M., Iorizzo, M., Simon, P.W., Cavagnaro, P.F., Dissecting the genetic control of root and leaf tissue specific anthocyanin pigmentation in carrot (*Daucus carota* L.). *Theor. Appl. Genet.*, 132, 2485–2507, 2019.
120. Kumar, R., Sharma, M.K., Kapoor, S., Tyagi, A.K., Sharma, A.K., Transcriptome analysis of rin mutant fruit and *in silico* analysis of promoters of differentially regulated genes provides insight into LeMADS-RIN-regulated ethylene-dependent as well as ethylene-independent aspects of ripening in tomato. *Mol. Genet. Genomics*, 287, 189–203, 2012.

121. Zhou, X., Fei, Z., Thannhauser, T.W., Li, L., Transcriptome analysis of ectopic chloroplast development in green curd cauliflower (*Brassica oleracea* L. var. botrytis). *BMC Plant Biol.*, 11, 169, 2011.
122. Tao, X., Gu, Y.H., Wang, H.Y., Zheng, W., Li, X., Zhao, C.W., and Zhang, Y.Z., Digital gene expression analysis based on integrated *de novo* transcriptome assembly of sweet potato [*Ipomoea batatas* (L.) Lam]. *PLoS One*, 7, 4, e36234 2012.
123. Trick, M., Long, Y., Meng, J., Bancroft, I., Single nucleotide polymorphism (SNP) discovery in the polyploid Brassica napus using Solexa transcriptome sequencing. *Plant Biotechnol. J.*, 7, 4, 334–346, 2009.
124. Mammadov, J., Aggarwal, R., Buyyarapu, R., Kumpatla, S., SNP markers and their impact on plant breeding. *Int. J. Plant Genomics*, 1, 728398, 2012.
125. Blanca, J., Cañizares, J., Roig, C., Ziarsolo, P., Nuez, F., Picó, B., Transcriptome characterization and high throughput SSRs and SNPs discovery in *Cucurbita pepo* (Cucurbitaceae). *BMC Genomics*, 12, 104, 2011.
126. Nicolaï, M., Pisani, C., Bouchet, J.P., Vuylsteke, M., Palloix, A., Discovery of a large set of SNP and SSR genetic markers by high-throughput sequencing of pepper (*Capsicum annuum*). *Genet. Mol. Res.*, 11, 2295–2300, 2012.
127. Lee, S. and Choi, D., Comparative transcriptome analysis of pepper (*Capsicum annuum*) revealed common regulons in multiple stress conditions and hormone treatments. *Plant Cell Rep.*, 32, 1351–1359, 2013.
128. Kondrak, M., Marincs, F., Antal, F., Juhasz, Z., Banfalvi, Z., Effects of yeast trehalose-6-phosphate synthase 1 on gene expression and carbohydrate contents of potato leaves under drought stress conditions. *BMC Plant Biol.*, 12, 74, 2012.
129. Wang, Y., Xu, L., Chen, Y., Shen, H., Gong, Y., Limera, C., Liu, L., Transcriptome profiling of radish (*Raphanus sativus* L.) root and identification of genes involved in response to lead (Pb) stress with next generation sequencing. *PLoS One*, 8, 6, e66539, 2013.
130. Kaur, P. and Gaikwad, K., From Genomes to GENE-omes: Exome Sequencing Concept and Applications in Crop Improvement. *Front. Plant Sci.*, 8, 2164, 2017.
131. Mertes, F., Elsharawy, A., Sauer, S., van Helvoort, J.M., van der Zaag, P.J., Franke, A., Nilsson, M., Lehrach, H., Brookes, A.J., Targeted enrichment of genomic DNA regions for next-generation sequencing. *Brief Funct. Genomics*, 10, 6, 374–86, 2011.
132. Pootakham, W., Shearman, J.R., Ruang-Areerate, P., Sonthirod, C., Sangsrakru, D., Jomchai, N., Yoocha, T., Triwitayakorn, K., Tragoonrung, S., Tangphatsornruang, S., Large-scale SNP discovery through RNA sequencing and SNP genotyping by targeted enrichment sequencing in cassava (*Manihot esculenta* Crantz). *PLoS One*, 9, 12, e116028, 2014.
133. Jupe, F., Witek, K., Verweij, W., Sliwka, J., Pritchard, L., Etherington, G.J., Maclean, D., Cock, P.J., Leggett, R.M., Bryan, G.J., Cardle, L., Hein, I., Jones, J.D., Resistance gene enrichment sequencing (RenSeq) enables reannotation

of the NB-LRR gene family from sequenced plant genomes and rapid mapping of resistance loci in segregating populations. *Plant J.*, 76, 3, 530–44, 2013.

134. Uitdewilligen, J., Wolters, A., D'hoop, B.B., Borm, T.J.A., Visser, R.G.F., van Eck, H.J., A Next-Generation Sequencing Method for Genotyping-by-Sequencing of Highly Heterozygous Autotetraploid Potato. *PLoS One*, 8, 5, 62355, 2016.
135. Terracciano, I., Cantarella, C., D'Agostino, N., Hybridization-based enrichment and next generation sequencing to explore genetic diversity in plants, in: *Dynamics of Mathematical Models in Biology*, A. Rogato, V. Zazzu, M. Guarracino, (Eds.), Springer, Cham, 2016.
136. Sharma, S.K., McLean, K., Colgan, R.J., Rees, D., Young, S. *et al.*, Combining conventional QTL analysis and whole-exome capture-based bulk-segregant analysis provides new genetic insights into tuber sprout elongation and dormancy release in a diploid potato population. *Heredity*, 127, 3, 253–265, 2021.
137. Jouet, A., Saunders, D.G.O., McMullan, M., Ward, B., Furzer, O., Jupe, F., Cevik, V., Hein, I., Thilliez, G.J.A., Holub, E., van Oosterhout, C., Jones, J.D.G., *Albugo candida* race diversity, ploidy and host-associated microbes revealed using DNA sequence capture on diseased plants in the field. *New Phytol.*, 221, 3, 1529–1543, 2019.

2

Application of Bioinformatics Tools in Rice Genomics Research

Dhanawantari L. Singha[1†], Debajit Das[2†] and Jitendra Maharana[3*]

[1]*Department of Botany, Rabindranath Tagore University, Hojai, Assam, India*
[2]*DBT-North East Centre for Agricultural Biotechnology, Assam Agricultural University, Jorhat, Assam, India*
[3]*Distributed Information Centre (DIC), Department of Agricultural Biotechnology, Assam Agricultural University, Jorhat, India*

Abstract
Rice, a type of monocotyledon cereal grain, is a model species for genomic studies. The release of initial draft rice genome sequences, coupled with significant advancements in molecular biotechnology and genomics tools, has sparked a remarkable surge in rice research. The advancements in genomics technology, in particular, has resulted in a substantial augmentation of biological information related to rice. Efficient bioinformatics tools have facilitated the organization of this vast information into well-structured databases. The detailed sequence information now available has streamlined the application of molecular breeding, mutation breeding, and CRISPR/Cas-based editing of target genome in rice. This chapter emphasizes the implications of bioinformatics tools on rice functional genomics, including transcriptomics, mutation studies, and CRISPR/Cas-based genome editing, all aimed at enhancing the resilience of rice crops against both biotic and abiotic stresses. Additionaly, this chapter also highlights the existing limitations and current challenges in implementing bioinformatics applications in rice research.

***Keywords*:** Rice, genomics, bioinformatics, databases, CRISPR/Cas9 system

**Corresponding author*: jitued@gmail.com
†Authors contributed equally.

Hrudayanath Thatoi, Sonali Mohapatra, Swagat Kumar Das and Sukanta Kumar Pradhan (eds.) Applied Biotechnology and Bioinformatics: Agriculture, Pharmaceutical Research and Environment, (37–72) © 2025 Scrivener Publishing LLC

2.1 Introduction

Bioinformatics is a cross-disciplinary field of study within the biological sciences that merges biological science, computer science, and statistics to develop methodologies and tools for acquiring, manipulating, and interpreting a wide array of biological information. Its significance was underscored in 2002 when the first rice genomes were decoded, representing a major milestone in biological research. Within rice genomics, bioinformatics tools are indispensable, aiding in tasks like comparing gene sequences, assembling genomes, predicting elements, and conducting comparative genetic analyses across various rice species. Bioinformatics is critical to the area of rice genetics and breeding since it allows for the identification of genes linked to agronomic features, genotype-based phenotypic result prediction, and genetic impact research. Rice breeding might undergo a radical change in synthetic genomics and genome modification with the combination of bioinformatics and cutting-edge genome editing technology. Nearly two decades ago, scientists at The Institute for Genomic Research (TIGR) in the United States first used the phrase "rice bioinformatics" to describe their studies of sequencing data from rice, a model monocot species [1].

This chapter presents a comprehensive summary of the accomplishments in rice genome sequencing and assembly. Moreover, it highlights the remarkable strides made in the creation of databases that includes rice genomic information and computational methods for genome exploration and breeding research. The integration of bioinformatics with cutting-edge tools such as ability for single-cell sequencing and artificial intelligence has emerged as a promising avenue for advancing rice research. In the age of abundant big omics data, it is postulated that the significance of bioinformatics will be unparalleled in driving advancements in rice research and agricultural studies, while also enhancing the efficacy of rice breeding.

2.2 Role of Genomics in Rice Research

Genomics refers to the study of genomes, which encompasses the complete genetic composition of an organism, including the arrangement of genes and genetic information. This arrangement plays a pivotal role in determining the functionality of specific genes within an organism spanning across plants and animals. Every organism possesses a genome that contains the biological information development and sustenance of a viable member within its species. Hence, the comprehensive examination of all coding and non-coding sequences through mapping, sequencing, and functional analysis is regarded as genomic research across all species. The fundamental objectives of genomics can be catalogued into three overarching categories. Structural genomics focuses on studying the organization of genetic material within genomes and analyzing the physical properties of all proteins generated by DNA sequences. Functional genomics is a scientific field that explores the function and interaction of genes within their organizational context, including the temporal and spatial expression of genes and their effects. This field enables scientists to understand the specific behavior of genes under specific circumstances. By studying functional genomics, scientists can understand the specific behavior of a gene in a given circumstance. Proteomics, a subset of functional genomics, explores the functionality of an organism's proteins. Comparative genomics, on the other hand, examines the synteny linkages among genomes of closely related species to identify potential opportunities for gene exploration and exchange between these groups.

Genomic research has significantly transformed our understanding of plant systems. It focuses on studying the structure, function, and evolutionary history of genes [2]. Through complex calculations, genomic research transforms a vast pool of genetic information into a usable form through bioinformatics. The study of genomics as a discipline bridges diverse kingdoms, establishes connections across gaping genetic chasms, and opens up new avenues across domains that are specific to particular species. The intent of scientific analysis has switched to genomic research as a consequence of the depleting biophysical resource base and the resulting excessive demand on the already-limited natural resources for agriculture. Our understanding of genome information has diversified, and has greatly influenced crop improvement, particularly in the context of breeding. The application of genomics has revolutionized breeding techniques in every aspect. Instead of solely researching phenotypes and predicting genotypes, we now delve into the intricate details of genetic makeup to predict plant phenotypes accurately.

2.3 Model Plant for Genomic Research: Rice

The suitability of an organism for impactful genomic studies is contingent on a variety of parameters, including genome size (the amount of DNA that makes up the entire genome), the total chromosome count, the cumulative distance of recombination in the genome, the type and magnitude of repetitive DNA sequences, the number of active genes, and the feasibility of genetic modification through genetic transformation. Advancements in sequencing techniques over the past two decades have significantly propelled genomic and genetic research in rice. Several cutting-edge genomic technologies have facilitated the establishment and enhancement of cutting-edge genomic techniques and approaches, revolutionizing the field of rice research. For instance, the acquisition of a whole genome of rice-specific resources has become simpler and quicker thanks to modern long-read sequencing technology, scaffolding approaches, and recently developed assembly algorithms. The accessibility of full genomes made it possible to uncover genomic differences, particularly structural changes, which are notoriously challenging to define without whole genome sequences. The advent of innovative genomic insights, which profoundly altered the paradigm of rice research, was also made possible by these technical breakthroughs. As an illustration, the concept of a pan-genome, along with associated technique, has been established based on recent genomic research findings, supported by breakthrough sequencing technology and advanced algorithms. The pan-genome approach, which replaces a single or limited number of reference genomes with a broader catalog representing greater genetic variation within a species, has revolutionized the paradigm for genomic study [3]. These state-of-the-art sequencing and genomic approaches have transformed practically all aspects of fundamental and breeding research on rice: (1) Facilitating the assembly of multiple superior rice reference genomes can provide a cohesive groundwork for further genomic and genetic studies on rice. (2) Giving rice gene banks accessibility to technologies that are affordable and effective for retrieving the sequencing of accessions. (3) Offering tools for a comprehensive assessment of genetic resources and the molecular processes underpinning notable characteristics, as well as making it easier to acquire genome sequences. (4) Supporting innovative concepts and approaches for exploring and harnessing genetic differences in rice germplasms. (5) Making it possible to use cutting-edge methods to make use of genetic variations in standard breeding procedures and strengthen rice breeding's success and precision. In the coming years, we anticipate a substantial integration and widespread adoption of these sequencing and genomic technologies in rice genomic research and breeding programs. This will be driven by factors such as additional cost reductions, standardization of sequencing technologies, and a rapid advancement of genomic and bioinformatics. These advancements will facilitate a more comprehensive and efficient utilization of genomic tools, leading to enhanced precision and success in rice breeding and research. The evolution of genomic-assisted breeding in rice can be attributed to the remarkable progress achieved through the application of these technologies [4].

2.4 High-Throughput Sequencing

The next-generation sequencing (NGS) technologies were not yet developed when the Rice Genome Sequencing Project began. The rice reference genome, however, was substantially upgraded and refined following the development of NGS technology. Nearly 4000 oversights were discovered and rectified in the previous version of the Nipponbare rice genome, which was updated and confirmed in 2013 using data from two short-read sequencing technologies.

A high-quality reference genome encompasses both the assembly accuracy and the comprehensive annotation, which involves identifying the functional significance of specific regions within the genome. However, despite the availability of rapid advancement in the rice genome sequencing, the genome annotation has faced delays. Novel sequencing techniques and algorithms are now being utilized for rice genome annotation. For instance, an RNA-seq made possible by NGS technology offers an abundance of data regarding the expression of genes at the whole-genome level, which may be used to pinpoint the genome's transcriptional component. Sang *et al.* utilized an extensive dataset of RNA-seq data to improve the prediction of gene models in the Nipponbare genome. This effort led to a more coherent and higher-quality genome annotation compared to previous methodologies [5].

Rice research has achieved a significant milestone with the effective assembly of the high-quality reference genome. This accomplishment establishes a structured foundation for comprehensive genomic investigations and yielding valuable insights into rice genetics, evolution, and biology. The area of functional gene cloning has experienced the most expansion because of the accessibility of the reference rice genome and comparable resources. During the period from 1990 to 1994, around 20–30 research articles focused on rice gene cloning were published annually. However, with the release of the first rice BAC library in 1995, the number of publications increased ranging from 40 to 80 per year. Subsequently, with the whole genome became accessible, the annual publication rate surged to well over 100.

Over the past decade, resequencing techniques utilizing NGS have gained significant prominence as a valuable tool for identifying genomic variations within rice germplasms, especially notable given the existence of a comprehensive reference genome. The resequencing approach entails the alignment of NGS data to the reference genome and subsequently identifying variants by comparing the aligned data with the reference sequence. It should be emphasized that relying solely on this resequencing method may not be enough to accurately detect most structural variations (SVs) in rice germplasms. Furthermore, an increasing amount of evidence suggests that relying on a single reference genome is insufficient to capture the extensive diversity present in the rice gene pool. This restriction is a result of notable differences in the genetic makeup that may be specific to certain individuals and thus go unnoticed in the genomes of others. For example, when comparing the indica rice lines Zhenshan97 and Minghui63, researchers found significant differences in their genetic structures. These differences include inversions, translocations, presence/absence variations (PAVs), and segmental duplications [6]. Considering these observations, it is crucial to expand initiatives aimed at obtaining and understanding genetic variations, particularly SVs, from a broader selection of rice gene banks. Furthermore, it is crucial to generate reference genomes of superior quality that more accurately represent the wide range of genetic variation found in rice germplasms [7]. Thankfully, the progress made in sequencing technologies, such as Next-Generation and Third-Generation Sequencing, has greatly simplified the process of assembling genomes from different rice accessions maintained in gene banks.

Prior to the emergence of third-generation sequencing technologies, scientists embarked on ambitious endeavors to decode the genetic blueprint of rice using cutting-edge Next-Generation Sequencing (NGS) methods. Among these endeavors, Gao *et al.* undertook a resequencing initiative aimed at refining the genomes of PA64 and 93-11, two crucial parental strains integral to the development of the LYP9 super hybrid rice [8]. The advent of third-generation sequencing platforms marked a pivotal shift in rice genome research, ushering in a new era characterized by the capability to generate sequencing data with significantly elongated reads. Unlike the conventional NGS technologies, which typically churn out reads spanning around 150 base pairs, third-generation sequencing platforms boast the remarkable ability to produce and analyze reads exceeding 10 kilobases in length. This extended read length bears immense potential in overcoming the formidable challenges associated with genome assembly, particularly in navigating through the intricate landscape of repetitive regions, which can constitute up to 45% of the rice genome. Concurrently, a plethora of scaffolding techniques has emerged to facilitate the assembly of genomes into comprehensive chromosome-scale representations. Optical mapping, for instance, stands out as a notable method capable of amalgamating assembled contigs into cohesive super-scaffolds. By leveraging the creation of DNA sequence fingerprints and the construction of a consensus optical map, optical mapping holds promise in seamlessly merging fragmented genomic segments into coherent and contextually relevant structures [9].

The advancement of cutting-edge technologies has revolutionized our understanding of intricate genomic structures within rice genomes, streamlining the process of assembling rice genomes. Zhang *et al.* harnessed the power of third-generation sequencing technology, specifically PacBio sequencing, to sequence Bacterial Artificial Chromosomes (BACs), culminating in the assembly of genomes for two indica cultivars [6]. In 2017, Du and colleagues carried out an extensive examination of the genome of the indica rice variety Shuhui498. Their groundbreaking research combined third-generation sequencing with mapping data, a genetic map, and fosmid sequence tags, resulting in a nearly complete representation of the genome, covering more than 99% of the estimated genome size. This significant achievement in genome assembly became a crucial benchmark for identifying genes and structural changes in rice, as emphasized by reference [10]. Since this breakthrough, a plethora of rice genome assemblies have been unveiled at an unprecedented pace, with more than 100 newly generated rice genomes accessible through *de novo* assembly [11]. Since then, an incredible number of rice genome assemblies have been released, with *de novo* assembly making over 100 newly created rice genomes available [11].

The realm of genomic studies has witnessed a remarkable breakthrough with the advent of the Pan-Genome concept, particularly in the domain of rice genetics. This innovative approach has revolutionized our capacity to delve deep into the genetic makeup of rice, paving the way for the development of a comprehensive reference genome. Unlike traditional methods constrained by the utilization of a single reference genome, the Pan-Genome methodology embraces the diversity within rice germplasms by incorporating *de novo* genome assemblies from multiple accessions. Conventional analytical techniques such as resequencing analysis are inherently limited, as they rely on a singular reference genome, unable to capture the full spectrum of genetic variations across populations or species. Recognizing this limitation, the scientific community has embarked on a quest to identify optimal methodologies capable of encapsulating the myriad genetic variations present within rice populations or species. In response to these challenges, novel sequencing technologies have emerged alongside the adoption of the pan-genome approach as promising solutions. By leveraging these innovative tools, researchers aim to capture, depict, and elucidate the maximum breadth of genetic variations within rice populations or species. This transformative shift promises to propel genomic studies to unprecedented heights, offering invaluable insights into the intricate genetic landscape of rice.

The term "pan-genome" refers to the totality of genetic sequences existing within a biological group, such as a species or a particular subset of a population, rather than within an individual organism [12]. The original formulation of this concept was introduced by Tettelin *et al.* [13]. By scrutinizing bacterial genomes comparatively, the researchers unearthed a notable diversity of genes demonstrating fluctuations in their presence or absence among various bacterial strains. They defined the pan-genome as the collective ensemble of genes found across all bacterial strains, as cited in reference [13]. Following this pioneering work, the concept and its methodology were extended to encompass studies involving human, animal, and plant subjects [14, 15]. The pan-genome introduces a multitude of approaches for assessing genomic diversity within a given dataset and estimating the substantial number of additional whole-genome sequences required to thoroughly elucidate such diversity [16].

Presently, there are three well-established methods for developing a pan-genome [12]. One of these methods involves utilizing iterative construction or map-to-pan strategies. This method involves the process of aligning Next-Generation Sequencing (NGS) reads to a reference genome, followed by collecting the unaligned reads for *de novo* assembly. This method was used to compile the expendable genome of rice by analyzing low-coverage sequencing data from 1483 cultivated rice populations [17], leading to the generation of the initial rice pan-genome. The study successfully compiled a wide range of protein-coding genes, including several genes that are of great importance in agriculture and were not found in the Nipponbare reference genome. The pan-genome analysis provided valuable insights into the dispensable genome of rice and enhanced the reference genome by uncovering potential genes linked to phenotypic diversity. A comprehensive genome of rice was constructed using 3000 rice accessions, following a similar approach as described in previous studies [18, 19].

Another approach involves the identification and gathering of genomic variations among distinct accessions through comprehensive analysis of the entire genome. The accessibility of several *de novo* assembled genomes is an essential for implementing this strategy. The use of *de novo* assembled genomes significantly enhances the quality of the pan-genome by improving the depth and a pattern of genetic variations. In the year 2018, Zhao and colleagues implemented the approach and developed a pan-genome for rice through extensive sequencing and *de novo* assembly of 66 distinct rice accessions [20]. Developments in sequencing technologies, specifically the long-read sequencing technologies of the third generation, have facilitated the process of high-quality *de novo* assembly. Consequently, this has resulted in an increased preference for utilizing this strategy for constructing pan-genomes [21].

A novel approach involving bi-directed variation graphs (VGs) has been established to construct a graph-based pan-genome, as a third strategy [22]. The methodology employed by VG involves the use of sequencing reads that are corroborated by linear genome references, which results in the generation of a graph illustration of the variable regions. This approach effectively circumvents the issue of assembly bias that may arise due to reliance on a reference genome [23, 24]. Qin and co-workers employed this methodology to perform *de novo* assembly of 31 genomes of high quality, which were representative of the genetic diversity present in a population of rice. They subsequently developed a pan-genome of rice based on a graph-based approach. The study employed the pan-genome to look into the distribution of structural variations. Furthermore, an assessment of the mechanisms underlying the formation of structural variations and their impact on gene expression was carried out [25]. The creation of rice pan-genomes has enhanced comprehension of genomic diversity in rice while offering a more insightful framework for scientists to acquire significantly more data than relying on a single or limited reference genomes. In addition, they provide significant assistance for analyses that rely on sequencing, like RNA sequencing or identification of structural variations.

2.5 Genome-Wide Association Study (GWAS)

The advent of ultra-high throughput genotyping technologies has heralded a new era in agricultural research, empowering scientists with a potent tool known as Genome-Wide Association Study (GWAS). This innovative approach offers a window into the intricate world of quantitative traits in plants, particularly in vital crops like rice and other cereals [26]. Through GWAS, researchers can delve deeper into understanding the intricate mechanisms of transcriptional regulation and metabolic responses within these crops under various environmental conditions, including challenging abiotic stresses [27]. Despite the requisite establishment of a mapping population, the application of genome-wide association mapping unveils crucial regions harboring Quantitative Trait Loci (QTLs), thus shedding light on pivotal genetic determinants.

The usage of GWAS marks a pioneering stride in the realm of rice genetics, offering a sophisticated avenue for delving into the intricate genetic architecture governing a plethora of traits across diverse rice varieties. By scrutinizing the rice genome for sequence polymorphisms and correlating them with the observed phenotypic variations, geneticists can pinpoint the precise loci and, in some instances, the very genes responsible for these traits. What sets GWAS apart from conventional biparental populations is twofold. Firstly, the pool of rice varieties enlisted in GWAS harbors a remarkable spectrum of genetic diversity, far surpassing the limited genetic scope inherent in typical segregation populations with only two parent lines. Secondly, the richness of historical recombination events within GWAS populations often translates to a higher mapping resolution, offering finer insights into the genetic architecture of traits [28]. The advent of second-generation sequencing technologies, notably the widespread adoption of platforms like Illumina sequencing [29, 30], coupled with the development of robust statistical models tailored for plant populations, particularly the mixed model approach [31–33], has catalyzed the integration of GWAS into rice genetic research in recent years. Since its inception, GWAS has emerged as a powerful tool, enabling the genetic dissection of numerous complex traits [2, 34]. Despite lingering challenges, the past decade has witnessed a surge in GWAS endeavors within the rice research community, yielding a wealth of associated loci and advancing our understanding of rice genetics.

Rice stands out as a prime candidate for Genome-Wide Association Studies (GWAS) owing to its remarkable array of genetic resources. With an unparalleled genetic diversity surpassing that of many other plant species, rice provides an ideal platform for identifying functional genes through GWAS. Its distinctive characteristics, including self-pollination and inbreeding, further amplify the efficacy of this method. One notable advantage lies in the ability to conduct genotyping just once on a population, thus facilitating the perpetual maintenance of inbred lines within seed banks. These lines serve as invaluable resources for ongoing and future research initiatives, allowing for the evaluation of various traits across diverse environments [35].

2.6 Bioinformatics Approach to Study Stress Conditions in Rice

Bioinformatics stands as a multidisciplinary domain utilizing computational methods to document and explore the functionalities of various genes. With the advent of genomics, biological science across all realms has undergone a profound transformation. The pressing necessity for enhanced methodologies to analyze, interpret, and manage scientific data has become evident due to the sheer volume of available data [36, 37]. A plethora of databases and repositories have been established to cater to a wide array of datasets, encompassing the genome sequences, gene and amino acid sequences, gene expression profiles, coexpression patterns, as well as genomic variations such as SNPs and InDels [38]. The evolution of sequencing methods, resulting in increased efficiency and affordability. Following the initial publication of the whole-genome sequence of rice in 2005, numerous varieties were subsequently sequenced as part of the 3000 Rice Genomes Project [39, 40]. The project's data have facilitated the creation of rice variation databases, which have yielded significant

Table 2.1 Widely used bioinformatics databases and resources for rice genomics research.

Databases	Description	References
TIGR	Genome browser	[43]
IRRI	Rice germplasm	
RiceGE	Rice mutant database	
RPAN	Rice pan-genome database	[18]
RiceNavi	Molecular breeding	[44]
PlantCARE	Promoter analysis	[45]
KEGG_rice	Metabolite analysis	[46]
Gramene	Comparative genomics	[47]
RiceXPro	Expression profile	[48]
TENOR	Expression profile	[49]
RiceFREND	Coexpression analysis	[50]
ATTED-II	Coexpression analysis	[51]
RiceVarMap V2.0	GWAS	[52]
Rice SNP-Seek	GWAS	[53]
IC4R	Comprehensive database	[5]
CRISPR-P	CRISPR/Cas9 tool	[54]

understanding into the evolutionary process and domestication of rice [41]. Moreover, easily accessible data serve as a valuable resource, aiding breeders in making informed decisions pertaining to cultivar selection and marker utilization for breeding various traits across different cultivars. The techniques employed for scoring gene expression have undergone significant changes over time. Earlier methods such as serial analysis of gene expression (SAGE) have been replaced by more advanced techniques like microarrays and RNA-sequencing, as highlighted by Perez-de-Castro *et al.* [42]. Besides genome repositories, databases documenting gene and protein expression patterns play a pivotal role in unraveling the intricate mechanisms governing specific traits, encompassing phenomena like flowering time, developmental stages, resilience to environmental stressors, and resistance to diseases. Table 2.1 shows the available bioinformatics databases and resources for rice genomics research.

2.6.1 Rice Stress Responsive Transcription Factor Database (RiceSRTFDB)

Through the power of genetic engineering, scientists can manipulate specific genes involved in transcriptional circuits, paving the way for the creation of transgenic organisms with enhanced stress tolerance. This approach not only sheds light on the intricate mechanisms governing stress responses but also holds promise for developing resilient crops capable of withstanding various environmental challenges. Within this realm of genetic manipulation, a plethora of stress-related genes and transcription factors (TFs) have been unearthed, offering valuable insights into how organisms react to abiotic stressors. These TFs, acting as pivotal regulators, orchestrate the expression of a myriad of target genes [55–58], thereby influencing the organism's ability to cope with adverse conditions. However, despite significant progress, the full spectrum of rice TFs' roles during abiotic stress remains elusive, necessitating further exploration. To bridge this knowledge gap, initiatives such as the Stress-Responsive Transcription Factor Database (STIFDB) have emerged, furnishing researchers with a wealth of information on Arabidopsis genes and potential TFs induced in response to abiotic stress [59, 60]. Similarly, the Rice Stress Responsive Transcription Factor Database (RiceSRTFDB) serves as a comprehensive repository, offering insights into rice TFs' characteristics, including their cis-regulatory elements and expression patterns across diverse stress conditions, tissues, and developmental stages. This database not only obviates the need for laborious updates to genomic information but also facilitates the prioritization of TFs for further functional studies. Coupled with cutting-edge technological tools, researchers can delve deeper into understanding the nuances of each TF's action and response to abiotic stress events. As new gene expression data continually enrich our understanding, the database undergoes regular updates to ensure its relevance and comprehensiveness. Moreover, as the database continues to expand, it holds the promise of encompassing a broader spectrum of rice TFs, accompanied by detailed analyses of their expression patterns across various cultivars and genetic mutants. Such advancements promise to propel our understanding of stress responses in rice and pave the way for the development of more resilient crop varieties tailored to withstand the challenges of a changing environment.

2.6.2 RiceMetaSys

The RiceMetaSys platform is a readily accessible and freely available web-based interface that provides extensive information on genes present in rice genotypes that exhibit susceptibility to salt and drought stress. The acquisition of these data is facilitated by diverse microarray methodologies. The platform has been meticulously crafted to ensure a user-friendly experience while adhering to the principles of open-source software. The execution of a "Physical position search" involves employing a QTL-derived methodology to acquire discernment into the functioning of genes that respond to the effects of salt and drought stress, and their corresponding functions in augmenting resilience to these stress-inducing factors. This particular tool has the potential to be utilized in the identification of genes that are efficacious and may potentially facilitate advancements in crop enhancement. The incorporation of SSR analysis into the fine-mapping analysis and marker-assisted selection (MAS) of genes related to salt and drought tolerance is predicated on its capacity to furnish data pertaining to primers for polymorphism surveys. This particular tool not only enables the creation of an extrinsic connection to markers that are exclusive to introns, but also allows for the gathering of extensive data without any limitations. The application of SSR and ISM-ILP markers, which are accessible through the RiceMetaSys Database [61], can effectively aid molecular biologists and breeders in the implementation of fine-mapping methodologies. This has the potential to confer significant benefits to their respective domains. During the forthcoming period, plans are in place to upgrade the database to include supplementary attributes, such as increased tolerance to elevated temperatures and enhanced resistance to rice blast disease.

Furthermore, the integration of RNAseq data will serve to provide additional clarification regarding these characteristics.

2.6.3 Rice Oligonucleotide Array Database (ROAD 2.0)

Established in 2012, the Rice Oligonucleotide Array Database (ROAD) was designed to facilitate the monitoring of gene expression across different spatial and temporal contexts, providing valuable insights into specific gene functions. While microarray-based studies have been instrumental in discovering new genes, ROAD has faced limitations in accessing comprehensive metadata. Its primary aim is to conduct meta-analyses on microarray data from various anatomical specimens and developmental stages. Recognizing the shortcomings of the original ROAD, researchers have developed an upgraded version, ROAD 2.0, which includes a broader range of profiles covering biotic and abiotic stresses as well as hormone-induced responses. Additionally, RNAseq datasets have been integrated into the new database, enhancing our understanding of gene expression in rice tissues under different stress conditions and developmental stages [62].

2.6.4 Plant Stress Protein Database (PSPDB)

Plants exhibit a wide range of responses to various environmental stressors, both biotic and abiotic, often involving the production of specific stress proteins. One invaluable resource in this realm is the Plant Stress Protein Database (PSPDB), meticulously curated to encompass 2064 stress proteins derived from 134 plant species. These proteins, originating from diverse sources, have been subjected to over 30 different stress stimuli. Notably, PSPDB prioritizes the inclusion of proteins with well-established functional roles and confirmed experimental associations with both biotic and abiotic stress conditions. Accessible online, PSPDB stands as a comprehensive repository facilitating research into plant stress responses [63].

2.6.5 Plant Stress Gene Database (PSGD)

The Plant Stress Gene Database (PSGD) stands as a comprehensive reservoir of genetic insights, encompassing a rich compendium of 259 genes spanning 11 distinct plant species, among them the esteemed *Oryza sativa*. These genetic elements have garnered recognition for their pivotal roles in diverse responses to both biotic and abiotic stressors. Delving into the intricacies of each gene, the website furnishes elaborate details encompassing species delineation, gene identification codes, functional annotations, and the discernment of paralogous or orthologous counterparts across species (http://ccbb.jnu.ac.in/stressgenes/).

2.6.6 Drought Stress Gene Database (Drought DB)

The Drought DB serves as one of the significant resource and information tool to identify and analyze the genes that are accountable for drought stress tolerance in rice. Through meticulous curation, it serves as an indispensable tool for identifying and scrutinizing these genetic elements crucial for the development of drought-resistant crop varieties. Central to its functionality is the meticulous identification of orthologs across various species, a process fundamental in enhancing our understanding of drought tolerance mechanisms. Each ortholog is intricately linked to its corresponding reference entry in the MIPS/PGSB Plants DB, thereby facilitating access to supplementary information encompassing genome context and sequence data. Such integration of resources underscores the manual compilation process undertaken to assemble genes responsive to drought stress, ultimately contributing to the advancement of agricultural resilience [64].

2.7 Application of Bioinformatics Tools in Advanced Rice Genomics Research

In the year 2002, the rice genome was first uncovered by making use of diverse bioinformatics algorithms and tools. Subsequently, the utilization of sequence information has facilitated the identification of genes accountable for the coveted agronomic characteristics. The utilization of advanced bioinformatics tools has facilitated the estimation of the phenotypic manifestation that corresponds to the genotype data. The amalgamation of these tools with state-of-the-art techniques like genome editing yields remarkable outcomes in the field of rice crop improvement.

2.7.1 CRISPR/Cas-Based Study in Rice

In recent years, there has been a remarkable surge in the advancement of crop improvement through the utilization of cutting-edge genetic techniques known as Clustered Regularly Interspaced Short Palindromic Repeats (CRISPR) and CRISPR Associated protein nucleases (Cas). These techniques, along with their derivatives, have evolved significantly from their traditional counterparts. Field experiments for various crops, such as rice, have been greenlit thanks to the revolutionary CRISPR technology, which allows precise gene editing at specific locations in the genome without the need for introducing foreign genes. CRISPR/Cas has outshined other genome editing tools like zinc finger nucleases (ZFNs), meganucleases, and transcription activator-like effector nucleases (TALENs), showcasing its unparalleled effectiveness and efficiency in editing crop genomes, including rice [65–68]. Consequently, it has emerged as an incredibly promising and potent genome editing system. Furthermore, the rapid advancements in CRISPR derivatives make it even more enticing for future applications. The CRISPR RNA, in tandem with the Cas protein, demonstrates an astonishing ability to pinpoint the desired target region within the genome and induce double-strand breaks (DSBs). Subsequent to cleaving the double helix, the cell's innate repair mechanism kicks in. This repair process can occur through either non-homologous end joining (NHEJ) or homology-directed repair (HDR) techniques, as elucidated by Jinek *et al.* [69] and Zetsche *et al.* [70]. The repair process results in highly specific and targeted alterations within the genome, brought about by mutations such as deletions, insertions, and substitutions.

2.7.2 The Application of Derivatives of CRISPR/Cas Tools in Rice

The conventional CRISPR/Cas tools are subject to significant limitations due to their propensity to generate DSBs [71]. According to Zhang *et al.* [72], the repair mechanism of double-strand breaks (DSBs) is associated with a higher likelihood of off-target effects. The occurrence of off-target effects has the potential to cause lethality in the plant. Additionally, the process of repairing through HDR necessitates the availability of a donor template [73]. Cas9 is a widely used nuclease enzyme with a larger protein size, while other derivatives of Cas protein, like Cas12a (formerly Cpf1), have a limited PAM size. In order to surmount these constraints, contemporary technologies, namely, base editing and prime editing, have been conceived. Both of these tools have the capability to introduce precise mutations without generating DSBs. Base editing has the potential to introduce precise modifications to individual nucleotides within a specific region of the rice genome. The Adenosine/Cytosine Base Editor (A/CBE) is composed of a CRISPR/Cas9 domain that is inactive, as well as an adenosine or cytosine deaminase domain. Prime editing, alternatively referred to as the search and paste gene manipulation technique, employs a lengthier guide RNA

Table 2.2 A list of application of advanced CRISPR/Cas tools in rice.

Gene name	Specifications	Tool used	Reference
OsCDC48	Regulates senescence and cell death; editing resulted substitution of mutation of 43.48% with no off targets and no indels.	CBE	[74]
OsALS	Conferred herbicide imazamox resistance.	CBE	[75]
OsSPL14	Responsible for ideal plant architecture, mutation resulted in enhanced rice plant architecture and creates 59.1% mutation frequency.	ABE	[76]
OsWx	Amylose synthesis gene, mutation creates low amylose synthesis in rice endosperm.	ABE	[77]
GL2/OsGRF4 and *OsGRF3*	Responsible for determining grain size and yield, mutations lead to increase in grain weight and higher yield.	ABE	[78]
OsMPK6	The construct targeted the pathogen responsive phosphorylation site.	rBE14 system, modified ABE	[79, 80]
OsSPL14	Responsible for ideal plant architecture in rice.	pRABEsp-OsU6, modified ABE	[76]
OsACC-T1, OsALS-T1, OsCDC48-T3, OsDEP1-T1, OsDEP1-T2, and *OsNRT1.1B-T1*	Targeted six rice genomic loci; substitution efficiencies from 15.8 to 59.1%.	PABE-7 system, modified ABE	[81]

termed as prime editing guide RNA (pegRNA) in conjunction with Cas9 H840 nickase that is fused with reverse transcriptase (RT) enzyme. The prime editing technique surpasses the need for template DNA, which is typically associated with conventional CRISPR/Cas.

2.7.3 Base Editing and Prime Editing in Rice

The implementation of CRISPR/Cas tools has yielded favorable outcomes in the rice crop with respect to targeted traits associated with biotic and abiotic stresses. Within the realm of biotic stresses affecting rice, laboratory-based efforts have yielded cultivars that exhibit resistance to both disease and herbicides. This was achieved through targeted manipulation of various negative regulatory genes. Numerous trials have been efficaciously carried out to improve abiotic stress resilience in rice, encompassing drought, cold, salinity, herbicide resistance, and heavy metal stress, via the CRISPR/Cas mechanism. A comprehensive overview of the diverse applications of advanced CRISPR/Cas tools in the context of rice is presented in Table 2.2.

2.7.4 Computational Tools for CRISPR/Cas Based Study in Rice

The significance of bioinformatics tools and algorithms in advancing CRISPR/Cas technology cannot be overstated. The efficacy of genome editing, with the aim of minimizing off-target effects, heavily relies on the meticulous selection of target region and the design of an appropriate guide RNA, facilitated by a suitable gRNA design tool. Several gRNA design tools

Table 2.3 Popular CRISPR guide RNA designing tools in rice.

Tool name	Organism	Reference
CRISPR-Cereal	Rice, wheat, maize	[82]
E-CRISP	Rice, other plants	[83]
CRISP direct	Rice, other plants	[84]
CCTop	Rice, other plants, vertebrates, and invertebrates	[85]
CHOPCHOP	Rice, other plants, vertebrates, and invertebrates	[86]
Benchling	Rice, other plants	[87]

are available to facilitate efficient targeted gene knockout. Herein, Table 2.3 presents a selection of commonly employed gRNA design tools.

2.7.5 Advanced Machine Learning Approaches for CRISPR/Cas

Zhu *et al.* [88] have introduced an ensemble predictor (Pre), PreAcrs, utilizing machine learning techniques for the identification of anti-CRISPR proteins (Acrs). Bacteriophages and mobile genetic elements are responsible for synthesizing anti-CRISPR proteins (ACRs) with the aim of inhibiting the bacterial CRISPR/Cas immune system. Identification of such Acrs has the potential to unveil novel avenues for regulating the CRISPR/Cas machinery, leading to the advancement of enhanced CRISPR/Cas tools. Several additional servers employed for the prediction and illustration of Acrs include PaCRISPR [89], DeepAcr [90], and AcRanker [91]. The PreAcrs, PaCRISPR, and DeepAcr algorithms have demonstrated successful predictions of various Acrs that exhibit effectiveness across various CRISPR/Cas systems. For a comprehensive reference, Table 2.4 presents an extensive compilation of servers utilized for the identification of Acr proteins, along with their respective specifications.

Table 2.4 The servers used to identify the Acr proteins and their specifications.

Server name	**Identified Acrs**	**Specifications**	**Reference**
PreAcrs	AcrIIA7-980	Using the PreAcrs algorithm, it gave the highest score than PaCRISPR and AcRanker	[88]
PaCRISPR	AcrIIA16, AcrIIA17, AcrIIA18, AcrIIA19	Class II-A CRISPR–Cas systems	[89]
DeepAcr	AcrVIB1	Cas13b	[90]
AcRanker	AcrllA20 (ML1) and AcrIIA21 (ML8)	Inhibit Cas9 (SinCas9 and SpyCas9)	[90]

2.7.6 Bioinformatics Approach for CRISPR/Cas Off Target Detection

Various studies have reported that the CRISPR/Cas mechanism can lead to unintended off-target effects within the genome [92]. These off-target effects can have significant implications, including lethality induction or hindered organismal growth, through the initiation of extensive gene deletions and genomic rearrangements in crucial genes [93, 94]. Consequently, it is imperative to conduct thorough off-target assessments during the selection of gRNA. Several bioinformatics tools are accessible for identifying and evaluating off target sites (Table 2.5).

2.7.7 Detection of Direct Off Target Mutation Detection Through NGS Analysis

Genome sequencing serves as a sensitive and viable method for detecting off-target sites in genomes edited using CRISPR/Cas. Zhang *et al.* performed whole-genome sequencing (WGS) to examine potential off-target effects in the CRISPR/Cas9-edited rice. The WGS process involves creating a library using the total DNA of the samples, followed by high-throughput

Table 2.5 List of *in silico* tools available to find most probable offtarget sites in rice genome.

Tool used	Specification	References
Benchling	Off target scores are calculated based on the algorithm developed by Hsu *et al.* 2013	[95]
CHOPCHOP	Bowtie	[96]
Cas-OFFinder	-	[97]
CRISPR-Cereal	FlashFry: enables a genome-wide scan for off-targets	[98]
GT-Scan	Bowtie	[96]
E-CRISP	Bowtie2	[99]
Crisflash	FlashFry, scans a genomic FASTA file for PAM-adjacent protospacers, adding identified gRNAs to an N-ary tree.	[100]
CCTop	Bowtie	[96]

sequencing technologies, such as Illumina, for comprehensive sequencing. The sequenced genome was then analyzed using the BLASTN algorithm to search for the off-target sites within the 20-bp target region [101]. Single-nucleotide polymorphisms (SNPs) or insertions/deletions (indels) were eventually regarded as potential off-targets. WGS was also employed by Tang *et al.* [102] to study Cas12a (Cpf1)-edited rice plants, leveraging the Illumina platform. While various advanced sequencing-based platforms and bioinformatics algorithms have been developed, accurately identifying off-target sites without false negatives remains a challenge due to errors associated with PCR and next-generation sequencing (NGS). Nevertheless, minimizing background noise and false negatives can be achieved by implementing negative controls, increasing the number of replicates, and employing rigorous sequencing protocols.

2.8 Current Challenges of Bioinformatics Tools for Rice Genomics Research

2.8.1 Estimation of Heterosis

The reduction in sequencing expenses, the widespread availability of bioinformatics tools, and growing numbers of superior/advanced hybrid rice cultivars offer substantial opportunities for investigating the genomic structure of rice heterosis. Nevertheless, uncovering the genetic basis underlying rice heterosis via genetic mapping may encounter challenges due to non-random recombination events and insufficient population sizes. To address this, it is crucial to increase the number of populations and enhance their sizes for a comprehensive understanding of the genetic basis of heterosis in rice hybrids. Furthermore, there exists considerable variation in the significance of epistasis in heterosis among reports on hQTLs. Properly evaluating the impact of epistasis on heterosis requires the utilization of substantial population sizes.

The detection of a majority of hQTLs can be accomplished through the implementation of GWAS or QTL mapping techniques on various populations [103, 104]. Ideally speaking, the identification of a greater number of hQTLs is expected in populations of indica-japonica hybrid crosses, when population size and number are taken into consideration. Assessing hQTL effects presents a favorable occasion to quantify heterosis associated with yield. Advancements in statistical computations and prediction models within the field of statistical genetics are essential for enhancing the measurement of heterosis, as well as for improving genome-enabled prediction and genomic selection (GS) in rice breeding [105, 106].

2.8.2 *De Novo* Domestication of Rice

Rice domestication process presents a paradigmatic framework for investigating adaptation and holds significant intrigue for enhancing crop amelioration [107]. Rice domestication has led to the improvement of its yield-related characteristics, including but not limited to high yield, ideotype, and ease of harvest. Conversely, it has engendered a genetic bottleneck that has culminated in a diminution of genetic heterogeneity and stress resilience. Furthermore, it is noteworthy that while maize domestication often exhibited standing regulatory variation resulting in a gain of function, rice domestication was characterized by *de novo* coding variation leading to a loss of function [107]. Therefore, it can be inferred that the utilization of genome modification for the *de novo* domestication of wild rice holds significant potential as a viable method for enhancing the quality of rice cultivars. An exemplary instance can be mentioned in the *de novo* domestication of wild allotetraploid *Oryza alta* (CCDD), wherein the manipulation of genes facilitated rapid enhancement of six crucial agronomical attributes [108]. Given the conditions, the establishment of a proficient transgenic and genome editing mechanism, alongside an excellent genome assembly, will prove to be indispensable for the purpose of initiating *de novo* domestication.

2.8.3 Using Genome Editing to Revolutionize Rice Breeding

The groundbreaking technique of genome editing has opened new doors in the realm of rice breeding, allowing for precise and targeted modifications to the genetic makeup of this staple crop. This innovation has sparked a revolution, transforming traditional breeding methods and offering unprecedented opportunities for enhancing rice varieties. By combining traditional crossbreeding with advanced genome manipulation techniques, researchers have unlocked the potential to revolutionize rice breeding. Studies suggest that inter-(sub)specific hybrid rice, achieved through genome editing, could boast a 15%–30% increase in productivity compared to conventional intraspecific hybrid indica rice. However, one challenge hindering the effective utilization of distant heterosis is hybrid sterility (HS). Recent research by Xie *et al.* demonstrated successful generation of hybrid-compatible lines (*indica-japonica*) in rice by eliminating *SaF* and *SaM* alleles in indica rice [109]. Similarly, in African rice, the elimination of *S1* alleles has shown promise in mitigating hybrid sterility [110]. Moreover, genome editing techniques have been employed to facilitate clonal reproduction, a critical aspect for maintaining hybrids through clonal means [111, 112]. The use of CRISPR-mediated base editors has shown tremendous potential in expediting targeted evolution of plant genes, offering new avenues for rice cultivation. In the pursuit of enhancing crop quality, researchers have developed innovative techniques such as saturated targeted endogenous mutagenesis editors (STEMEs) and base-editing-mediated gene evolution (BEMGE). These approaches aim to introduce significant genetic variations in target genes, as demonstrated by the creation of novel mutations in *OsALS1* and *OsACC* genes, leading to varying degrees of herbicide resistance [113, 114]. Furthermore, the integration of genome editing with rapid cultivation methods holds promise in addressing the challenge of feeding the growing global population [115]. Thus, the convergence of genome editing with cutting-edge technologies and recent discoveries in rice genomics is essential for accelerating the breeding process and meeting the demands of an ever-expanding human population.

2.8.4 Uncovering the Wide Range of Diversity in Rice Varieties

The significance of advancing rice breeding lies in the rich tapestry of allelic variations associated with pivotal agronomic traits. To date, approximately 20 genomes of *Oryza* species, alongside numerous cultivated varieties, have undergone comprehensive sequencing efforts. However, it is noteworthy that the bulk of genomic data stems from the *indica* and *japonica* subspecies, inadvertently neglecting crucial variations found in alternative *Oryza* species and distinct subgroups of *O. sativa*. One noteworthy example is the haplotype *OsTCP19-H*, correlated with heightened tillering response to nitrogen and enhanced nitrogen use efficiency (NUE), predominantly observed in aus and aromatic rice varieties. Conversely, the *OsTCP19-L* haplotype, associated with diminished NUE, prevails in japonica and indica rice strains [116]. Moreover, *Oryza glaberrima*, endowed with genes conferring resilience to both biotic and abiotic stresses, emerges as a promising candidate for agricultural pursuits in the western reaches of Africa. Recent investigations by Li *et al.* [117] and Kan *et al.* [118] pinpoint *Thermo-tolerance 1* (*TT1*) and *Thermotolerance 2* (TT2) as pivotal genetic determinants contributing to rice's ability to withstand heat stress. Intriguingly, these quantitative trait loci (QTLs) trace their origins back to African rice, rather than *O. sativa*. To ensure a comprehensive understanding of rice genetic diversity, it becomes imperative to pivot towards exploring variations present in other *Oryza* species and diverse subgroups of *O. sativa*. This necessitates fortifying the integration of all-encompassing pan-genomic studies, encompassing a broader spectrum of varieties from disparate *O. sativa* subgroups or their ancestral progenitors. In tandem, bolstering the annotation precision of the rice pan-genome through robust algorithmic methodologies and expanded data integration stands as a crucial step forward.

2.9 Conclusion

The use of bioinformatics in the field of plant biotechnology signifies a fundamental paradigm shift in the approach of scientific inquiry concerning living entities. Bioinformatics holds considerable importance in advancing the agricultural industry by enabling the analysis of stress tolerance and plant pathogens, which are crucial factors in the advancement of plant breeding process. The utilization of NGS in conjunction with various sequencing techniques will facilitate the accessibility of additional plant genome information in all public domains. This will additionally allow the detection of genetic variations and the prediction of protein structure and function [119]. Moreover, GWAS have facilitated the detection of loci and allelic variations associated with desirable traits, thereby facilitating crop alteration and improvement [120]. In summary, the incorporation of bioinformatics in the fields of plant biotechnology has empowered scientists to gain a comprehensive and structured understanding of economically significant plants. Notwithstanding the remarkable progress made in plant biotechnology through the utilization of bioinformatics, the attainment of cost-effective, fully automated genome sequencing and assembly remains a formidable challenge. The exigency for proficient bioinformatic tools that can furnish longer reads with impartial coverage is of vital importance to surmount the intricacy of the plant genome. To address this challenge, the development of advanced algorithms that facilitate data extraction, analysis, and comparison is imperative. Individuals proficient in bioinformatics, mathematics, and programming will assume a crucial responsibility in introducing innovative methodologies and insights into this field. This advancement will not only drive progress in plant biotechnology and agriculture but also hold great significance for the future of humanity.

Conflict of Interest

The authors declare that they have no conflict of interest.

References

1. Yuan, Q., Quackenbush, J., Sultana, R., Pertea, M., Salzberg, S.L., Buell, C.R., Rice Bioinformatics. Analysis of Rice Sequence Data and Leveraging the Data to Other Plant Species. *Plant Physiol.*, 125, 3, 1166–74, 2001, doi: 10.1104/pp.125.3.1166.
2. McCouch, S.R., Wright, M.H., Tung, C.W., Maron, L.G., McNally, K.L., Fitzgerald, M. *et al.*, Open access resources for genome-wide association mapping in rice. *Nat. Commun.*, 7, 10532, 2016, Epub 20160204, doi: 10.1038/ncomms10532, PubMed PMID: 26842267, PubMed Central PMCID: PMCPMC4742900.
3. Bayer, P.E., Golicz, A.A., Scheben, A., Batley, J., Edwards, D., Plant pan-genomes are the new reference. *Nat. Plants*, 6, 8, 914–20, 2020, doi: 10.1038/s41477-020-0733-0, PubMed PMID: WOS:000550646600001.
4. Varshney, R.K., Bohra, A., Yu, J.M., Graner, A., Zhang, Q.F., Sorrells, M.E., Feature Review Designing Future Crops: Genomics-Assisted Breeding Comes of Age. *Trends Plant Sci.*, 26, 6, 631–49, 2021, doi: 10.1016/j.tplants.2021.03.010, PubMed PMID: WOS:000651146800011.
5. Sang, J., Zou, D., Wang, Z., Wang, F., Zhang, Y., Xia, L. *et al.*, IC4R-2.0: Rice Genome Reannotation Using Massive RNA-seq Data. *Genom. Proteom. Bioinf.*, 18, 2, 161–72, 2020, Epub 20200716, doi: 10.1016/j.gpb.2018.12.011, PubMed PMID: 32683045, PubMed Central PMCID: PMCPMC7646092.
6. Zhang, J.W., Chen, L.L., Xing, F., Kudrna, D.A., Yao, W., Copetti, D. *et al.*, Extensive sequence divergence between the reference genomes of two elite indica rice varieties Zhenshan 97 and Minghui 63. *Proc. Natl. Acad. Sci. U. S. A.*, 113, 35, E5163–E71, 2016, doi: 10.1073/pnas.1611012113, PubMed PMID: WOS:000383090700012.
7. Zhou, Y., Chebotarov, D., Kudrna, D., Llaca, V., Lee, S., Rajasekar, S. *et al.*, A platinum standard pan-genome resource that represents the population structure of Asian rice. *Sci. Data*, 7, 113, 2020, doi: 10.1038/s41597-020-0438-2, PubMed PMID: WOS:000525726600001.
8. Gao, Z.Y., Zhao, S.C., He, W.M., Guo, L.B., Peng, Y.L., Wang, J.J. *et al.*, Dissecting yield-associated loci in super hybrid rice by resequencing recombinant inbred lines and improving parental genome sequences. *Proc. Natl. Acad. Sci. U. S. A.*, 110, 35, 14492–7, 2013, doi: 10.1073/pnas.1306579110. PubMed PMID: WOS:000323564600082.
9. Shelton, J.M., Coleman, M.C., Hemdon, N., Lu, N.Y., Lam, E.T., Anantharaman, T. *et al.*, Tools and pipelines for BioNano data: molecule assembly pipeline and FASTA super scaffolding tool. *BMC Genomics*, 16, 734, 2015, doi: 10.1186/s12864-015-1911-8. PubMed PMID: WOS:000361869500002.
10. Du, H.L., Yu, Y., Ma, Y.F., Gao, Q., Cao, Y.H., Chen, Z. *et al.*, Sequencing and *de novo* assembly of a near complete indica rice genome. *Nat. Commun.*, 8, 15324, 2017, doi: 10.1038/ncomms15324. PubMed PMID: WOS:000400562200001.

11. Jia, L., Xie, L.J., Lao, S.T., Zhu, Q.H., Fan, L.J., Rice bioinformatics in the genomic era: Status and perspectives. *Crop J.*, 9, 3, 609–21, 2021, doi: 10.1016/j.cj.2021.03.003, PubMed PMID: WOS:000661096400012.
12. Golicz, A.A., Bayer, P.E., Bhalla, P.L., Batley, J., Edwards, D., Pangenomics Comes of Age: From Bacteria to Plant and Animal Applications. *Trends Genet.*, 36, 2, 132–45, 2020, doi: 10.1016/j.tig.2019.11.006. PubMed PMID: WOS:000506882100006.
13. Tettelin, H., Masignani, V., Cieslewicz, M.J., Donati, C., Medini, D., Ward, N.L. *et al.*, Genome analysis of multiple pathogenic isolates of Streptococcus agalactiae: Implications for the microbial 'pan-genome'. *Proc. Natl. Acad. Sci. U. S. A.*, 102, 45, 16530, 2005, doi: 10.1073/pnas.0508532102, PubMed PMID: WOS:000233283700072.
14. Tian, X.M., Li, R., Fu, W.W., Li, Y., Wang, X.H., Li, M. *et al.*, Building a sequence map of the pig pan-genome from multiple *de novo* assemblies and Hi-C data. *Sci. China Life Sci.*, 63, 5, 750–63, 2020, doi: 10.1007/s11427-019-9551-7, PubMed PMID: WOS:000527942100010.
15. Zhao, J.L., Bayer, P.E., Ruperao, P., Saxena, R.K., Khan, A.W., Golicz, A.A. *et al.*, Trait associations in the pangenome of pigeon pea (Cajanus cajan). *Plant Biotechnol. J.*, 18, 9, 1946–54, 2020, doi: 10.1111/pbi.13354, PubMed PMID: WOS:000557374700013.
16. Vernikos, G., Medini, D., Riley, D.R., Tettelin, H., Ten years of pan-genome analyses. *Curr. Opin. Microbiol.*, 23, 148–54, 2015, doi: 10.1016/j.mib.2014.11.016, PubMed PMID: WOS:000349879800023.
17. Yao, W., Li, G.W., Zhao, H., Wang, G.W., Lian, X.M., Xie, W.B., Exploring the rice dispensable genome using a metagenome-like assembly strategy. *Genome Biol.*, 16, 187, 2015, doi: 10.1186/s13059-015-0757-3. PubMed PMID: WOS:000361798700001.
18. Sun, C., Hu, Z.Q., Zheng, T.Q., Lu, K.C., Zhao, Y., Wang, W.S. *et al.*, RPAN: rice pan-genome browser for similar to 3000 rice genomes. *Nucleic Acids Res.*, 45, 2, 597–605, 2017, doi: 10.1093/nar/gkw958, PubMed PMID: WOS:000396576300013.
19. Wang, W.S., Mauleon, R., Hu, Z.Q., Chebotarov, D., Tai, S.S., Wu, Z.C. *et al.*, Genomic variation in 3,010 diverse accessions of Asian cultivated rice. *Nature*, 557, 7703, 43–+, 2018, doi: 10.1038/s41586-018-0063-9, PubMed PMID: WOS:000431234500027.
20. Zhao, Q., Feng, Q., Lu, H.Y., Li, Y., Wang, A.H., Tian, Q.L. *et al.*, Pan-genome analysis highlights the extent of genomic variation in cultivated and wild rice. *Nat. Genet.*, 50, 8, 278, 2018, doi: DOI 10.1038/s41588-018-0136-6, PubMed PMID: WOS:000440423400022.
21. Tao, Y.F., Zhao, X.R., Mace, E., Henry, R., Jordan, D., Exploring and Exploiting Pan-genomics for Crop Improvement. *Mol. Plant*, 12, 2, 156–69, 2019, doi: 10.1016/j.molp.2018.12.016, PubMed PMID: WOS:000457568200012.

22. Tao, Y.F., Jordan, D.R., Mace, E.S., A Graph-Based Pan-Genome Guides Biological Discovery. *Mol. Plant*, 13, 9, 1247–9, 2020, doi: 10.1016/j.molp.2020.07.020, PubMed PMID: WOS:000566895400010.
23. Garrison, E., Siren, J., Novak, A.M., Hickey, G., Eizenga, J.M., Dawson, E.T. *et al.*, Variation graph toolkit improves read mapping by representing genetic variation in the reference. *Nat. Biotechnol.*, 36, 9, 875–+, 2018, doi: 10.1038/nbt.4227, PubMed PMID: WOS:000443986000030.
24. Danilevicz, M.F., Fernandez, C.G.T., Marsh, J.I., Bayer, P.E., Edwards, D., Plant pangenomics: approaches, applications and advancements. *Curr. Opin. Plant Biol.*, 54, 18–25, 2020, doi: 10.1016/j.pbi.2019.12.005, PubMed PMID: WOS:000558338400004.
25. Qin, P., Lu, H.W., Du, H.L., Wang, H., Chen, W.L., Chen, Z. *et al.*, Pan-genome analysis of 33 genetically diverse rice accessions reveals hidden genomic variations. *Cell*, 184, 13, 3542–+, 2021, doi: 10.1016/j.cell.2021.04.046. PubMed PMID: WOS:000665547300018.
26. Alqudah, A.M., Sallam, A., Baenziger, P.S., Borner, A., GWAS: Fast-forwarding gene identification and characterization in temperate Cereals: lessons from Barley - A review. *J. Adv. Res.*, 22, 119–35, 2020, doi: 10.1016/j.jare.2019.10.013, PubMed PMID: WOS:000514801100012.
27. Muthuramalingam, P., Krishnan, S.R., Pandian, S., Mareeswaran, N., Aruni, W., Pandian, S.K. *et al.*, Global analysis of threonine metabolism genes unravel key players in rice to improve the abiotic stress tolerance. *Sci. Rep.*, 8, 1, 9270, 2018, Epub 20180618, doi: 10.1038/s41598-018-27703-8, PubMed PMID: 29915249, PubMed Central PMCID: PMCPMC6006157.
28. Takeda, S. and Matsuoka, M., Genetic approaches to crop improvement: responding to environmental and population changes. *Nat. Rev. Genet.*, 9, 6, 444–57, 2008, doi: 10.1038/nrg2342, PubMed PMID: 18475268.
29. Huang, X., Feng, Q., Qian, Q., Zhao, Q., Wang, L., Wang, A. *et al.*, High-throughput genotyping by whole-genome resequencing. *Genome Res.*, 19, 6, 1068–76, 2009, Epub 20090506, doi: 10.1101/gr.089516.108, PubMed PMID: 19420380, PubMed Central PMCID: PMCPMC2694477.
30. Elshire, R.J., Glaubitz, J.C., Sun, Q., Poland, J.A., Kawamoto, K., Buckler, E.S. *et al.*, A robust, simple genotyping-by-sequencing (GBS) approach for high diversity species. *PLoS One*, 6, 5, e19379, 2011, Epub 20110504. doi: 10.1371/journal.pone.0019379. PubMed PMID: 21573248; PubMed Central PMCID: PMCPMC3087801.
31. Zhou, X. and Stephens, M., Genome-wide efficient mixed-model analysis for association studies. *Nat. Genet.*, 44, 7, 821–4, 2012, Epub 20120617. doi: 10.1038/ng.2310. PubMed PMID: 22706312; PubMed Central PMCID: PMCPMC3386377.
32. Listgarten, J., Lippert, C., Kadie, C.M., Davidson, R.I., Eskin, E., Heckerman, D., Improved linear mixed models for genome-wide association studies. *Nat. Methods*, 9, 6, 525–6, 2012, Epub 20120530, doi: 10.1038/nmeth.2037. PubMed PMID: 22669648, PubMed Central PMCID: PMCPMC3597090.

33. Wang, Q., Tian, F., Pan, Y., Buckler, E.S., Zhang Z. A SUPER powerful method for genome wide association study. *PLoS One*, 9, 9, e107684, 2014, Epub 20140923. doi: 10.1371/journal.pone.0107684, PubMed PMID: 25247812, PubMed Central PMCID: PMCPMC4172578.
34. Huang, X. and Han, B., Natural variations and genome-wide association studies in crop plants. *Annu. Rev. Plant Biol.*, 65, 531–51, 2014, Epub 20131120. doi: 10.1146/annurev-arplant-050213-035715. PubMed PMID: 24274033.
35. Wang, H., Xu, X., Vieira, F.G., Xiao, Y., Li, Z., Wang, J. *et al.*, The Power of Inbreeding: NGS-Based GWAS of Rice Reveals Convergent Evolution during Rice Domestication. *Mol. Plant*, 9, 7, 975–85, 2016, Epub 20160512, doi: 10.1016/j.molp.2016.04.018, PubMed PMID: 27179918.
36. Bayat, A., Science, medicine, and the future: Bioinformatics. *BMJ*, 324, 7344, 1018–22, 2002, doi: 10.1136/bmj.324.7344.1018, PubMed PMID: 11976246, PubMed Central PMCID: PMCPMC1122955.
37. Vassilev, D., Leunissen, J., Atanassov, A., Nenov, A., Dimov, G., Application of Bioinformatics in Plant Breeding. *Biotechnol. Biotechnol. Equip.*, 19, sup3, 139–52, 2014, doi: 10.1080/13102818.2005.10817293.
38. Garg, P. and Jaiswal, P., Databases and bioinformatics tools for rice research. *Curr. Plant Biol.*, 7-8, 39–52, 2016, doi: 10.1016/j.cpb.2016.12.006.
39. Matsumoto, T., Wu, J.Z., Kanamori, H., Katayose, Y., Fujisawa, M., Namiki, N. *et al.*, The map-based sequence of the rice genome. *Nature*, 436, 7052, 793–800, 2005, doi: 10.1038/nature03895, PubMed PMID: WOS:000231116500034.
40. Li, J.Y., Wang, J., Zeigler, R.S., The 3,000 rice genomes project: new opportunities and challenges for future rice research. *Gigascience*, 3, 8, 2014, doi: 10.1186/2047-217x-3-8. PubMed PMID: WOS:000365652900002.
41. Chen, E.W., Huang, X.H., Tian, Z.X., Wing, R.A., Han, B., The Genomics of Oryza Species Provides Insights into Rice Domestication and Heterosis. *Annu. Rev. Plant Biol.*, 70, 639–65, 2019, doi: 10.1146/annurev-arplant-050718-100320, PubMed PMID: WOS:000482700100023.
42. Perez-de-Castro, A.M., Vilanova, S., Canizares, J., Pascual, L., Blanca, J.M., Diez, M.J. *et al.*, Application of Genomic Tools in Plant Breeding. *Curr. Genomics*, 13, 3, 179–95, 2012. PubMed PMID: WOS:000303477600002.
43. Kawahara, Y., de la Bastide, M., Hamilton, J.P., Kanamori, H., McCombie, W.R., Ouyang, S. *et al.*, Improvement of the Oryza sativa Nipponbare reference genome using next generation sequence and optical map data. *Rice (N Y)*, 6, 1, 4, 2013, Epub 20130206. doi: 10.1186/1939-8433-6-4. PubMed PMID: 24280374; PubMed Central PMCID: PMCPMC5395016.
44. Wei, X., Qiu, J., Yong, K.C., Fan, J.J., Zhang, Q., Hua, H. *et al.*, A quantitative genomics map of rice provides genetic insights and guides breeding. *Nat. Genet.*, 53, 2, 243–+, 2021, doi: 10.1038/s41588-020-00769-9, PubMed PMID: WOS:000613615200003.
45. Lescot, M., Dehais, P., Thijs, G., Marchal, K., Moreau, Y., Van de Peer, Y. *et al.*, PlantCARE, a database of plant cis-acting regulatory elements and a portal to tools for *in silico* analysis of promoter sequences. *Nucleic Acids*

Res., 30, 1, 325–7, 2002, doi: DOI 10.1093/nar/30.1.325, PubMed PMID: WOS:000173077100087.

46. Kanehisa, M., Furumichi, M., Sato, Y., Ishiguro-Watanabe, M., Tanabe, M., KEGG: integrating viruses and cellular organisms. *Nucleic Acids Res.*, 49, D1, D545–D51, 2021, doi: 10.1093/nar/gkaa970. PubMed PMID: WOS:000608437800070.
47. Gupta, P., Naithani, S., Tello-Ruiz, M.K., Chougule, K., D'Eustachio, P., Fabregat, A. *et al.*, Gramene Database: Navigating Plant Comparative Genomics Resources. *Curr. Plant Biol.*, 7-8, 10–5, 2016, doi: 10.1016/j.cpb.2016.12.005, PubMed PMID: 28713666, PubMed Central PMCID: PMCPMC5509230.
48. Sato, Y., Antonio, B.A., Namiki, N., Takehisa, H., Minami, H., Kamatsuki, K. *et al.*, RiceXPro: a platform for monitoring gene expression in japonica rice grown under natural field conditions. *Nucleic Acids Res.*, 39, Database issue, D1141–8, 2011, Epub 20101102. doi: 10.1093/nar/gkq1085. PubMed PMID: 21045061; PubMed Central PMCID: PMCPMC3013682.
49. Kawahara, Y., Oono, Y., Wakimoto, H., Ogata, J., Kanamori, H., Sasaki, H. *et al.*, TENOR: Database for Comprehensive mRNA-Seq Experiments in Rice. *Plant Cell Physiol.*, 57, 1, e7, 2016, Epub 20151116, doi: 10.1093/pcp/pcv179, PubMed PMID: 26578693.
50. Sato, Y., Namiki, N., Takehisa, H., Kamatsuki, K., Minami, H., Ikawa, H. *et al.*, RiceFREND: a platform for retrieving coexpressed gene networks in rice. *Nucleic Acids Res.*, 41, Database issue, D1214–21, 2013, doi: Epub 20121124, doi: 10.1093/nar/gks1122, PubMed PMID: 23180784, PubMed Central PMCID: PMCPMC3531108.
51. Obayashi, T., Hibara, H., Kagaya, Y., Aoki, Y., Kinoshita, K., ATTED-II v11: A Plant Gene Coexpression Database Using a Sample Balancing Technique by Subagging of Principal Components. *Plant Cell Physiol.*, 63, 6, 869–81, 2022, doi: 10.1093/pcp/pcac041, PubMed PMID: 35353884.
52. Zhao, H., Yao, W., Ouyang, Y., Yang, W., Wang, G., Lian, X. *et al.*, RiceVarMap: a comprehensive database of rice genomic variations. *Nucleic Acids Res.*, 43, Database issue, D1018–22, 2015, doi: Epub 20141001. doi: 10.1093/nar/gku894. PubMed PMID: 25274737; PubMed Central PMCID: PMCPMC4384008.
53. Mansueto, L., Fuentes, R.R., Borja, F.N., Detras, J., Abriol-Santos, J.M., Chebotarov, D. *et al.*, Rice SNP-seek database update: new SNPs, indels, and queries. *Nucleic Acids Res.*, 45, D1, D1075–D81, 2017, Epub 20161129. doi: 10.1093/nar/gkw1135, PubMed PMID: 27899667, PubMed Central PMCID: PMCPMC5210592.
54. Lei, Y., Lu, L., Liu, H.Y., Li, S., Xing, F., Chen, L.L., CRISPR-P: a web tool for synthetic single-guide RNA design of CRISPR-system in plants. *Mol. Plant*, 7, 9, 1494–6, 2014, Epub 20140409. doi: 10.1093/mp/ssu044. PubMed PMID: 24719468.

55. Nakashima, K., Ito, Y., Yamaguchi-Shinozaki, K., Transcriptional Regulatory Networks in Response to Abiotic Stresses in Arabidopsis and Grasses. *Plant Physiol.*, 149, 1, 88–95, 2009, doi: 10.1104/pp.108.129791, PubMed PMID: WOS:000262261500012.
56. Urano, K., Kurihara, Y., Seki, M., Shinozaki, K., 'Omics' analyses of regulatory networks in plant abiotic stress responses. *Curr. Opin. Plant Biol.*, 13, 2, 132–8, 2010, doi: 10.1016/j.pbi.2009.12.006, PubMed PMID: WOS:000278303200004.
57. Yang, S.J., Vanderbeld, B., Wan, J.X., Huang, Y.F., Narrowing Down the Targets: Towards Successful Genetic Engineering of Drought-Tolerant Crops. *Mol. Plant*, 3, 3, 469–90, 2010, doi: 10.1093/mp/ssq016, PubMed PMID: WOS:000278819900001.
58. Golldack, D., Luking, I., Yang, O., Plant tolerance to drought and salinity: stress regulating transcription factors and their functional significance in the cellular transcriptional network. *Plant Cell Rep.*, 30, 8, 1383–91, 2011, doi: 10.1007/s00299-011-1068-0, PubMed PMID: WOS:000293795100002.
59. Shameer, K., Ambika, S., Varghese, S.M., Karaba, N., Udayakumar, M., Sowdhamini, R., STIFDB-Arabidopsis Stress Responsive Transcription Factor DataBase. *Int. J. Plant Genomics*, 2009, 583429, 2009, Epub 20091018, doi: 10.1155/2009/583429, PubMed PMID: 19841686, PubMed Central PMCID: PMCPMC2763139.
60. Priya, P. and Jain, M., RiceSRTFDB: a database of rice transcription factors containing comprehensive expression, cis-regulatory element and mutant information to facilitate gene function analysis. *Database (Oxford)*, 2013, bat027, 2013, Epub 20130509, doi: 10.1093/database/bat027, PubMed PMID: 23660286, PubMed Central PMCID: PMCPMC3649641.
61. Sandhu, M., Sureshkumar, V., Prakash, C., Dixit, R., Solanke, A.U., Sharma, T.R. *et al.*, RiceMetaSys for salt and drought stress responsive genes in rice: a web interface for crop improvement. *BMC Bioinf.*, 18, 1, 432, 2017, Epub 20170930, doi: 10.1186/s12859-017-1846-y, PubMed PMID: 28964253, PubMed Central PMCID: PMCPMC5622590.
62. Cao, P., Jung, K.H., Choi, D., Hwang, D., Zhu, J., Ronald, P.C., The Rice Oligonucleotide Array Database: an atlas of rice gene expression. *Rice (N Y)*, 5, 1, 17, 2012, Epub 20120719, doi: 10.1186/1939-8433-5-17, PubMed PMID: 24279809, PubMed Central PMCID: PMCPMC4883718.
63. Kumar, S. and Shanker, A., Bioinformatics Resources for the Stress Biology of Plants, in: *Biotic and Abiotic Stress Tolerance in Plants*, pp. 367–86, 2018.
64. Alter, S., Bader, K.C., Spannagl, M., Wang, Y., Bauer, E., Schon, C.C. *et al.*, DroughtDB: an expert-curated compilation of plant drought stress genes and their homologs in nine species. *Database-Oxford*, 2015:bav046, 2015, doi: 10.1093/database/bav046, PubMed PMID: WOS:000361051200001.
65. Shan, Q.W., Wang, Y.P., Li, J., Zhang, Y., Chen, K.L., Liang, Z. *et al.*, Targeted genome modification of crop plants using a CRISPR-Cas system. *Nat.*

Biotechnol., 31, 8, 686–8, 2013, doi: 10.1038/nbt.2650, PubMed PMID: WOS:000322911000010.

66. Wood, A.J., Lo, T.W., Zeitler, B., Pickle, C.S., Ralston, E.J., Lee, A.H. *et al.*, Targeted Genome Editing Across Species Using ZFNs and TALENs. *Science*, 333, 6040, 307, 2011, doi: 10.1126/science.1207773, PubMed PMID: WOS:000292732000032.
67. Li, T., Huang, S., Zhao, X.F., Wright, D.A., Carpenter, S., Spalding, M.H. *et al.*, Modularly assembled designer TAL effector nucleases for targeted gene knockout and gene replacement in eukaryotes. *Nucleic Acids Res.*, 39, 14, 6315–25, 2011, doi: 10.1093/nar/gkr188, PubMed PMID: WOS:000293919600049.
68. Zegeye, W.A., Tsegaw, M., Zhang, Y.X., Cao, L.Y., CRISPR-Based Genome Editing: Advancements and Opportunities for Rice Improvement. *Int. J. Mol. Sci.*, 23, 8, 4454, 2022, doi: 10.3390/ijms23084454, PubMed PMID: WOS:000787957400001.
69. Jinek, M., Chylinski, K., Fonfara, I., Hauer, M., Doudna, J.A., Charpentier, E., A Programmable Dual-RNA-Guided DNA Endonuclease in Adaptive Bacterial Immunity. *Science*, 337, 6096, 816–21, 2012, doi: 10.1126/science.1225829, PubMed PMID: WOS:000307535600036.
70. Zetsche, B., Gootenberg, J.S., Abudayyeh, O.O., Slaymaker, I.M., Makarova, K.S., Essletzbichler, P. *et al.*, Cpf1 Is a Single RNA-Guided Endonuclease of a Class 2 CRISPR-Cas System. *Cell*, 163, 3, 759–71, 2015, doi: 10.1016/j.cell.2015.09.038, PubMed PMID: WOS:000364828900027.
71. Antoniou, P., Miccio, A., Brusson, M., Base and Prime Editing Technologies for Blood Disorders. *Front. Genome Ed.*, 3, 618406, 2021, Epub 20210128, doi: 10.3389/fgeed.2021.618406, PubMed PMID: 34713251, PubMed Central PMCID: PMCPMC8525391.
72. Zhang, X.H., Tee, L.Y., Wang, X.G., Huang, Q.S., Yang, S.H., Off-target Effects in CRISPR/Cas9-mediated Genome Engineering. *Mol. Ther.-Nucleic Acids*, 4, e264, 2015, doi: 10.1038/mtna.2015.37, PubMed PMID: WOS:000368913300001.
73. Aird, E.J., Lovendahl, K.N., St Martin, A., Harris, R.S., Gordon, W.R., Increasing Cas9-mediated homology-directed repair efficiency through covalent tethering of DNA repair template. *Commun. Biol.*, 1, 54, 2018, doi: 10.1038/s42003-018-0054-2, PubMed PMID: WOS:000461126500054.
74. Zong, Y., Wang, Y.P., Li, C., Zhang, R., Chen, K.L., Ran, Y.D. *et al.*, Precise base editing in rice, wheat and maize with a Cas9-cytidine deaminase fusion. *Nat. Biotechnol.*, 35, 5, 438–+, 2017, doi: 10.1038/nbt.3811, PubMed PMID: WOS:000400809800017.
75. Shimatani, Z., Kashojiya, S., Takayama, M., Terada, R., Arazoe, T., Ishii, H. *et al.*, Targeted base editing in rice and tomato using a CRISPR-Cas9 cytidine deaminase fusion. *Nat. Biotechnol.*, 35, 5, 441–+, 2017, doi: 10.1038/nbt.3833, PubMed PMID: WOS:000400809800018.

76. Hua, K., Tao, X.P., Yuan, F.T., Wang, D., Zhu, J.K., Precise A.T to G.C Base Editing in the Rice Genome. *Mol. Plant*, 11, 4, 627–30, 2018, doi: 10.1016/j.molp.2018.02.007, PubMed PMID: WOS:000428983100012.
77. Sato, H., Suzuki, Y., Sakai, M., Imbe, T., Molecular characterization of Wx-mq, a novel mutant gene for low-amylose content in endosperm of rice (Oryza sativa L.). *Breed. Sci.*, 52, 2, 131–5, 2002, doi: DOI 10.1270/jsbbs.52.131, PubMed PMID: WOS:000176692800009.
78. Che, R., Tong, H., Shi, B., Liu, Y., Fang, S., Liu, D. *et al.*, Control of grain size and rice yield by GL2-mediated brassinosteroid responses. *Nat. Plants*, 2, 15195, 2015, Epub 20151221, doi: 10.1038/nplants.2015.195, PubMed PMID: 27250747.
79. Hou, Y., Tong, X., Wang, Y., Qiu, J., Li, Z., Zhang, W. *et al.*, Data set from a comprehensive phosphoproteomic analysis of rice variety IRBB5 in response to bacterial blight. *Data Brief.*, 6, 282–5, 2016, Epub 20151217, doi: 10.1016/j.dib.2015.11.071, PubMed PMID: 26862573, PubMed Central PMCID: PMCPMC4706621.
80. Yan, F., Kuang, Y.J., Ren, B., Wang, J.W., Zhang, D.W., Lin, H.H. *et al.*, Highly Efficient A.T to G.C Base Editing by Cas9n-Guided tRNA Adenosine Deaminase in Rice. *Mol. Plant*, 11, 4, 631–4, 2018, doi: 10.1016/j.molp.2018.02.008, PubMed PMID: WOS:000428983100013.
81. Li, C., Zong, Y., Wang, Y.P., Jin, S., Zhang, D.B., Song, Q.N. *et al.*, Expanded base editing in rice and wheat using a Cas9-adenosine deaminase fusion. *Genome Biol.*, 19, 59, 2018, doi: 10.1186/s13059-018-1443-z. PubMed PMID: WOS:000433985400001.
82. He, C., Liu, H., Chen, D.J., Xie, W.Z., Wang, M.X., Li, Y.Q. *et al.*, CRISPR-Cereal: a guide RNA design tool integrating regulome and genomic variation for wheat, maize and rice. *Plant Biotechnol. J.*, 19, 11, 2141–3, 2021, doi: 10.1111/pbi.13675, PubMed PMID: WOS:000684878800001.
83. Heigwer, F., Kerr, G., Boutros, M. E-CRISP, fast CRISPR target site identification. *Nat. Methods*, 11, 2, 122–4, 2014, doi: 10.1038/nmeth.2812. PubMed PMID: WOS:000331141600006.
84. Naito, Y., Hino, K., Bono, H., Ui-Tei, K., CRISPRdirect: software for designing CRISPR/Cas guide RNA with reduced off-target sites. *Bioinformatics*, 31, 7, 1120–3, 2015, doi: 10.1093/bioinformatics/btu743, PubMed PMID: WOS:000352269500018.
85. Stemmer, M., Thumberger, T., Keyer, M.D., Wittbrodt, J., Mateo, J.L., CCTop: An Intuitive, Flexible and Reliable CRISPR/Cas9 Target Prediction Tool. *PLoS One*, 12, 4, e0176619, 2017, doi: 10.1371/journal.pone.0176619, PubMed PMID: WOS:000399875900143.
86. Montague, T.G., Cruz, J.M., Gagnon, J.A., Church, G.M., Valen, E., CHOPCHOP: a CRISPR/Cas9 and TALEN web tool for genome editing. *Nucleic Acids Res.*, 42, W1, W401–W7, 2014, doi: 10.1093/nar/gku410, PubMed PMID: WOS:000339715000067.

87. Doench, J.G., Fusi, N., Sullender, M., Hegde, M., Vaimberg, E.W., Donovan, K.F. *et al.*, Optimized sgRNA design to maximize activity and minimize off-target effects of CRISPR-Cas9. *Nat. Biotechnol.*, 34, 2, 184–+, 2016, doi: 10.1038/nbt.3437, PubMed PMID: WOS:000369619100021.
88. Zhu, L., Wang, X.Y., Li, F.Y., Song, J.N., PreAcrs: a machine learning framework for identifying anti-CRISPR proteins. *BMC Bioinf.*, 23, 1, 444, 2022, doi: 10.1186/s12859-022-04986-3. PubMed PMID: WOS:000874128800002.
89. Wang, J.W., Dai, W., Li, J.H., Xie, R.P., Dunstan, R.A., Stubenrauch, C. *et al.*, PaCRISPR: a server for predicting and visualizing anti-CRISPR proteins. *Nucleic Acids Res.*, 48, W1, W348–W57, 2020, doi: 10.1093/nar/gkaa432, PubMed PMID: WOS:000562474100048.
90. Wandera, K.G., Alkhnbashi, O.S., Bassett, H.V.I., Mitrofanov, A., Hauns, S., Migur, A. *et al.*, Anti-CRISPR prediction using deep learning reveals an inhibitor of Cas13b nucleases. *Mol. Cell*, 82, 14, 2714–+, 2022, doi: 10.1016/j.molcel.2022.05.003, PubMed PMID: WOS:000841749300005.
91. Eitzinger, S., Asif, A., Watters, K.E., Iavarone, A.T., Knott, G.J., Doudna, J.A. *et al.*, Machine learning predicts new anti-CRISPR proteins. *Nucleic Acids Res.*, 48, 9, 4698–708, 2020, doi: 10.1093/nar/gkaa219, PubMed PMID: WOS:000573914600014.
92. Alkan, F., Wenzel, A., Anthon, C., Havgaard, J.H., Gorodkin, J., CRISPR-Cas9 off-targeting assessment with nucleic acid duplex energy parameters. *Genome Biol.*, 19, 177, 2018, doi: 10.1186/s13059-018-1534-x, PubMed PMID: WOS:000448669600001.
93. Guilinger, J.P., Pattanayak, V., Reyon, D., Tsai, S.Q., Sander, J.D., Joung, J.K. *et al.*, Broad specificity profiling of TALENs results in engineered nucleases with improved DNA-cleavage specificity. *Nat. Methods*, 11, 4, 429–+, 2014, doi: 10.1038/nmeth.2845, PubMed PMID: WOS:000333749900026.
94. Cai, Y.P., Chen, L., Sun, S., Wu, C.X., Yao, W.W., Jiang, B.J. *et al.*, CRISPR/Cas9-Mediated Deletion of Large Genomic Fragments in Soybean. *Int. J. Mol. Sci.*, 19, 12, 3835, 2018, doi: 10.3390/ijms19123835, PubMed PMID: WOS:000455323500137.
95. Hsu, P.D., Scott, D.A., Weinstein, J.A., Ran, F.A., Konermann, S., Agarwala, V. *et al.*, DNA targeting specificity of RNA-guided Cas9 nucleases. *Nat. Biotechnol.*, 31, 9, 827–+, 2013, doi: 10.1038/nbt.2647, PubMed PMID: WOS:000324306300023.
96. Langmead, B., Trapnell, C., Pop, M., Salzberg, S.L., Ultrafast and memory-efficient alignment of short DNA sequences to the human genome. *Genome Biol.*, 10, 3, R25, 2009, Epub 20090304, doi: 10.1186/gb-2009-10-3-r25, PubMed PMID: 19261174, PubMed Central PMCID: PMCPMC2690996.
97. Bae, S., Park, J., Kim, J.S., Cas-OFFinder: a fast and versatile algorithm that searches for potential off-target sites of Cas9 RNA-guided endonucleases. *Bioinformatics*, 30, 10, 1473–5, 2014, doi: 10.1093/bioinformatics/btu048, PubMed PMID: WOS:000336530000021.

98. McKenna, A. and Shendure, J., FlashFry: a fast and flexible tool for large-scale CRISPR target design. *BMC Biol.*, 16, 1, 74, 2018, Epub 20180705, doi: 10.1186/s12915-018-0545-0, PubMed PMID: 29976198, PubMed Central PMCID: PMCPMC6033233.
99. Langdon, W.B., Performance of genetic programming optimised Bowtie2 on genome comparison and analytic testing (GCAT) benchmarks. *BioData Min.*, 8, 1, 1, 2015, Epub 20150108, doi: 10.1186/s13040-014-0034-0, PubMed PMID: 25621011, PubMed Central PMCID: PMCPMC4304608.
100. Jacquin, A.L.S., Odom, D.T., Lukk, M., Crisflash: open-source software to generate CRISPR guide RNAs against genomes annotated with individual variation. *Bioinformatics*, 35, 17, 3146–7, 2019, doi: 10.1093/bioinformatics/btz019, PubMed PMID: 30649181, PubMed Central PMCID: PMCPMC6735888.
101. Zhang, H., Zhang, J., Wei, P., Zhang, B., Gou, F., Feng, Z. *et al.*, The CRISPR/Cas9 system produces specific and homozygous targeted gene editing in rice in one generation. *Plant Biotechnol. J.*, 12, 6, 797–807, 2014, doi: Epub 20140523, doi: 10.1111/pbi.12200, PubMed PMID: 24854982.
102. Tang, X., Liu, G.Q., Zhou, J.P., Ren, Q.R., You, Q., Tian, L. *et al.*, A large-scale whole-genome sequencing analysis reveals highly specific genome editing by both Cas9 and Cpf1 (Cas12a) nucleases in rice. *Genome Biol.*, 19, 84, 2018, doi: 10.1186/s13059-018-1458-5, PubMed PMID: WOS:000437331000001.
103. Huang, X.H., Yang, S.H., Gong, J.Y., Zhao, Y., Feng, Q., Gong, H. *et al.*, Genomic analysis of hybrid rice varieties reveals numerous superior alleles that contribute to heterosis. *Nat. Commun.*, 6, 6258, 2015, doi: 10.1038/ncomms7258, PubMed PMID: WOS:000350202800008.
104. Huang, X.H., Yang, S.H., Gong, J.Y., Zhao, Q., Feng, Q., Zhan, Q.L. *et al.*, Genomic architecture of heterosis for yield traits in rice. *Nature*, 537, 7622, 629–+, 2016, doi: 10.1038/nature19760, PubMed PMID: WOS:000384331700039.
105. Wallace, J.G., Rodgers-Melnick, E., Buckler, E.S., On the Road to Breeding 4.0: Unraveling the Good, the Bad, and the Boring of Crop Quantitative Genomics. *Annu. Rev. Genet.*, 52, 421–44, 2018, doi: 10.1146/annurev-genet-120116-024846, PubMed PMID: WOS:000453889500020.
106. Cui, Y.R., Li, R.D., Li, G.W., Zhang, F., Zhu, T.T., Zhang, Q.F. *et al.*, Hybrid breeding of rice via genomic selection. *Plant Biotechnol. J.*, 18, 1, 57–67, 2020, doi: 10.1111/pbi.13170, PubMed PMID: WOS:000530370800010.
107. Chen, Q., Chen, C., Zhang, C., Li, T., Chen, S., Wu, D. *et al.*, Study on flood management of the middle reach of Huaihe River based on Sobek. *International Conference on Artificial Intelligence, Virtual Reality, and Visualization (AIVRV 2021)*, 2021.
108. Yu, W., Zhou, J., Yin, Y., Feng, X., Nan, H., Lin, J. *et al.*, Effects of Hot Isostatic Pressing and Heat Treatment on the Microstructure and Mechanical Properties of Cast TiAl Alloy. *Metals*, 11, 8, 3423, 2021, doi: 10.3390/met11081156.

109. Xie, Y., Niu, B., Long, Y., Li, G., Tang, J., Zhang, Y. *et al.*, Suppression or knockout of SaF/SaM overcomes the Sa-mediated hybrid male sterility in rice. *J. Integr. Plant Biol.*, 59, 9, 669–79, 2017, Epub 20170823, doi: 10.1111/jipb.12564, PubMed PMID: 28636095.
110. Xie, Y., Tang, J., Xie, X., Li, X., Huang, J., Fei, Y. *et al.*, An asymmetric allelic interaction drives allele transmission bias in interspecific rice hybrids. *Nat. Commun.*, 10, 1, 2501, 2019, Epub 20190607, doi: 10.1038/s41467-019-10488-3, PubMed PMID: 31175302, PubMed Central PMCID: PMCPMC6555797.
111. Wang, C., Liu, Q., Shen, Y., Hua, Y., Wang, J., Lin, J. *et al.*, Clonal seeds from hybrid rice by simultaneous genome engineering of meiosis and fertilization genes. *Nat. Biotechnol.*, 37, 3, 283–6, 2019, Epub 20190104, doi: 10.1038/s41587-018-0003-0, PubMed PMID: 30610223.
112. Khanday, I., Skinner, D., Yang, B., Mercier, R., Sundaresan, V., A male-expressed rice embryogenic trigger redirected for asexual propagation through seeds. *Nature*, 565, 7737, 91–+, 2019, doi: 10.1038/s41586-018-0785-8, PubMed PMID: WOS:000454708600033.
113. Kuang, Y.J., Li, S.F., Ren, B., Yan, F., Spetz, C., Li, X.J. *et al.*, Base-Editing-Mediated Artificial Evolution of OsALS1 In Planta to Develop Novel Herbicide-Tolerant Rice Germplasms. *Mol. Plant*, 13, 4, 565–72, 2020, doi: 10.1016/j.molp.2020.01.010, PubMed PMID: WOS:000523654200006.
114. Li, C., Zhang, R., Meng, X.B., Chen, S., Zong, Y., Lu, C.J. *et al.*, Targeted, random mutagenesis of plant genes with dual cytosine and adenine base editors. *Nat. Biotechnol.*, 38, 7, 875–U66, 2020, doi: 10.1038/s41587-019-0393-7, PubMed PMID: WOS:000508165600005.
115. Hickey, L.T., Hafeez, A.N., Robinson, H., Jackson, S.A., Leal-Bertioli, S.C.M., Tester, M. *et al.*, Breeding crops to feed 10 billion. *Nat. Biotechnol.*, 37, 7, 744–54, 2019, doi: 10.1038/s41587-019-0152-9, PubMed PMID: WOS:000478028700019.
116. Liu, Y., Wang, H., Jiang, Z., Wang, W., Xu, R., Wang, Q., *et al.*, Genomic basis of geographical adaptation to soil nitrogen in rice. *Nature,* 590 600–605. 2021, doi: 10.1038/s41586-020-03091-w.
117. Li, X.M., Chao, D.Y., Wu, Y., Huang, X., Chen, K., Cui, L.G. *et al.*, Natural alleles of a proteasome alpha2 subunit gene contribute to thermotolerance and adaptation of African rice. *Nat. Genet.*, 47, 7, 827–33, 2015, Epub 20150518, doi: 10.1038/ng.3305, PubMed PMID: 25985140.
118. Kan, Y., Mu, X.R., Zhang, H., Gao, J., Shan, J.X., Ye, W.W. *et al.*, TT2 controls rice thermotolerance through SCT1-dependent alteration of wax biosynthesis. *Nat. Plants*, 8, 1, 53–67, 2022, doi: 10.1038/s41477-021-01039-0, PubMed PMID: WOS:000739789100001.
119. Kyriakidou, M., Tai, H.H., Anglin, N.L., Ellis, D., Strömvik, M.V., Current Strategies of Polyploid Plant Genome Sequence Assembly. *Front. Plant Sci.*, 9, 2018, doi: 10.3389/fpls.2018.01660.
120. Claros, M.G., Bautista, R., Guerrero-Fernández, D., Benzerki, H., Seoane, P., Fernández-Pozo, N., Why Assembling Plant Genome Sequences Is So Challenging. *Biology*, 1, 2, 439–59, 2012, doi: 10.3390/biology1020439.

3

Computer-Aided Vaccine Design: Applications in Agriculture

Tanmaya Kumar Sahu[1]* and Atmakuri Ramakrishna Rao[2]

[1]*School of Bioinformatics, ICAR-Indian Institute of Agricultural Biotechnology Garhkhatanga, Ranchi, Jharkhand, India*
[2]*Indian Council of Agricultural Research, Krishi Bhavan, New Delhi, India*

Abstract
The ongoing demand for sustainable and efficient agricultural practices has led to an increased focus on the development of innovative technologies for disease management in livestock and crops. Computer-aided vaccine design (CAVD) has emerged as a promising approach to enhance the efficiency and precision of vaccine development in the agricultural sector. Further, the advancements in CAVD offer innovative solutions to address the challenges associated with disease outbreaks in agriculturally important organisms. Looking at the importance of CAVD in the agriculture sector, this chapter has been composed to provide an overview of the various computational tools and techniques used in the design and development of vaccines for mitigating infectious diseases in agricultural settings. It outlines animal diseases, vaccination in agriculture, and integration of bioinformatics and structural biology approaches in the rational design of vaccines, facilitating the identification of potential antigenic targets and the prediction of their immunogenicity. The chapter also emphasizes the application of machine learning and artificial intelligence in predicting potential vaccine candidates. Further, the prospective of this interdisciplinary approach to accelerate the production of cost-effective, targeted, and sustainable vaccines, thereby promoting disease control and enhancing agricultural productivity, has been elaborated in this chapter.

Keywords: Antigen prediction, epitope, vaccine formulation, sustainable agriculture, disease resistance, livestock health

**Corresponding author*: tanmayabioinfo@gmail.com

Hrudayanath Thatoi, Sonali Mohapatra, Swagat Kumar Das and Sukanta Kumar Pradhan (eds.) Applied Biotechnology and Bioinformatics: Agriculture, Pharmaceutical Research and Environment, (73–96)

3.1 Introduction

Agriculture involves crop cultivation, animal husbandry, fish farming, and other forms of practices to obtain plant and animal products for human use. The agricultural industry plays a significant role in the global economy in providing food, raw materials, and employment for millions of people worldwide. It has been a fundamental part of human civilization for thousands of years. Animal agriculture is an important part of the global food system, and it provides a significant portion of the world's protein and other nutrients. Commonly raised animals include cattle, pigs, chickens, sheep, and fish, among others. Infectious diseases in agricultural animals are a significant concern for the livestock industry, as they can cause high economic loss having a considerable impact on animal welfare. There are many infectious diseases that affect agricultural animals, and they are caused by an array of pathogens, including viruses, bacteria, fungi, etc. Common infectious diseases in agricultural animals include foot-and-mouth disease (FMD) in cloven-hoofed animals; avian influenza in birds; Newcastle disease (ND) in chickens, ducks, and turkeys; bluetongue disease in sheep and cattle; porcine epidemic diarrhea (PED) in pigs; and many more. Veterinary professionals play a critical role in diagnosing and treating infectious diseases in agricultural animals, as well as providing guidance and advice to farmers and other stakeholders in the agriculture industry. Efforts to control and prevent infectious diseases in agricultural animals include the use of vaccines, bio-security measures, surveillance and monitoring programs.

Vaccination is an effective measure in controlling many diseases affecting the agriculturally important animals. In animal agriculture, vaccines are currently being used to protect livestock from various diseases such as protecting chickens from avian influenza and ND, pigs from PRRS, and cattle FMD. Use of vaccines in protecting plant disease is also being explored in the present era. Damage-associated molecular patterns (DAMPs) have been reported as a potential and novel class of plant vaccines [1]. The vaccine molecules are derived from the plants and trigger the immune response either by detecting the damage when they are present in the extracellular space or detecting the volatile compounds produced due to damage in the surrounding air where the plant is located. Therefore, vaccine designing is important in agriculture for disease prevention, animal health, addressing the evolving pathogens, and enhancing the preparedness in the event of a pandemic.

In last few decades, the time for vaccine designing is being reduced by the emergence of computer-aided vaccine design (CAVD), making it a cost-effective process of disease prevention [2]. The CAVD has the potential to revolutionize vaccine development, as it uses advanced computer algorithms and modeling techniques to accelerate and improve the process of vaccine development. Computational methods for vaccine designing are appreciated nowadays due to speed, efficiency, precision, efficacy, cost-effectiveness, safety, and scalability. Thus, this chapter provides an overview of vaccine, its computational design, and its applications in agriculture to address infectious diseases.

3.2 Agriculturally Important Animals

Agricultural animals are mainly raised for food, fiber, or other agricultural purposes. These animals are raised, and managed by farmers, and other agricultural producers to produce meat, milk, eggs, wool, leather, and other products. The major agricultural animals are cattle, goat, pig, buffalo, sheep, horses, donkeys, rabbits, chicken, ducks, turkeys, etc. Cattle and buffalo are among the major agricultural animals that are generally raised for dairy products, ploughing fields, and pulling carts. Other animals like goats, sheep, and pigs are reared for milk, meat, and fiber. Among the agriculturally important birds, chickens, ducks, and turkeys are reared generally for meat, eggs, and feather. These animals have a vital role in the global food supply and provide important products that are used in a variety of industries.

3.3 Diseases Affecting Animal Health in Agriculture

Agricultural animals are susceptible to a variety of diseases that can have a significant impact on animal health and productivity, as well as the livelihoods of farmers and the overall food supply. Among the animal diseases, some of the most common agricultural animal diseases that badly affect animal health and farmers' livelihood are FMD, bovine tuberculosis (TB), bovine respiratory disease (BRD), bluetongue disease, brucellosis, African swine fever, porcine reproductive and respiratory syndrome (PRRS), PED, classical swine fever (CSF), rift valley fever, Johne's disease, avian influenza, Newcastle disease (ND), and Salmonellosis.

Foot-and-mouth disease (FMD) is a rapidly transmittable viral infection that affects cloven-hoofed animals like, cattle, sheep, and pigs. It can cause fever, blisters spread over the mouth and feet, and weight loss. FMD can spread quickly and have serious economic consequences for the agricultural industry. FMD outbreaks can have significant economic impacts on the livestock industry [3]. Among the infectious bacterial diseases, BRD affects cattle and causes respiratory illness, reduced weight gain, and death. It is a common disease in feedlot cattle and can lead to significant economic losses in the beef industry. Bovine tuberculosis (TB) is another bacterial disease that affects cattle, and can also be transmitted to other animals and humans. It also causes weight loss, fever, and coughing, and can have serious consequences for animal health and productivity. Further, brucellosis is caused by bacteria that affect cattle, goats, and sheep, and can also be transmitted to humans. It can cause fever, miscarriage, and reduced milk production. Similarly, Johne's disease is a chronic bacterial infection that affects ruminant animals such as cattle, sheep, and goats resulting in weight loss, diarrhea, and reduced milk production in infected animals. Johne's disease is difficult to diagnose and can have significant economic impacts on the dairy and beef industries [4].

Another viral disease called African Swine Fever (ASF) affects pigs, and is characterized by fever, loss of appetite, and internal bleeding. It is highly contagious and can have devastating effects on the pig population, resulting in economic losses for farmers and the pork industry [5]. The disease is currently spreading in various parts of the globe, and is considered a major threat to global pork production [6]. PRRS is another viral disease in pigs that causes respiratory illness and reproductive problems [7]. It is a major concern for pig farmers due to high mortality rates in piglets, and can therefore result in significant economic losses to the pig farmers in many countries [8].

Avian influenza (bird flu) and ND are two highly contagious viral diseases that primarily affect birds, including chickens, ducks, and turkeys, and are a major threat to poultry production in many parts of the world. The diseases cause severe respiratory illness, and reduced egg production, and high mortality rates in infected birds and have serious consequences for the poultry industry [9]. Specifically, ND is also characterized with nervous system disorders. Avian influenza can also be transmitted to humans, and has the potential to cause a pandemic across the globe [10].

Apart from the above-discussed agriculturally important animal diseases, many other infectious diseases exist that can have significant impacts on animal health and productivity. These diseases are managed, controlled, and prevented by many measures. However, vaccination has been an effective measure for controlling infectious microbial diseases in agricultural animals.

3.4 Vaccination in Agriculture

Vaccines have been widely used in human medicine to prevent or/and control the spread of infectious diseases. The use of vaccines has been well explored in the agriculture sector to prevent and control the spread of infectious diseases in livestock that can cause illness or death. Livestock vaccines have been developed for FMD, avian influenza and ND, PRRS, and brucellosis. Researchers at Iowa State University developed a PRRS virus vaccine using a modified live virus, which was shown to be effective in protecting pigs from the disease [11]. Further, the PED virus vaccine was demonstrated to be effective in preventing the disease in piglets and reducing mortality and clinical signs of infection [12].

Vaccines specifically designed for domestic and wild animals to prevent the spread of zoonotic diseases are called veterinary vaccines like Rabies vaccine. Further, the vaccines used to control and prevent the spread of diseases in a particular environment are called environmental vaccines [13]. An example of environmental vaccine is the vaccine for botulism in fish that reduces mortality and the number of fish carrying the botulism toxin [14]. Many researchers across the globe reviewed the current vaccination strategies followed to protect the livestock animals, fishes, and birds from infectious diseases. Ma *et al.* [15] described the use of conventional aquaculture vaccines and provided an overview of promising molecular approaches and strategies for new aquaculture vaccine development. Ravikumar *et al.* [16] reviewed various vaccines for viral diseases and vaccination protocols available for most of the avian viral infections, and strategies for developing safer and more effective viral vaccines for poultry. Brun [17] presented an overview of current vaccine technologies that can be approached for virus pathogens affecting the livestock population.

3.5 Vaccine

A vaccine is a biological substance designed to induce active acquired immunity against a specific disease. It is usually composed of weakened or dead pathogens or their parts, which stimulate the immune system to produce antibodies against the disease-causing agent such as a virus or bacterium without causing the disease itself. The body's immune system recognizes and remembers the antigen, allowing it to quickly recognize and destroy the pathogen if it is encountered in the future. Vaccines are an effective way to prevent infectious diseases and have been instrumental in reducing the incidence of many infectious diseases affecting mankind. There are various types of vaccines available, including live attenuated vaccines, subunit vaccines, and inactivated vaccines, each designed to elicit an immune response in a slightly different way. Vaccines typically contain small amounts of the virus or bacterium, or a protein or other component of the pathogen, that have been weakened or killed so that they do not cause disease, but are still able to trigger an immune response in the body. Vaccination has been widely applied in human health; however, its application in agricultural animal health is yet to be fully explored.

3.5.1 Types of Vaccines

Various types of vaccines exist, including live attenuated vaccines, subunit vaccines, inactivated or killed vaccines, recombinant, or conjugate vaccines, vector vaccines, and mRNA/DNA vaccines, among others. Though all kinds of vaccines can be developed for every disease, the specific type of vaccine is often designed depending on the target pathogen and its specific characteristics.

Inactivated vaccines employ a dead or inactivated virus or bacteria to trigger an immune response. This approach has proven effective in creating vaccines for diseases such as polio. The safety and efficacy of the inactivated poliovirus vaccine were validated by researchers, who utilized inactivated poliovirus to elicit an immune response against polio [18]. On the other hand, live attenuated vaccines contain weakened viruses or bacteria that can still replicate but are less virulent than the original pathogen. Though these vaccines are capable of establishing long-lasting immunity, they are not advised for individuals with compromised immune systems. Subunit, recombinant, or conjugate vaccines, on the other hand, utilize a fragment of the pathogen, such as a protein or sugar molecule, to evoke an immune response. They are often used when it is difficult to grow or inactivate the pathogen. Subunit vaccines have been used to develop vaccines against human papillomavirus (HPV). The successful use of subunit vaccines for HPV has been demonstrated in Paavonen *et al.* [19].

The mRNA vaccines include a short fragment of mRNA that directs cells to generate a protein located on the virus's surface. This protein triggers an immune response, leading to the development of immunity. The mRNA vaccines contain a small piece of genetic material (mRNA) that instructs cells to produce a protein found on the surface of the virus. This protein stimulates an immune response and creates immunity. The mRNA has been applied recently for COVID-19, which uses mRNA to encode the spike protein of SARS-CoV-2 [20]. Like mRNA, DNA is also used as a vaccine containing a small piece of DNA that encodes a protein located on the virus surface. The DNA is taken up by cells, which produce the protein and stimulate an immune response.

Vector vaccines are developed using a modified virus or bacterium to deliver a piece of the pathogen's genetic material to cells. This genetic material stimulates an immune response and creates immunity. Adenovirus vector vaccine is an example of vector vaccines that use a weakened adenovirus for delivering antigenic materials to the immune system. Adenovirus vector vaccines have been used to develop vaccines against Ebola virus where a vesicular stomatitis virus (VSV) backbone has been used to express the glycoprotein of Ebola virus [21].

3.5.2 Steps of Vaccine Designing

Vaccine designing involves the identification of key antigens and vaccine development to prevent and manage infectious diseases. It is a complex process that requires collaboration between scientists, healthcare providers, and regulatory agencies. The general steps involved in vaccine designing are given in Figure 3.1 and described in the subsequent subsections.

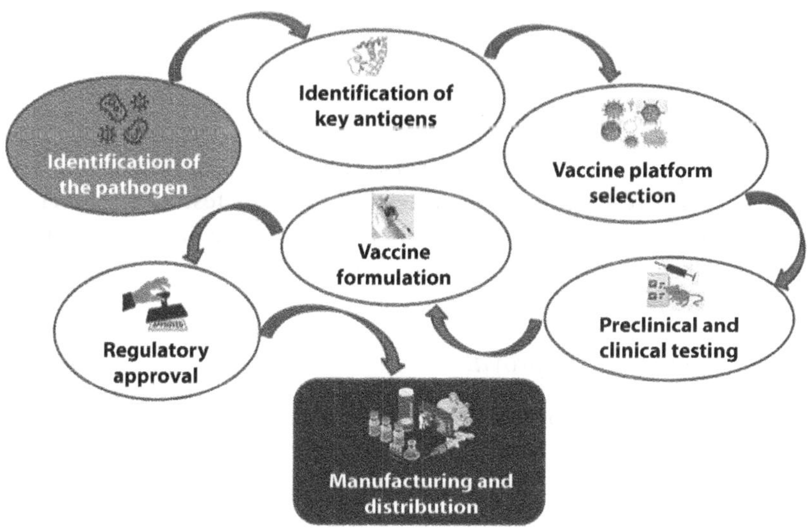

Figure 3.1 The general steps involved in vaccine designing.

3.5.2.1 Identification of the Pathogen

The first step in vaccine designing is to identify the pathogen that causes the disease. This can involve laboratory tests, clinical samples, or epidemiological studies.

3.5.2.2 Identification of Key Antigens

Once the pathogen is identified, the next step is to recognize the potential immunogenic antigens that can stimulate an immune response. Antigens can be proteins, carbohydrates, or nucleic acids that are unique to the pathogen.

3.5.2.3 Vaccine Platform Selection

A vaccine platform is the method used for antigen delivery to the immune system. Selecting an appropriate vaccine platform often depends on factors such as the type of antigen, the age of the target population, and the cost and availability of the vaccine.

3.5.2.4 Vaccine Formulation

The vaccine formulation involves combining the antigen with adjuvant, stabilizers, and other ingredients to enhance the immune response and ensure stability during storage and transportation.

3.5.2.5 Preclinical and Clinical Testing

Before a vaccine can be approved for use, it must undergo rigorous preclinical and clinical testing. Preclinical testing involves testing the safety and efficacy of the vaccine in animals, while clinical testing involves testing the safety and efficacy of the vaccine in humans.

3.5.2.6 Regulatory Approval

Once the vaccine has been shown to be safe and effective, it must be approved by regulatory agencies like, European Medicines Agency (EMA) or the US Food and Drug Administration (FDA).

3.5.2.7 Manufacturing and Distribution

After regulatory approval, the vaccine can be manufactured and distributed to healthcare providers and the general public.

3.6 Intervention of Computer in Vaccine Designing

Computers have revolutionized the field of vaccine designing by enabling the analysis of complex data and the development of predictive models. As the field of vaccine designing continues to evolve, the use of computers is likely to become even more important. Computers allow scientists to analyze large amount of data, simulate complex biological processes, and predict the efficacy of different vaccine candidates. These interventions have the potential to accelerate the development of safe and effective vaccines against a wide range of infectious diseases. Computational models are being used to predict the efficacy of a vaccine by analyzing the binding affinity between a vaccine antigen and a host receptor, and optimizing the dose and route of administration of a vaccine. Computational modeling can be used to simulate the interaction between the vaccine and the immune system, and to predict the optimal vaccine dose and schedule and optimize the design of a vaccine adjuvant [22]. Another computational technique, the high-throughput screening has been used to identify compounds/adjuvant that stimulate an immune response, or enhance the efficacy of existing vaccine [23]. Through bioinformatics approaches, the genome of a pathogen can be analyzed for the potential vaccine targets [24]. Further, using the bioinformatics tools, the 3D structure of a pathogen protein can be predicted to identify candidate epitopes for developing a vaccine. In vaccine designing, data mining and machine learning techniques are used to identify patterns and correlations in clinical trial data, or to predict the safety and efficacy of vaccine candidates.

3.6.1 Computer-Aided Vaccine Design

Computer-aided vaccine design (CAVD) is the use of computational methodologies and tools to aid in the discovery and design of new vaccines. CAVD has the potential to accelerate the development of safe and effective vaccines, by allowing researchers to analyze large amounts of data, predict the efficacy of different vaccine candidates, and optimize vaccine design parameters. CAVD is used in reverse and structural vaccinology.

Reverse vaccinology is a computational method used to identify potential vaccine targets by analyzing the genome of a pathogen. This approach involves the use of bioinformatics tools to identify genes that code for antigens, which are proteins that stimulate an immune response. By using this approach, researchers can identify potential vaccine targets quickly and efficiently [25]. Structural vaccinology is the use of computational methods to predict the structure of antigens and antibodies. By using this approach, researchers can identify specific regions of antigens that are recognized by antibodies, and can design vaccines that target these regions. This approach can help improve the efficacy of vaccines and reduce the risk of side effects [26]. Immune response to a designed vaccine is analyzed and predicted through Immunoinformatics studies where researchers can predict the efficacy of different vaccine candidates and optimize vaccine design parameters such as the dose and schedule. Immunoinformatics can also be used to identify potential adjuvants, which are substances that enhance the immune response to a vaccine [27]. Use of machine learning is being popularized in CAVD to predict potential epitopes as vaccine candidates and also to analyze clinical trial data to identify subgroups of patients

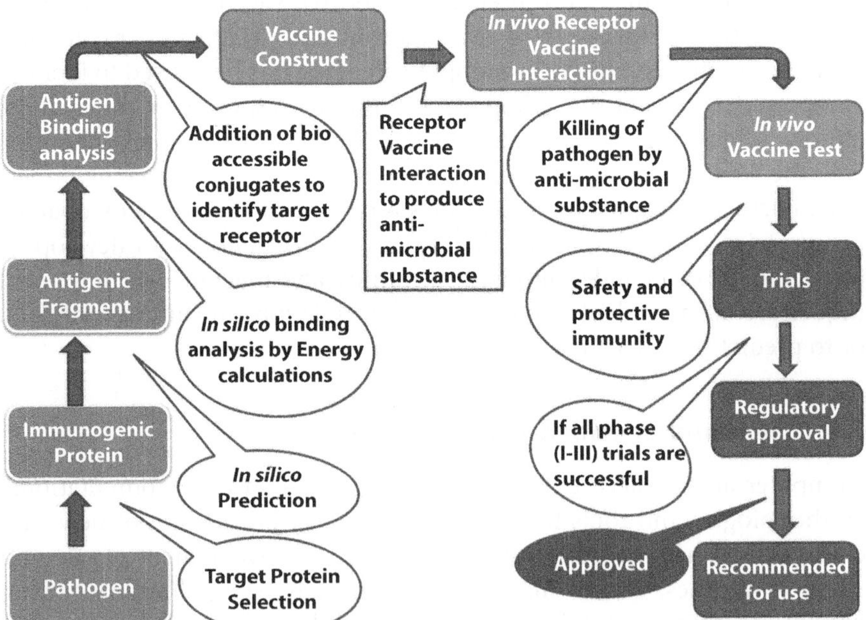

Figure 3.2 Basic steps to develop a vaccine through computer-aided vaccine design.

who are more likely to benefit from a particular vaccine. Machine learning can also be used to predict the safety and efficacy of vaccine candidates, and to optimize vaccine design parameters [28]. The basic steps to develop a vaccine through CAVD are given in Figure 3.2.

3.6.2 Artificial Intelligence for Vaccine Design

Artificial intelligence (AI) has emerged as a powerful tool for vaccine design in recent years. AI algorithms have the ability to analyze voluminous data and to identify significant patterns by allowing the rapid identification of potential vaccine candidates and can considerably accelerate the process of vaccine development. AI is applied to vaccine designing for epitope prediction, antigen selection, vaccine formulation, clinical trial design, etc. Machine learning algorithms being a part of AI are used to predict potential epitopes (both B-cells and T-cells) in a pathogen's protein sequences, which can then be used to design a vaccine. In addition to epitope prediction, AI is also used to predict the interactions of epitopes with host immune receptors for identifying new vaccine targets and optimizing vaccine efficacy. It helps predict the most effective combination of antigens and adjuvants, optimal vaccine formulation, and delivery method based on the specific characteristics of the antigen and the host. AI for optimizing the clinical trial design, including selecting the most appropriate patient populations and dosing regimens, has also been established [29]. Further, AI has a crucial role in identifying potential side effects and toxicity of vaccine components before they are tested in animals and humans. The rapid development of COVID-19 mRNA vaccines is an undeniable evidence of AI-supported vaccine development for combating various pathogens [20, 30, 31]. Demonstrating the potential of AI in this arena, Rahman *et al.* [32] utilized *in silico* techniques to identify 22 membrane proteins within the *Helicobacter pylori* proteome as promising antigens. Similarly, for *Acinetobacter baumannii*, Singh *et al.* [33] employed AI-driven methodologies to suggest and subsequently validate the potential candidate vaccines from an outer membrane protein named FilF (a putative pilus assembly protein). Another approach to using AI in vaccine design is through the use of molecular docking simulations, which can predict the binding of antigens to specific host immune receptors. These simulations can be used to predict the binding of potential epitopes to MHC molecules, which can help researchers to identify novel T-cell epitopes for vaccine development.

3.6.3 Epitope Prediction in Vaccine Designing

The discovery of antigenic sites on proteins holds significant importance in the development of synthetic peptide vaccines, immunodiagnostic tests, antibody production, and allergy research. Traditional experimental approaches for identifying these sites are both time-consuming and resource-intensive. Hence, there is a pressing need for computational tools capable of accurately predicting these sites. Epitope prediction, the process of computationally identifying short protein segments or peptides recognized by antibodies or T-cell receptors, addresses this need. These prediction techniques fall under two main categories: sequence-based and structure-based methods. Sequence-based methods analyze the primary amino acid sequence of the protein or peptide, while structure-based methods utilize the 3D structure of the protein or peptide for epitope prediction.

3.6.3.1 Epitopes: The Antigenic Determinants

An epitope, alternatively referred to as an antigenic determinant, denotes a distinct area on the surface of an antigen, like a protein, virus, or bacterium that triggers recognition by the immune system, particularly by T-cells, B-cells, or antibodies. Upon encountering a foreign entity, the immune system generates antibodies that identify and attach to the epitope(s) located on the antigen's surface. This binding triggers a series of immune responses that can lead to the destruction or elimination of the foreign substance. Understanding the structure and location of epitopes is essential for the development of vaccines, diagnostic tests, and therapies that target specific pathogens or diseases. Basically, these are short sequences of amino acids (peptides) having antigenic and immunogenic properties. Specifically, antibodies or T-cell receptors within the adaptive immune system are responsible for recognizing epitopes. These epitopes are situated on the surface proteins of pathogens like viruses or bacteria. There are two distinct types of epitopes: linear and conformational. Linear epitopes consist of a continuous sequence of amino acids, while conformational epitopes result from the 3D folding of a protein or protein complex. Identifying and designing vaccines for linear epitopes is generally more straightforward, as they can be predicted from the primary amino acid sequence of a protein. On the other hand, conformational epitopes present a greater challenge as their identification requires knowledge of the protein's 3D structure, which can be more challenging to acquire. Epitopes are essential for the development of vaccines, as they are the parts of a pathogen that elicit an immune response. By designing vaccines that contain epitopes from a pathogen, the immune system can be primed to recognize and respond to the pathogen if it is encountered in the future.

T-cell epitopes consist of shot peptide sequences recognized by T-cells, a category of white blood cells involved in cell-mediated immunity. These epitopes are present on the surfaces of pathogenic microorganisms such as viruses, bacteria, parasites, and cancer cells. Major histocompatibility complex (MHC) molecules facilitate the presentation of T-cell epitopes to T-cells. MHC molecules form complexes with T-cell epitopes, and display them on the surfaces of infected cells. This display enables T-cells to identify and target infected cells for elimination. T-cell epitopes play a crucial role in vaccine development by inducing a T-cell-mediated immune response against pathogens. Researchers employ various computational methods to predict the peptides that can bind with MHC molecules and be recognized by T-cells. A B-cell epitope is a precise region on the antigenic surface that is recognized and bound by antibodies produced by B-cells. B-cell epitopes can be either linear or conformational, and they play a pivotal role in the immune response against pathogens. The B-cell produces an antibody as soon as it recognizes a B-cell epitope and the antibody specifically binds to that epitope. This binding triggers a series of immune responses, including the activation of other immune cells that can lead to the elimination of the foreign substance. B-cell epitopes can be identified using various methods, such as peptide mapping, phage display, and protein microarrays. Understanding the location and structure of B-cell epitopes is important for the design of vaccines and immunotherapies that can target specific pathogens or diseases.

3.6.3.2 Sequence-Based Epitope Prediction

Vaccines designed based on epitope sequences can prompt the immune system to recognize specific antigenic regions, thereby triggering the production of antibodies [34]. The process of predicting epitopes from pathogen protein sequences involves employing diverse algorithms and computational tools to analyze the amino acid sequence of the pathogen protein. Many current epitope prediction tools utilize machine learning algorithms like random forests, support vector machines, artificial neural networks, deep learning, etc. In sequence-based epitope prediction, it is not feasible to directly input sequences into machine learning algorithms. Consequently, sequences are typically encoded into numeric feature vectors using various amino acid encoding schemes and are then employed to train the model [35]. Numerous successful epitope prediction methods have been proposed in the past, demonstrating efficacy in identifying potential epitopes from pathogenic protein sequences. Table 3.1 provides an overview of various tools available for sequence-based epitope prediction.

Table 3.1 Major tools for sequence-based epitope prediction.

Tool	Epitope type	Descriptors used	MLA algorithm	Link
BepiPred [36]	B-cell epitopes	Physicochemical properties and other sequence-based features	Random forest	https://services.healthtech.dtu.dk/services/BepiPred-2.0/
ABCpred [37]	B-cell epitopes	Physicochemical properties of amino acids, as well as their local and global interactions within the protein	Artificial Neural Networks	http://crdd.osdd.net/raghava/abcpred/
NetMHC [38]	T-cell epitopes for MHC class I binding peptides	Analyzing peptide–MHC binding	Artificial Neural Networks	https://services.healthtech.dtu.dk/services/NetMHC-4.0/
NetMHCIIPan [39]	T-cell epitopes for MHC class II binding peptides	Analyzing peptide-MHC binding	Artificial Neural Networks	https://services.healthtech.dtu.dk/services/NetMHCIIpan-4.0/
MHCPred [40]	T-cell epitopes	Analyzing peptide–MHC binding	Deep learning algorithm	http://www.ddg-pharmfac.net/mhcpred/MHCPred/

(Continued)

Table 3.1 Major tools for sequence-based epitope prediction. (*Continued*)

Tool	Epitope type	Descriptors used	MLA algorithm	Link
SVMTriP [41]	B-cell and T-cell epitopes	Tri-peptide similarity and propensity scores	Support Vector Machine (SVM)	http://sysbio.unl.edu/SVMTriP/
EpiMatrix [42]	T-cell epitopes	Uses a library of known epitopes sequence similarity	Matrix-based method	https://epivax.com/immunogenicity-assessment/ivax-web-based-vaccine-design
ProPred1 [43]	T-cell epitopes	Binding of peptides	Matrix-based method	http://www.imtech.res.in/raghava/propred1/

3.6.3.3 Structure-Based Epitope Prediction Methods

Structure-based epitope prediction methods utilize the 3D structure of the antigen and/or its complex with an antibody or receptor to predict epitopes. Table 3.2 shows the major tools for structure-based epitope prediction along with the algorithm and access link. The methods for structure-based epitope prediction are based on several concepts like computational alanine scanning, mapping of electrostatic potential, protein–protein docking, molecular dynamics simulation, conformational epitope prediction, and epitope mapping by Nuclear Magnetic Resonance (NMR) spectroscopy, which are briefly described in the following subsections.

Computational Alanine Scanning (CAS): CAS is a method that predicts the energetic contribution of each residue in the antigen to the antibody–antigen interaction [51] where each residue in the antigen structure is mutated to alanine and the change in binding affinity of the antigen–antibody/receptor complex is calculated [52]. The residues with the largest energy contribution are predicted to be part of the epitope. CAS has been used to predict epitopes for various antigens, including influenza hemagglutinin, Ebola virus glycoprotein, HIV-1 envelope protein, and SARS-CoV-2 spike protein.

Mapping of Electrostatic Potential: This method predicts epitopes by identifying regions on the antigen surface with complementary electrostatic

Table 3.2 Major tools for structure-based epitope prediction.

Tool	Epitope type	Method	Algorithm	Link
DiscoTope [44]	B-cell epitopes	By analyzing three-dimensional structure of protein antigens	An algorithm that combines energy-based and spatial-based methods to predict epitopes	http://tools.iedb.org/discotope/
EpiDock [45]	MHC Class II T- Cell Epitopes	By analyzing the three-dimensional structure of a protein and predicts antibody-antigen interactions	Docking algorithm	https://www.ddg-pharmfac.net/epidock/
PatchDock [46]	B-cell epitopes (predicts the binding mode between an antibody and antigen)	Geometric matching of antigen–antibody 3D structures to identify potential binding orientations	Docking algorithm and energy-based scoring function	https://bioinfo3d.cs.tau.ac.il/PatchDock/
ElliPro [47]	B-cell epitopes	Analyzing the protein antigen structure	An algorithm that calculates the protrusion index of each residue on the protein surface and identifies the residues that are most exposed and solvent-accessible.	http://tools.iedb.org/ellipro/

(*Continued*)

Table 3.2 Major tools for structure-based epitope prediction. (*Continued*)

Tool	Epitope type	Method	Algorithm	Link
SEPPA (Spatial Epitope Prediction of Protein Antigens) [48]	B-cell epitopes	By analyzing the three-dimensional structure of protein antigens	An algorithm that identifies potential epitope residues based on the geometric and physicochemical properties of the protein surface	http://bidd2.nus.edu.sg/SEPPA3/ or at http://www.badd-cao.net/seppa3/index.html
COBEpro [49]	B-cell epitopes	Uses a combination of accessible surface area and residue conservation to identify potential epitopes	Support vector machine	http://scratch.proteomics.ics.uci.edu.
BEpro [50]	B-cell epitope	Amino-acid propensity scores and half sphere exposure values at multiple distances	Epitope scoring calculated using a linear combination of terms	https://pepito.proteomics.ics.uci.edu/

potential to the antibody/receptor [51]. The method uses Coulomb's law to calculate the electrostatic potential of the antigen and the antibody/receptor. The method has been used to predict epitopes in various antigens, including hepatitis C virus E2 protein, dengue virus envelope protein, and HIV-1 gp120 protein.

Protein–Protein Docking: This method involves docking the antigen and the antibody/receptor structures to predict their binding mode and the epitope [53]. The docking algorithm predicts the binding orientation of the two proteins, and the residues at the interface are predicted to be part of the epitope. This method has been used to predict epitopes in various antigens, including the human papillomavirus E6 protein, the zika virus envelope protein, and the Ebola virus glycoprotein.

Molecular Dynamics Simulation: Molecular dynamics simulation is used to predict the conformational changes in the antigen and the antibody/receptor upon binding [54]. It predicts the residues that show large conformational changes and are located at the interface of the antigen and the antibody/receptor. This method has been used to predict epitopes in various antigens, including the SARS-CoV-2 spike protein, the influenza hemagglutinin protein, and the hepatitis B virus surface antigen.

Conformational Epitope Prediction (CEP): CEP predicts epitopes based on the antigen's 3D structure and binding site of the antibody [55]. It uses a scoring function that considers various physicochemical properties of the binding site residues, such as hydrophobicity, charge, and solvent accessibility, to predict potential epitopes. CEP has been applied in the recent past to predict epitopes for an array of antigens, including viral proteins and allergens.

Epitope Mapping by NMR Spectroscopy (EMNS): EMNS uses NMR spectroscopy to map the epitope on an antigen [56]. It involves labeling the antigen with stable isotopes and measuring the NMR spectra of the antigen in the presence and absence of the antibody. The difference in the spectra provides information about the epitope. EMNS has been used to map epitopes on various proteins, including HIV-1 gp120 and influenza hemagglutinin.

3.7 *In Silico* Vaccine Designing: Agricultural Applications

In silico vaccine designing is being implemented to develop vaccines for many infectious diseases occurring in agriculturally important animals and birds. Mugunthan and Mani Chandra [57] employed a computational reverse vaccinology strategy to develop a potential multi-epitope vaccine candidate. This vaccine consisted of cytotoxic and helper T-cell epitopes as well as B-cell epitopes, all predicted from the antigenic proteins of *M. gallisepticum*. This pathogen has become a significant challenge in the poultry industry, leading to substantial economic losses. Beyond the computational design of the multi-epitope vaccine, the researchers predicted its tertiary model and subjected it to refinement and validation using various bioinformatics tools. Furthermore, they conducted molecular docking experiments, confirming the effective binding between the multi-epitope vaccine construct and chicken TLR-2 and 5 receptors. Among various bovine diseases, Sahu *et al.* [58] have predicted a total of 78 potential MHC class I T-cell epitopes for four diseases, of which 27 were for FMD, 25 were for bovine viral diarrhea, 12 were for calf scour, and 14 were for infectious bovine rhinotracheitis. Further, they have also proposed a principal component analysis-based epitope score (PbES) that estimates a consensus score for seven different existing epitope prediction algorithms. In crop agriculture, vaccines work by priming the plant's immune system for recognizing and responding to specific pathogens. Researchers at the University of Florida developed a plant-based vaccine using the bacterium that causes citrus greening, which was shown to be effective in protecting citrus trees from the disease [59]. Apart from these examples, few more applications of CAVD in agriculture exists; however, the area is yet to be explored fully for many contagious diseases of animals and birds that cause huge economic loss to farmers.

3.8 Conclusion and Future Prospects

The field of *in silico* vaccine design is rapidly advancing, and it is expected to play a significant role in the development of new vaccines in the future. One of the main advantages of *in silico* vaccine design is that it can significantly reduce the time and cost of vaccine development. By using computational methods to predict the immune response, researchers can narrow down the list of potential vaccine candidates, which can then be tested in the animal models and eventually in the clinical trials. Another advantage of *in silico* vaccine design is that it can allow for the development of vaccines that are tailored to specific populations or even individuals. The progress made in the field of machine learning and artificial intelligence has contributed to the creation of sophisticated computational techniques for *in silico* vaccine design. Recent advancements in machine learning and artificial intelligence have also led to the development of more sophisticated computational methods for *in silico* vaccine design. These methods can analyze large amounts of data and predict the immune response with greater accuracy, which could lead to the development of more effective vaccines.

The future scope of CAVD in agriculture is highly promising, offering several potential avenues for advancement and development. CAVD can facilitate the identification of specific pathogenic targets in agriculturally important organisms, enabling the development of vaccines that precisely target and neutralize these threats. By utilizing advanced computational models and simulations, computer-aided vaccine design can expedite the vaccine development process. This acceleration can significantly reduce the time and resources required for traditional vaccine development, allowing for the rapid deployment of effective vaccines to combat emerging agricultural diseases. Through computational analysis of various host-pathogen interactions, CAVD can optimize the efficacy of vaccines in agriculture. By considering the unique characteristics of different farming systems, CAVD can maximize the efficacy of vaccines and minimize the risks associated with prevalent diseases, contributing to sustainable and efficient agricultural practices. The future of computer-aided vaccine design in agriculture holds tremendous potential to revolutionize disease management practices, foster sustainable agricultural production, and contribute to global food security. As advancements in technology continue to unfold, this approach is poised to play a pivotal role in shaping the future of agricultural innovation and resilience. However, it is important to note that CAVD is still in its early stages, and further research and development are needed to fully realize its potential in agriculture.

References

1. Quintana-Rodriguez, E., Duran-Flores, D., Heil, M., Camacho-Coronel, X., Damage associated molecular patterns (DAMPs) as future plant vaccines that protect crops from pests. *Sci. Hortic.*, 237, 207–220, 2018, https://doi.org/10.1016/j.
2. Flower, D.R., Computer-Aided Vaccine Design. *Hum. Vaccin. Immunother.*, 10, 2, 241–243, 2014, doi: 10.4161/hv.26687.
3. FAO (Food and Agriculture Organization of the United Nations), Foot-and-Mouth Disease, 2021, http://www.fao.org/animal-health/animal-diseases/foot-and-mouth-disease/en/.
4. USDA (United States Department of Agriculture), Johne's Disease, 2021, https://www.aphis.usda.gov/aphis/ourfocus/animalhealth/animal-disease-information/cattle-disease-information/johnes-disease/ct_johnes_disease.
5. OIE (World Organisation for Animal Health), African Swine Fever, 2020, https://www.oie.int/en/animal-health-in-the-world/animal-diseases/African-swine-fever/.
6. OIE (World Organisation for Animal Health), Animal Diseases, 2021, https://www.oie.int/en/animal-health-in-the-world/animal-diseases/.
7. Benfield, D.A., Collins, J.E., Dee, S.A., Halbur, P.G., Joo, H.S., Lager, K.M., Mengeling, W.L., Murtaugh, M.P., Rossow, K.D., Stevenson, G.W., Zimmerman, J.J., Porcine reproductive and respiratory syndrome, in: *Diseases of Swine*, vol. 18, pp. 201–232, 1999.

8. Perez, A.M., Davies, P.R., Goodell, C.K., Holtkamp, D.J., Mondaca-Fernandez, E., Poljak, Z. *et al.*, Lessons learned and knowledge gaps about the epidemiology and control of porcine reproductive and respiratory syndrome virus in North America. *J. Am. Vet. Med. Assoc.*, 246, 1304–1317, 2015.
9. Rahman, M.S., Rabbani, M.G., Uddin, M.J., Chakrabartty, A., Her, M., Prevalence of Avian Influenza and Newcastle Disease Viruses in poultry in selected areas of Bangladesh using rapid antigen detection kit. *Arch. Clin. Microbiol.*, 3, 1:3, 1–8, 2012, doi: 10.3823/248.
10. CDC (Centers for Disease Control and Prevention), Avian Influenza, 2020, Retrieved from https://www.cdc.gov/flu/avianflu/index.htm.
11. Wang, H., Xu, Y., Feng, W., Porcine Reproductive and Respiratory Syndrome Virus: Immune Escape and Application of Reverse Genetics in Attenuated Live Vaccine Development. *Vaccines*, 9, 5, 480, 2021, https://doi.org/10.3390/vaccines9050480.
12. Gerdts, V. and Zakhartchouk, A., Vaccines for porcine epidemic diarrhea virus and other swine coronaviruses. *Vet. Microbiol.*, 206, 45–51, 2017, doi: 10.1016/j.vetmic.2016.11.029.
13. Smith, L.A. and Rusnak, J.M., Botulinum neurotoxin vaccines: past, present, and future. *Crit. Rev. Immunol.*, 27, 4, 303–18, 2007, doi: 10.1615/critrevimmunol.v27.i4.20.
14. Sundeen, G. and Barbieri, J.T., Vaccines against Botulism. *Toxins (Basel)*, 9, 9, 268, 2017, doi: 10.3390/toxins9090268.
15. Ma, J., Bruce, T.J., Jones, E.M., Cain, K.D., A Review of Fish Vaccine Development Strategies: Conventional Methods and Modern Biotechnological Approaches. *Microorganisms*, 7, 11, 569, 2019, doi: 10.3390/microorganisms7110569.
16. Ravikumar, R., Chan, J., Prabakaran, M., Vaccines against Major Poultry Viral Diseases: Strategies to Improve the Breadth and Protective Efficacy. *Viruse*, 14, 6, 1195, 2022, doi: 10.3390/v14061195.
17. Brun, A., Vaccines and Vaccination for Veterinary Viral Diseases: A General Overview. *Methods Mol. Biol.*, 1349, 1–24, 2016, doi: 10.1007/978-1-4939-3008-1_1.
18. Zaman, K., Bandyopadhyay, A.S., Hoque, M. *et al.*, Evaluation of the safety, immunogenicity, and faecal shedding of novel oral polio vaccine type 2 in healthy newborn infants in Bangladesh: a randomised, controlled, phase 2 clinical trial. *Lancet*, 401, 131–139, 2023.
19. Paavonen, J., Naud, P., Salmerón, J., Wheeler, C.M. *et al.*, Efficacy of human papillomavirus (HPV)-16/18 AS04-adjuvanted vaccine against cervical infection and precancer caused by oncogenic HPV types (PATRICIA): final analysis of a double-blind, randomised study in young women. *Lancet*, 374, 9686, 301–314, 2009.
20. Polack, F.P., Thomas, S.J., Kitchin, N., Absalon, J., Gurtman, A., Lockhart, S. *et al.*, Safety and efficacy of the BNT162b2 mRNA covid-19 vaccine. *N. Engl. J. Med.*, 383, 27, 2603–2615, 2020, doi: &]nbsp;10.1056/NEJMoa2034577.

21. Henao-Restrepo, A.M., Longini, I.M., Egger, M. *et al.*, Efficacy and effectiveness of an rVSV-vectored vaccine in preventing Ebola virus disease: final results from the Guinea ring vaccination, open-label, cluster-randomised trial (Ebola ÇaSuffit!). *Lancet*, 389, 10068, 505–518, 2015.
22. Del Giudice, G., Rappuoli, R., Didierlaurent, A.M., Correlates of adjuvanticity: A review on adjuvants in licensed vaccines. *Semin. Immunol.*, 39, 14–21, 2018.
23. O'Hagan, D.T., Ott, G.S., De Gregorio, E., Seubert, A., The mechanism of action of MF59-an innately attractive adjuvant formulation. *Vaccine*, 29, 27, 4439–4442, 2018.
24. Sunita, Sajid, A., Singh, Y., Shukla, P., Computational tools for modern vaccine development. *Hum. Vaccin. Immunother.*, 16, 3, 723–735, 2020, doi: 10.1080/21645515.2019.1670035.
25. Rappuoli, R., Black, S., Lambert, P.H., Vaccine discovery and translation of new vaccine technology. *Lancet*, 378, 360–368, 2011.
26. Anasir, M., II and Poh, C.L., Structural Vaccinology for Viral Vaccine Design. *Front. Microbiol.*, 10, 738, 2019, doi: 10.3389/fmicb.2019.00738.
27. Flower, D.R., Immunoinformatics. Predicting immunogenicity *in silico*, Preface. *Methods Mol. Biol.*, 409, v–vi, 2007, doi: 10.1007/978-1-60327-118-9.
28. Kim, J.Y., Rosenberger, M.G., Rutledge, N.S., Esser-Kahn, A.P., Next-Generation Adjuvants: Applying Engineering Methods to Create and Evaluate Novel Immunological Responses. *Pharmaceutics*, 15, 6, 1687, 2023, doi: 10.3390/pharmaceutics15061687.
29. Askin, S., Burkhalter, D., Calado, G., El Dakrouni, S., Artificial Intelligence Applied to clinical trials: opportunities and challenges. *Health Technol. (Berl.)*, 13, 2, 203–213, 2023, doi: 10.1007/s12553-023-00738-2.
30. Feikin, D., Higdon, M.M., Abu-Raddad, L.J., Andrews, N., Araos, R., Goldberg, Y. *et al.*, Duration of effectiveness of vaccines against SARS-CoV-2 infection and COVID-19 disease: results of a systematic review and meta-regression. *Lancet*, 399, 10328, 924–944, 2022, doi: 10.1016/S0140-6736(22)00152-0.
31. Zheng, C., Shao, W., Chen, X., Zhang, B., Wang, G., Zhang, W., Real-world effectiveness of COVID-19 vaccines: a literature review and meta-analysis. *Int. J. Infect. Dis.*, 114, 252–260, 2022, doi: 10.1016/j.ijid.2021.11.009.
32. Rahman, N., Ajmal, A., Ali, F., Rastrelli, L., Core proteome mediated therapeutic target mining and multi-epitope vaccine design for Helicobacter pylori. *Genomics*, 112, 5, 3473–3483, 2020, doi: 10.1016/j.ygeno.2020.06.026.
33. Singh, R., Garg, N., Shukla, G., Capalash, N., Sharma, P., Immunoprotective Efficacy of Acinetobacter baumannii Outer Membrane Protein, FilF, Predicted *In silico* as a Potential Vaccine Candidate. *Front. Microbiol.*, 7, 158, 2016, doi: 10.3389/fmicb.2016.00158.
34. Alonso-Padilla, J., Lafuente, E.M., Reche, P.A., Computer-aided design of an epitope-based vaccine against epstein-barr virus. *J. Immunol. Res.*, 2017, 9363750, 2017.

35. Zhang, H., Berg, A.C., Maire, M. *et al.*, SVM-KNN: discriminative nearest neighbor classification for visual category recognition, in: *Proc IEEE Comput. Soc. Conf. Comput. Vis. Pattern Recognit*, New Jersey, United States, Institute of Electrical and Electronics Engineers (IEEE), pp. 2126–2136, 2006.
36. Jespersen, M.C., Peters, B., Nielsen, M., Marcatili, P., BepiPred-2.0: improving sequence-based B-cell epitope prediction using conformational epitopes. *Nucleic Acids Res.*, 45, W1, W24–W29, 2017, doi: 10.1093/nar/gkx346.
37. Saha, S. and Raghava, G.P.S., Prediction of continuous B-cell epitopes in an antigen using recurrent neural network. *Proteins: Struct. Funct. Bioinf.*, 65, 1, 40–48, 2006, doi: 10.1002/prot.21078.
38. Lundegaard, C., Lamberth, K., Harndahl, M., Buus, S., Lund, O., Nielsen, M., NetMHC-3.0: accurate web accessible predictions of human, mouse and monkey MHC class I affinities for peptides of length 8-11. *Nucleic Acids Res.*, 36, Web Server issue, W509–12, 2008, doi: 10.1093/nar/gkn202.
39. Nielsen, M., Lundegaard, C., Blicher, T., Peters, B., Sette, A., Justesen, S. *et al.*, Quantitative predictions of peptide binding to any HLA-DR molecule of known sequence: NetMHCIIpan. *PLoS Comput. Biol.*, 4, 7, e1000107, 2007, doi: 10.1371/journal.pcbi.1000107.
40. Guan, P., Hattotuwagama, C.K., Doytchinova, I.A., Flower, D.R., MHCPred 2.0: an updated quantitative T-cell epitope prediction server. *Appl. Bioinf.*, 5, 1, 55–6, 2006, doi: 10.2165/00822942-200605010-00008.
41. Yao, B., Zhang, L., Liang, S., Zhang, C., SVMTriP: A method to predict antigenic epitopes using support vector machine to integrate tri-peptide similarity and propensity. *PLoS One*, 7, 9, e45152, 2012.
42. Ardito, M., De Groot, A., McMurry, J.A., Moise, L., Yang, W., Martin, W., EpiMatrix: Tool for Accelerated Epitope Selection and Vaccine Design. *Hum. Vaccin. Immunother.*, 11, 9, 2312–2321, 2008, doi: 10.1080/21645515.2015.1061159.
43. Singh, H. and Raghava, G.P., ProPred1: prediction of promiscuous MHC Class-I binding sites. *Bioinformatics*, 19, 8, 1009–14, 2003, doi: 10.1093/bioinformatics/btg108.
44. Haste Andersen, P., Nielsen, M., Lund, O., Prediction of residues in discontinuous B-cell epitopes using protein 3D structures. *Protein Sci.*, 15, 11, 2558–67, 2006, doi: 10.1110/ps.062405906.
45. Ponomarenko, J. and Bourne, P.E., Antibody-protein interactions: benchmark datasets and prediction tools evaluation. *BMC Struct. Biol.*, 7, 1, 64, 2007, doi: 10.1186/1472-6807-7-64.
46. Schneidman-Duhovny, D., Inbar, Y., Nussinov, R., Wolfson, H.J., PatchDock and SymmDock: servers for rigid and symmetric docking. *Nucleic Acids Res.*, 33, Web Server issue, W363–7, 2005, doi: 10.1093/nar/gki481.
47. Ponomarenko, J., Bui, H.H., Li, W. *et al.*, ElliPro: a new structure-based tool for the prediction of antibody epitopes. *BMC Bioinf.*, 9, 1, 514, 2008.
48. Zhou, C., Chen, Z., Zhang, L., Yan, D., Mao, T., Tang, K., Qiu, T., Cao, Z., SEPPA 3.0-enhanced spatial epitope prediction enabling glycoprotein

antigens. *Nucleic Acids Res.*, 47, W1, W388–W394, 2019, doi: 10.1093/nar/gkz413.

49. Sweredoski, M.J. and Baldi, P., COBEpro: a novel system for predicting continuous B-cell epitopes. *Protein Eng. Des. Sel.*, 22, 3, 113–20, 2009, doi: 10.1093/protein/gzn075.
50. Sweredoski, M.J. and Baldi, P., PEPITO: Improved Discontinuous B-Cell Epitope Prediction Using Multiple Distance Thresholds and Half-Sphere Exposure. *Bioinformatics*, 24, 12, 1459–1460, 2008.
51. Kumari, R., Kumar, R., Lynn, A., g_mmpbsa - a GROMACS tool for high-throughput MM-PBSA calculations. *J. Chem. Inf. Model.*, 54, 7, 1951–1962, 2014.
52. Kortemme, T. and Baker, D., A simple physical model for binding energy hot spots in protein-protein complexes. *Proc. Natl. Acad. Sci. U. S. A.*, 99, 22, 14116–14121, 2002.
53. Gray, J.J., Moughon, S., Wang, C., Schueler-Furman, O., Kuhlman, B., Rohl, C.A., Baker, D., Protein-protein docking with simultaneous optimization of rigid-body displacement and side-chain conformations. *J. Mol. Biol.*, 331, 1, 281–299, 2003.
54. Silva, R.P., Huang, Y., Nguyen, A.W. *et al.*, Identification of a conserved S2 epitope present on spike proteins from all highly pathogenic coronaviruses. *Elife*, 12, e83710, 2023, doi: 10.7554/eLife.83710.
55. Kulkarni-Kale, U., Bhosle, S., Kolaskar, A.S., CEP: a conformational epitope prediction server. *Nucleic Acids Res.*, 33, Web Server issue, W168–71, 2005, doi: 10.1093/nar/gki460.
56. Bardelli, M., Livoti, E., Simonelli, L., Pedotti, M., Moraes, A., Valente, A.P., Varani, L., Epitope mapping by solution NMR spectroscopy. *J. Mol. Recognit.*, 28, 6, 393–400, 2015, doi: 10.1002/jmr.2454.
57. Mugunthan, S.P. and Harish, M.C., Multi-epitope-Based Vaccine Designed by Targeting Cytoadherence Proteins of Mycoplasma gallisepticum. *ACS Omega*, 6, 21, 13742–13755, 2021, doi: 10.1021/acsomega.1c01032.
58. Sahu, T.K., Rao, A.R., Meher, P.K., Sahoo, B.C., Gupta, S., Rai, A., Computational prediction of MHC class I epitopes for most common viral diseases in cattle (Bos taurus). *Indian J. Biochem. Biophys.*, 52, 1, 34–44, 2015.
59. Wang, N., Pierson, E., Setubal, J.C. *et al.*, The Candidatus Liberibacter-host interface: Insights into pathogenesis mechanisms and disease control. *Annu. Rev. Phytopathol.*, 54, 189–211, 2016.

4

Genomics to Phenomics: A Paradigm Shift in Crop Science Research

Biswajit Lenka[1], Manasi Dash[2] and Lakesh Muduli[2*]

[1]Department of Genetics and Plant Breeding, Faculty of Agricultural Sciences (IAS), Siksha 'O' Anusandhan (Deemed to be University), Bhubaneswar, Odisha, India
[2]Department of Genetics and Plant Breeding, College of Agriculture, Odisha University of Agriculture and Technology, Bhubaneswar, Odisha, India

Abstract

Novel tools, such as functional molecular markers and informatics, as well as updated statistics and inheritance phenomena are being produced through genomics research. Elucidating the fundamental mechanisms of heterosis and epigenetics and manipulating them have enormous potential. Large-scale data creation from the genome, transcriptome, proteome, metabolome, and epigenome, among other sources, has become common in a number of plant species in recent years. To comprehend the flow of biological information that underlies complex characters, a systems biology approach comprising integration of various omics data, modeling, and prediction of cellular processes is necessary. As more genomics, phenomics, and environmental information are available, advancements in the statistical and quantitative genetic models in multi-environment conditions have become more apparent. For better results, particularly in terms of prediction accuracy, an effective data-based integration through the establishment of linkage between genomics and phenomics is necessary.

Keywords: Functional molecular markers, heterosis, epigenetics, genomics, phenomics

**Corresponding author*: silulakesh@gmail.com

Hrudayanath Thatoi, Sonali Mohapatra, Swagat Kumar Das and Sukanta Kumar Pradhan (eds.) Applied Biotechnology and Bioinformatics: Agriculture, Pharmaceutical Research and Environment, (97–108) © 2025 Scrivener Publishing LLC

4.1 Introduction

Growing genetic gains are evidence of the effectiveness of well-established public and private plant breeding programs. However, over the past two decades, an increasing number of contemporary technologies with appealing advantages have been effectively integrated into the conventional breeding methods. Cultivars with robust yields will be delivered by incorporating these novel technologies in response to the anticipated unfavorable environmental conditions brought on by climate change. The novel technologies such as genomics rapid cycling [1], high-throughput phenotyping (HTP, phenomics), genome editing tools [2], and genome-wide association studies (GWAS) [3] play a pivotal role in modern-day crop improvement. A brief description on the implementation of such technologies, namely, genomics and phenomics, for cost-effective and time-saving plant breeding has been made in this chapter.

4.2 Genomics in Crop Improvement

Recent genetic and genomic developments have substantially improved our knowledge on the structural and functional characteristic of plant genomes and have integrated fundamental information in novel ways. It could improve our capacity to adapt crop plants to our advantage. Novel varieties can be developed through exercise of improved selection techniques or improved attributes caused by combinations of superior alleles at multiple loci. Thorough understanding of the biological principles may result in enhanced screening techniques that enable more effective selection of superior genotypes. The prediction of breeding values is one of the claims of genomics in plant breeding. Genomics and bioinformatics are rapidly expanding fields of research, being fueled by the continued growth and reducing cost of DNA sequencing and genotyping. Genomic selection (GS) shortens cycle time, improves estimation of breeding values, and enhances precision in selection. The number of crop improvement programs is switching from conventional to genomic selection. Genomic selection has been used in several crops, including rice, wheat, chickpea, and cassava [1, 4, 5].

4.3 Advances in Genomics-Assisted Breeding

Recent advancements in genetics and genomics have significantly enhanced our understanding of the structural and functional characteristics of plants.

Complete genome sequencing of many crops is completed. Two main strategies exist to benefit from genomics. First, a deeper comprehension of the biological principles may result in brand-new or enhanced screening techniques that enable more effective selection of superior genotypes. Second, new information can enhance decision-making for more effective breeding.

A deep understanding of the concepts and genomics-based tools is a need for genomics assisted crop improvement. Since the 1940s, researchers have researched artificial neural networks (ANNs), which were first inspired by the animal nervous system. Among the many different types of ANNs, deep learning network (DNN) is a still-emerging subfield. DNN, a subfield of machine learning, differs from ANN by having a significant number of extra hidden layers [6]. Convolutional Neural Networks (CNNs), a subset of DNNs having at least a convolutional layer, are capable of automatically extracting features from a continuous signal like plant image, DNA, or RNA sequence.

In plant breeding programs, the development of low-cost genetic markers through array-based genotyping and genotyping-by-sequencing [7] has opened the door for the implementation of genomic selection, also known as genomic assisted breeding. The strategy of employing genome-wide molecular markers for assessing an individual's overall success in breeding populations was first put forth by Meuwissen *et al.* [8] for use in animal breeding, but it has since been also proven across numerous plant species. The optimization of the training set composition is a crucial problem in genomic selection. Breeding initiatives must begin with early core parents (training populations) that reflect the genetic variety prevalent in the current progeny and to the maximum extent possible, and adhere to the testing population [9].

Over the past few decades, transcriptome analysis has progressed from Northern blotting to RNA sequencing (RNA-seq) through various techniques, the most well-known being microarray and real-time quantitative polymerase chain reaction (qRT PCR). Under specific conditions, the transcript pool of a plant component is a group of mRNAs that will pave for certain biochemical reaction of a plant to particular stimuli [10]. The basic physiology, developmental processes, and environmental stresses have been well studied with the aid of macro- and microarray techniques. Practical application of these technologies in plant breeding has been limited as the differential gene expression is affected by both trait of interest and background genetic diversity with an exception for near isogenic lines (NILs). Thus, the background effect should be eliminated in order to establish an association between level of gene expression and trait under study.

Genes that encode transcription factors (TFs), which play a key role in regulating proteins in all living cells, have the power to affect a wide range of biological processes.

The development of third-generation sequencing and NGS (next-generation sequencing) technologies has led to ever-increasing throughput and decreased sequencing prices [11]. The data produced by high-throughput genome sequencing also enable the simultaneous identification and sequencing of hundreds of genetic markers throughout entire genomes. These novel genome sequencing tools are useful for validation of genetic markers in different mapping populations [12]. Characterizing the genetic diversity seen in landraces and wild crop relatives preserved in gene banks is also a useful way to find novel genes that could increase yield and resilience to biotic and abiotic challenges. Further artificial selection has impacted hundreds of targets, including genes and nongenetic areas, resulting in an increase in the proportion of unusual alleles and a decrease in nucleotide diversity.

High-throughput resequencing methods have been applied in several crops to investigate genetic diversity and understand the molecular basis of important agronomic traits. High levels of sequence similarity are required to map the short reads onto the reference genome, which is necessary for the characterization of the genetic variations. Pangenomics investigations have therefore been conducted with the aim of capturing all potential variants within the germplasm of a certain species [13]. Genes or sequences that are core to all individuals and those that are accessory to some make up the pangenome. Pangenomics investigations have therefore been conducted with the aim of capturing all potential variants in a certain germplasm collection of a particular species [14]. The wheat (*Triticum aestivum* L.) pangenome was constructed using 18 cultivars, in which 128,656 genes were found, of which 64.3% were core genes and the rest were variable and showed presence–absence variation [15]. Similarly, eight top-notch rapeseed (*Brassica napus* L.) genomes used in a pangenome study revealed architecture and ecotype differences [16]. Pangenome offers thorough analysis of nonessential and species-specific genes discovered primarily from gene pools. Diverse gene stock derived from wild species is essential for achieving a crop's full genetic repertoire. Here, we introduce the idea of a super pangenome, which, by integrating several pangenomes from all the species within a given genus, reflects a whole genomic variation repertoire [13]. Utilizing the genetic diversity of wild species will aid in tapping the super-pangenome concept. It will speed up crop improvement by merging the wild side of a species with varied genetic stock.

High-density SNP arrays and DNA re-sequencing have shed light on most of the genotypic space for a variety of organisms over the past ten years, including humans, maize, rice, and *Arabidopsis*. Based on the two above tools, a new approach that is being frequently used is Genome-Wide Association Studies (GWAS). GWAS provides an opportunity to identify genomic variant across many individuals.

The fundamental strategy in GWAS is to decipher the relationship between each genotyped marker and an interesting phenotype that has been assessed over a wide range of population. This strategy was developed in human genetics around ten years ago [17]. Today, GWAS are often used in a variety of model species, such as Arabidopsis [18] and mice [19] as well as non-model systems like crops and cattle [20]. In GWAS, connections between phenotypic variation and genomic polymorphisms (such as SNPs, insertions and deletions, and structural variants) are looked for. The analysis of the related genomic structures is greatly facilitated by GWAS [21]. The following are the main benefits of GWAS that have been emphasized, as opposed to conventional gene mapping techniques that demand a lot of work to establish mapping groups. First, GWAS significantly shortens study time by using natural populations as research subjects. Second, it is highly effective with excellent resolution. Third, GWAS allows for the simultaneous detection of several alleles at a given site because it is carried out on a genome-wide scale. The genetic data and germplasm materials from a single natural population can be used for the genetic analysis of multiple traits, avoiding the need for repeated population establishment and significantly lowering the cost of gene sequencing. GWAS has been widely used in a variety of crops with high-throughput phenotyping (Table 4.1).

The above discussion highlighted how different genomics branches gain importance over traditional approach of crop improvement after emergence of molecular genetics. Also, at the same time, we have observed the integration of genomics methods with phenotyping for effective selection. Phenotyping was once the sole approach for selection when conventional plant breeding was the only method of crop improvement. However, later with the advent of molecular genetics, it was dominated by genomics and again it regained its equal value when reverse genetics came into existence. Thus, this reverse genetic approach again shifted the goal of improvement by giving equal importance to genotyping as well as phenotyping.

Table 4.1 Applications of high-throughput phenotyping integrated with GWAS in plants.

Crop	Methods	Traits	Number of associated loci/SNP/QTL/ candidate genes	References
Rice	X-ray CT, color-imaging Greenhouse 141 loci	15 morphological traits (e.g., plant height, tiller number, green leaf area)	141 loci	[22]
Rice	Visible light imaging	49 panicle traits (e.g., panicle length, rachis length, primary branch number)	10 candidate genes	[23]
Rice	Visible light/RGB imaging	Projected shoot area	7 QTLs	[24]
Winter wheat	RGB imaging	*Septoria tritici* blotch infected traits	26 chromosome segments	[25]
Durum wheat	Unmanned aerial vehicles integrated with multispectral imaging, tractor-based system integrated with Green Seeker spectral sensors	Normalized difference vegetation index (NDVI)	46 QTLs	[26]
Maize	Near-infrared, visible light/ RGB and fluorescence imaging	Plant fresh weight, plant dry weight, and biovolume estimation at 11 different developmental time points	12 MTAs, 6 pairs of epistatic interactions	[27]
Barley	Chlorophyll fluorescence	23 traits, including 19 chlorophyll fluorescence	205 markers	[28]

4.4 Phenotyping

Phenotyping always plays a crucial role in plant improvement as it is a measurement of the ultimate observable expression of a particular trait or group of traits. Thus, the accurate detection of a particular trait and its level of expression are otherwise known as phenotyping [29]. Accurate phenotyping of a crop plant for concerned characteristics is a tough task for a variety of reasons. The phenotype of a plant is the result of the interaction of the genotype of the plant, which is the stationary factor, and its surrounding environment, which is the variable factor [30]. In the entire life span as the plant encounters various environments, the phenotypic expression varies because of the genotype × environment interaction on trait expression. The above explanation regarding phenotypic variation is almost true for a monogenic or oligogenic trait where there is absence of gene × gene interaction on trait expression. However, when a character is governed by more than two genes (polygenic trait), interaction among various combinations of genes may also act as an addition factor on trait expression besides the environmental effect.

The effectiveness of phenotyping should consider all the factors and interactions affecting the trait. In the last two decades, efforts are being shifted towards high-throughput phenotyping to improve selection efficiency, which created a new branch in omics, i.e., phenomics.

4.5 Phenomics

Plant phenomics deals with the study of plant phenotype of a large population under a wide range of environmental conditions. Huge datasets from a large plant population across a diverse environment can capture all possible phenotypic variations and help in predicting the contribution of genetic components in overall expression of traits. Plant phenomics is a combined study of growth, varietal performance, and biochemical composition based on the huge data collected through high-throughput phenotyping approaches. Thus, it is a multi-disciplinary branch that needs expertise in different disciplines involving biology, genetics, molecular biology, mathematics, statistics, and information science.

4.6 Phenomics Approaches in Crop Improvement

All the phenomics approaches aim to evaluate the entire phenotype or traits of concern of large population grown under multiple environmental conditions. The important aspect of phenomics is the efficiency and accuracy in data collection that depend on the choice of appropriate strategies. The strategies should include the key factors like the developmental stages suitable for phenotyping, prevailing environmental condition at the time of sample collection, sufficient storage facility, and analysis pipelines. The classical phenomics approaches center on visual assays of the trait on individual plant and evaluation of biochemical constituents of the target trait. In this method, phenotyping is robust and accurate for few characters, but it is tedious, time-consuming, and expensive when data are recorded for multiple characters under a wide range of environmental conditions. The recent approaches are the refinements of the classical assay, i.e., high-throughput phenotyping (HTP). One of the widely used HTP is image based, in which acquisition of data is carried out by high-resolution cameras working on the visible, near-infrared, infrared, fluorescent, and hyper-spectral range. These cameras, when operated manually, are ether mounted on farm suitable vehicles or drones. In recent days, application of remote sensing technology has enhanced the HTP by gathering a large dataset of multiple areas for multiple traits in a few hours using an automated imaging system.

Phenotyping through HTP approaches is observed depending on plant types. Sometimes, phenotyping is done by keeping the plants fixed and moving the cameras in the case of *Arabidopsis*, a small plant with a simple architecture. However, in the case of plants with a complex architecture such as wheat, barley, and maize, observations are taken by keeping camera fixed and moving the plant parts [29]. This HTP system is also provided with appropriate software to regulate data acquisition as well as data processing such as conversion and normalization. These normalized processed data are always suitable for comparison and interpretation into valid results.

This HTP system has several advantages over the conventional methods. These approaches are non-destructive types as there is no tissue destroyed for analysis. Thus, observation can be taken for the same plant throughout its life cycle for several traits and also at different growth stages. These systems are easy to operate and highly sensitive with high-resolution cameras and multiple sensors. Even a single image data can be taken for several traits of a plant and also the whole plot or field can be analyzed in a single plot. Now, the application of cloud computing, artificial intelligence, and machine learning has enhanced the capability of data storage and processing by reducing the cost and saving time. Also, the HTP data are amenable to create a robust model for future prediction. However, this system needs proper standardization and validation software and models before data acquisition; otherwise, it will result in a biased estimate of the phenotype and biased selection, if any, which will lead to irregularity in standardization.

4.7 Conclusion

In the area of plant science research, selection of a potential plant is always a challenging task especially in the case of crop plants for fulfilling the demand of food security. During the time of green revolution when stabilizing the hunger index was the prime concern, selection was practiced mainly for increasing yield. Thus, a little effort in selection in those days particularly based on plant phenotype resulted in a better yield. As time passes by, the goal of selection was oriented towards climate resilience besides improving yield and related attributes. Although nutritional food security was almost achieved during the green revolution, the phenotypic basis of selection was ineffective in frequently changing environmental conditions, because the environment is a major factor that affects the trait expression and genetic gain in advance generation of selected individuals. Realizing this fact, the genetic basis of selection as an innate potential factor gained importance with the advancement in molecular biology and biotechnology. Various genomic technology and tools like molecular markers, genetic engineering, tissue culture, molecular cloning, next-generation sequencing, and genotyping-based sequencing enhanced the genomic selection for traits of interest directly from the population or selection after modifying the plant genetic architecture. There is no doubt about the phenomenal contribution of genomics in plant improvement, but tools like Qtl mapping and GWAS show how more efficient results are obtained when genomics is integrated with phenomics. Thus, research should prioritize genomics studies treated in a manner similar to phenomics investigations.

References

1. Crossa, J., Pérez-Rodríguez, P., Cuevas, J., Montesinos-López, O., Jarquín, D., De Los Campos, G., Burgueño, J., González-Camacho, J.M., Pérez-Elizalde,

S., Beyene, Y., Dreisigacker, S., Genomic selection in plant breeding: methods, models, and perspectives. *Trend Plant Sci.*, 22, 11, 961–975, 2017.
2. Zhang, K., Raboanatahiry, N., Zhu, B., Li, M., Progress in genome editing technology and its application in plants. *Front. Plant Sci.*, 8, 177, 2017.
3. Tam, V., Patel, N., Turcotte, M., Bossé, Y., Paré, G., Meyre, D., Benefits and limitations of genome-wide association studies. *Nat. Rev. Genet.*, 20, 8, 467–484, 2019.
4. Roorkiwal, M., Rathore, A., Das, R.R., Singh, M.K., Jain, A., Srinivasan, S., Gaur, P.M., Chellapilla, B., Tripathi, S., Li, Y., Hickey, J.M., Genome-enabled prediction models for yield related traits in chickpea. *Front. Plant Sci.*, 7, 1666, 2016.
5. Wolfe, M.D., Del Carpio, D.P., Alabi, O., Ezenwaka, L.C., Ikeogu, U.N., Kayondo, I.S., Lozano, R., Okeke, U.G., Ozimati, A.A., Williams, E., Egesi, C., Prospects for genomic selection in cassava breeding. *Plant Genome*, 2017.
6. Wang, H., Cimen, E., Singh, N., Buckler, E., Deep learning for plant genomics and crop improvement. *Curr. Opin. Plant Biol.*, 54, 34–41, 2020.
7. Peterson, B.K., Weber, J.N., Kay, E.H., Fisher, H.S., Hoekstra, H.E., Double digest RADseq: An inexpensive method for de novo SNP discovery and genotyping in model and non-model species. *PloS One*, 7, e37135, 2012.
8. Meuwissen, T.H.E., Hayes, B.J., Goddard, M.E., Prediction of total genetic value using genome-wide dense marker maps. *Genetics*, 157, 1819–1829, 2001.
9. Hickey, J.M., Crossa, J., Babu, R., de los Campos, G., Factors affecting the accuracy of genotype imputation in populations from several maize breeding programs. *Crop Sci.*, 52, 654–663, 2012.
10. Agarwal, P., Parida, S.K., Mahto, A., Das, S., Mathew, I.E., Malik, N., Tyagi, A.K., Expanding frontiers in plant transcriptomics in aid of functional genomics and molecular breeding. *Biotechnol. J.*, 9, 12, 1480–1492, 2014.
11. Kersey, P.J., Plant genome sequences: past, present, future. *Curr. Opin. Plant Biol.*, 48, 1–8, 2019.
12. Kim, M. and Tagkopoulos, I., Data integration and predictive modeling methods for multi-omics datasets. *Mol. Omics*, 14, 1, 8–25, 2018.
13. Khan, A.W., Garg, V., Roorkiwal, M., Golicz, A.A., Edwards, D., Varshney, R.K., Super-pangenome by integrating the wild side of a species for accelerated crop improvement. *Trend Plant Sci.*, 25, 2, 148–158, 2020.
14. Zhao, Q., Feng, Q., Lu, H., Li, Y., Wang, A., Tian, Q., Zhan, Q., Lu, Y., Zhang, L., Huang, T., Wang, Y., Pan-genome analysis highlights the extent of genomic variation in cultivated and wild rice. *Nat. Genet.*, 50, 2, 278–284, 2018.
15. Montenegro, J.D., Golicz, A.A., Bayer, P.E., Hurgobin, B., Lee, H., Chan, C.K.K., Visendi, P., Lai, K., Doležel, J., Batley, J., Edwards, D., The pangenome of hexaploid bread wheat. *Plant J.*, 90, 5, 1007–1013, 2017.
16. Song, J.M., Guan, Z., Hu, J., Guo, C., Yang, Z., Wang, S., Liu, D., Wang, B., Lu, S., Zhou, R., Xie, W.Z., Eight high-quality genomes reveal pan-genome

architecture and ecotype differentiation of Brassica napus. *Nat. Plants*, 6, 1, 34–45, 2020.
17. Hirschhorn, J.N. and Daly, M.J., Genome-wide association studies for common diseases and complex traits. *Nat. Rev. Genet.*, 6, 2, 95–108, 2005.
18. Atwell, S., Huang, Y.S., Vilhjálmsson, B.J., Willems, G., Horton, M., Li, Y., Meng, D., Platt, A., Tarone, A.M., Hu, T.T., Jiang, R., Genome-wide association study of 107 phenotypes in Arabidopsis thaliana inbred lines. *Nature*, 465, 7298, 627–631, 2010.
19. Flint, J. and Eskin, E., Genome-wide association studies in mice. *Nat. Rev. Genet.*, 13, 11, 807–817, 2012.
20. Ranc, N., Muños, S., Xu, J., Le Paslier, M.C., Chauveau, A., Bounon, R., Rolland, S., Bouchet, J.P., Brunel, D., Causse, M., Genome-wide association mapping in tomato (*Solanum lycopersicum*) is possible using genome admixture of *Solanum lycopersicum* var. cerasiforme. *G3: Genes Genomes Genet.*, 2, 8, 853–864, 2012.
21. Rebolledo, M.C., Peña, A.L., Duitama, J., Cruz, D.F., Dingkuhn, M., Grenier, C., Tohme, J., Combining image analysis, genome wide association studies and different field trials to reveal stable genetic regions related to panicle architecture and the number of spikelets per panicle in rice. *Front. Plant Sci.*, 7, 1384, 2016.
22. Yang, W., Guo, Z., Huang, C., Duan, L., Chen, G., Jiang, N., Fang, W., Feng, H., Xie, W., Lian, X., Wang, G., Combining high-throughput phenotyping and genome-wide association studies to reveal natural genetic variation in rice. *Nat. Commun.*, 5, 1, 5087, 2014.
23. Crowell, S., Korniliev, P., Falcao, A., Ismail, A., Gregorio, G., Mezey, J., and McCouch, S., Genome-wide association and high-resolution phenotyping link Oryza sativa panicle traits to numerous trait-specific QTL clusters. *Nat. Commun.*, 7, 1, 10527, 2016.
24. Campbell, M.T., Du, Q., Liu, K., Brien, C.J., Berger, B., Zhang, C., Walia, H., A comprehensive image-based phenomic analysis reveals the complex genetic architecture of shoot growth dynamics in rice (*Oryza sativa*). *Plant Genome*, 10, 2, 2017.
25. Yates, S., Mikaberidze, A., Krattinger, S.G., Abrouk, M., Hund, A., Yu, K., Studer, B., Fouche, S., Meile, L., Pereira, D., Karisto, P., Precision phenotyping reveals novel loci for quantitative resistance to *Septoria tritici*. *Plant Phenom.*, 2019.
26. Condorelli, G.E., Maccaferri, M., Newcomb, M., Andrade-Sanchez, P., White, J.W., French, A.N., Sciara, G., Ward, R., Tuberosa, R., Comparative aerial and ground based high throughput phenotyping for the genetic dissection of NDVI as a proxy for drought adaptive traits in durum wheat. *Front. Plant Sci.*, 9, 893, 2018.
27. Muraya, M.M., Chu, J., Zhao, Y., Junker, A., Klukas, C., Reif, J.C., and Altmann, T., Genetic variation of growth dynamics in maize (*Zea mays* L.)

revealed through automated non-invasive phenotyping. *Plant J.*, 89, 2, 366–380, 2017.
28. Rapacz, M., Wójcik-Jagła, M., Fiust, A., Kalaji, H.M., Kościelniak, J., Genome-wide associations of chlorophyll fluorescence OJIP transient parameters connected with soil drought response in barley. *Front. Plant Sci.*, 10, 78, 2019.
29. Singh, B.D. and Singh, A.K., *Marker-assisted plant breeding: principles and practices*, Springer Link, pp. 431–461, 2015.
30. Fasoula, D.A., Ioannides, I.M., Omirou, M., Phenotyping and plant breeding: overcoming the barriers. *Front. Plant Sci.*, 10, 1713, 2020.

Part II

PHARMACEUTICAL RESEARCH

5

Molecular Modeling and Drug Development

Howida A. Elseedy[1], Caroline Kiriacos[2] and Triveena M. Ramsis[3*]

[1]*Pharmaceutical Chemistry Department, Faculty of Pharmacy and Biotechnology, German University in Cairo (GUC), Cairo, Egypt*
[2]*Pharmaceutical Biology Department, Faculty of Pharmacy and Biotechnology, German University in Cairo (GUC), Cairo, Egypt*
[3]*Department of Pharmaceutical Chemistry, Faculty of Pharmacy, Sinai University – Kantara Branch, Ismailia, Egypt*

Abstract
Due to the utilization of conventional methods for drug design, the process of developing new drugs has been rigorous and time-consuming. *In silico* drug design offers a green way to design and develop drugs by avoiding many laboratory assays and chemicals, which is an important extra benefit that should not be neglected. The key *in silico* drug design methodologies include Ligand-Based Drug Design (LBDD), Structure-Based Drug Design (SBDD), homology modeling, and *in silico* pharmacokinetics. *In silico* drug design offers a green approach to design and develop drugs by avoiding many laboratory assays and chemicals. This can help minimize the time and cost associated with drug development by offering a more effective way to assess novel drugs before they are evaluated in clinical trials.

Keywords: *In silico*, include ligand-based drug design, structure-based drug design, computer-aided drug design techniques, docking

5.1 Introduction

In a world full of technological progress, the use of computational approaches and sophisticated tools has become inevitable in any scientific

**Corresponding author*: triveenaramsis@gmail.com; triveena.farid@su.edu.eg

Hrudayanath Thatoi, Sonali Mohapatra, Swagat Kumar Das and Sukanta Kumar Pradhan (eds.) Applied Biotechnology and Bioinformatics: Agriculture, Pharmaceutical Research and Environment, (111–138) © 2025 Scrivener Publishing LLC

research about drug development. Due to the utilization of conventional methods for drug design, the process of developing new drugs has been rigorous and time-consuming. Such a process could cost up to $1 billion USD and take 12 to 15 years, making it both expensive and time-consuming [1]. Nevertheless, drugs can still fail in clinical trials whether due to bad safety profiles or inactivity [1]. Computer-aided drug design (CADD) techniques are being used, not only for their efficiency, but also for overcoming many other hurdles such as the optimization of ligand binding strength, selectivity, or even the physicochemical and biological properties like bioavailability, solubility, and toxicity [1, 2]. Moreover, *in silico* drug design offers a green way to design and develop drugs by avoiding many laboratory assays and chemicals, which is an important extra benefit that could not be neglected [3]. In this chapter, we have highlighted important *in silico* drug design techniques including docking, pharmacophore modeling, quantitative structure activity relationship (QSAR), virtual screening, homology modeling, and *in silico* pharmacokinetics [4].

5.2 Structure-Based Drug Design

The necessity for the use of Structure-Based Drug Design (SBDD) has grown as a result of the increasing discovery of numerous novel targets in the disciplines of genomics and proteomics. To get the perfect target, first it should have a biological and biochemical basis and should be closely linked to human pathological diseases [5]. The good target should also have a distinct and defined pocket that will bind with a smaller molecule to perform a certain function or mechanism [5]. Clarifying the target types, they could be proteins (mostly used) or any genetic material such as the RNA. RNAs as targets are excellent since they are specific for the species and for the disease states [6, 7]. The smaller molecule that will bind to the target (ligand) could be a competitor with the natural endogenous one at certain potency levels to alter the function of the target [5]. The main step in structure-based drug design (SBDD) is to identify the possible binding sites of the ligand on its target. The binding pocket of the target could have a myriad of possible interactions like hydrogen bond donors, acceptors, and hydrophobic surfaces [5]. These binding sites could be active sites as in the case of enzymes, and a communication site is required for the mechanism of the molecule or an assembly site with another macromolecule [5]. The protein–protein interactions can experience difficulty in obtaining the target site location because it is usually large, flat, and hydrophobic, but these obstacles could be conquered [8–10]. SBDD highlights the inevitable role of co-crystallization studies because they give a crystallized structure of target macromolecule with an initial small molecule to specify the good target site upon which all the upcoming procedures are based on [5].

SBDD techniques consider the target 3D structure along with the estimation of compounds that can have an optimum binding interaction with the target binding site [5, 11]. Among the strategies used in SBDD for achieving this aim are Molecular Docking, Molecular Dynamics (MD), and Structure-Based Virtual Screening (SBVS) [12]. MD is a technique that simply allows the study of protein motion (conformations) with time. Protein motion is such an important characteristic because the function of the protein can still depend on it. Molecular dynamics can be used in studying any type of molecule whether in complexes or mixtures and at any state of matter at any temperature and pressure to give abundant information and understanding on a molecular level [13–15]. MD also gives information about the structural, energetic, and thermodynamic properties of the protein [16]. In other words, MD can give dynamic information about macromolecules and adequate knowledge on the mechanism of ligand and protein interactions [17]. MD simulations have been used extensively for studying mechanisms of drug resistance as a result of mutations in protein target or studying the mechanisms of pathogenic diseases, which resulted from protein misfolding as well [17]. In this chapter, we will be discussing molecular docking and structure-based virtual screening in detail.

Finding a lead compound is the next step after getting the 3D structure of the target. A lead compound is a molecule that shows binding affinity to the target. However, there are some problems that may appear afterwards that need refinement and optimization, for instance, weak affinity, high toxicity, instability, or any other pharmacokinetic problem. Lead compounds could be obtained by screening large databases of already synthesized molecules or naturally available ones [18].

To reach the lead optimization stage, various numbers of cycles should be done first. The whole process of SBDD starts by getting the purified and cloned protein target in solution and performing X-ray crystallography, nuclear magnetic resonance (NMR), or homology modeling to obtain the target's structure. The next step involves the docking of a database of compounds against the target site, where the docked poses can be scored and ranked based on steric and electrostatic interactions. The most promising candidates will undergo biochemical assays to see their ability to function as agonist/antagonist for the target in a solution. In the next stage, if the compounds possess the ability to activate or inhibit the target in the biological assay successfully (i.e., successfully bind to the target), the complex structure of both target with ligand should be established by X-ray crystallography or NMR. Consequently, the interactions between target and ligand can be analyzed. If the substrate is an inhibitor in the nM range by the potency tests, it will resume to further clinical trials. However, if the lead has no binding affinity in the nM range (usually the lead can have a dissociation constant of 10 μM or better) or even if it could not work as an inhibitor at all in the solution, modification and lead optimization should be taken into consideration and repeated for many cycles (i.e., an iterative process) [18]. However, if optimization trials are not successful, we will shift to the next compound in the list. After many cycles of optimization and testing, there is usually an improvement in the potency and compound specificity [5].

Screening tools will mostly get a list of defective leads such as those with weak affinity, low selectivity, lack of efficacy, and major or toxic side effects. Defective leads can also have unsuitable pharmacokinetic profiles as in absorption (which can be greatly affected by lipophilicity), distribution, metabolism, or excretion [19]. Indeed, this will require much optimization [18]. Optimization techniques that are used are based mainly on a hypothesis and a process of trial and error to prove or discard this hypothesis. Shedding light on some of the general optimization strategies, changes can be done in the lipophilicity or electronic character of the molecule by adding or removing hydrophilic or hydrophobic groups, such as changing the alicyclic or heterocyclic rings' size as well as inducing changes in the substituents of the rings. Other strategies to be used are opening of the rings, ring rigidification, heteroatoms addition, branching, addition or elimination of chiral atoms, and chain length variation. These changes could not only strengthen or weaken the binding but also influence the activity spectrum whether in a good or bad way. That is why after optimization, evaluation for this optimized lead must be done [20]. In an aim to find a way for calculating the binding affinity, electrostatic contribution to binding ΔG_{elec} can be used. ΔG_{elec} is strongly related to the total binding free energy, and when it gives a positive number, it means that electrostatic interaction opposes the binding [21]. Interaction scoring functions could also be based on geometric complementarity of surfaces, desolvation energy, explicit electrostatic interactions, and pairwise propensities of side chains of the amino acids to contact throughout the molecular interface [22].

Pharmacokinetic properties can be the reason behind failure of high-affinity drug, meaning that after optimization of the binding affinity, pharmacokinetic characteristics could be worse. As a result, optimization of these characteristics is fundamental. For example, absorption depends mainly on the lipophilicity level of the molecule; the higher the polarity level, the lower the absorption rate. Also, metabolism rate could be changed by substitutions or branches (steric hindrance), i.e., addition like substitution of aliphatic amino group with alkyl or branch can suppress the metabolism rate. Therefore, optimization of any of them can affect the other and all of these should be optimized before proceeding to clinical trials aiming to decrease the costs of possible failures [20]. This is when computational pharmacokinetics is employed to filter out drug candidates with bad pharmacokinetic profiles, as will be discussed later in this chapter.

5.3 Docking

Obtaining the 3D structures of potential biological targets has made the molecular docking an essential tool to be used during the last three decades [23]. Its significance appeared not only for its ability to search within large libraries of compounds but also its ability to evaluate the tightness of the fit by scoring and ranking the results as well as finding the binding mode [23, 24]. The main (known) utilization for docking is between a ligand and target, which could be protein, DNA, or RNA; however, there are other uses for docking as protein–protein docking [23]. The main aim of this part will be discussing the more common ligand–protein docking processes.

In order to proceed with docking, the biological target has to be first selected and prepared. The target structure is typically determined experimentally by X-ray crystallography or NMR. Many target structures are present in the Protein Data Bank (PDB) with an ID called PDB ID [25]. The higher the resolution of the structure, the better is the 3D structure quality. In order to prepare the protein structure prior to docking, hydrogen atoms are added, and to account for the absence of hydrogen atoms in the crystal structure, water atoms are deleted and the structure is refined by energy minimization [25].

Protein structures are typically not in a fixed position. In addition, the protein-bound state of the ligand can be different from the unbound one. Also, the protein conformations can differ with the different classes of the ligands. Therefore, protein flexibility is a very important aspect to consider. However, the resolved protein structures from X-ray do not represent flexibility [25]. Some docking software try to address this issue whether by rotamer inclusion in the side chains of binding sites residues, induced fit method, or ensemble docking [25, 26]. Ensemble docking is a method for docking that was introduced first in 1999 [27]. It is based on docking of a ligand to an ensemble of the protein target structure (molecular fragments only) because of the high degree of fluctuations and flexibility in the target binding site. These protein ensembles can be obtained using MD simulations [28]. This approach generates dynamic pharmacophores; thus, it is able to give more accurate prediction about binding of known ligands than the pharmacophores based on a single target structure [28]. However, induced fit model supposes that both ligand and target are partially compatible. Accordingly, when both become proximal to each other, conformational changes are induced to both according to the chemical forces created by interaction [29]. The induced fit model affects the local side chain along with inducing slight alterations in the backbone, in order to optimize the association of both [30].

Ligands, contrary to biological targets, are typically treated flexibly by docking tools except for ring conformations. Thus, as the number of rotatable bonds increase, the higher the number of torsions and conformations. The search space exponentially increases as well. The docking process, therefore, becomes more hectic and time-consuming. The same goes for the more complicated structures that are highly branched [23].

Digging deeper into docking, it requires two main algorithms. The first one is a conformational search algorithm that searches for the best poses for the ligand. The second one is an energy scoring function for evaluating ligand poses. Conformational search usually investigates millions of ligand conformations [23, 31]. It has two main types, which are systematic search and stochastic search [23]. Systematic search identifies all sterically allowed conformations of a molecule through the possible combinations of rotatable bonds and dihedral angle combinations [23, 32]. The stochastic method works by making random changes to the state of variables in a repetitive way until a user-defined cutoff is met. In the stochastic method, the experiment should run 50 times at least with changes in the initial conditions. Predicted binding modes from the stochastic method can be assessed for their similarity by RMSD values and clustering the docked conformations as claimed by the RMSD threshold, which is 2 Å [23, 33]. The results gained from systematic search is inevitable while for the stochastic method, the results can vary due to depending mainly on randomness (changing randomly all degrees of freedom for ligand as translational, rotational, and conformational). Therefore, a huge number of poses that will undergo evaluation is generated from the stochastic method [23, 31]. However, systematic search is mainly used for rigid protein–protein docking, as there are only six degrees of freedom, and is employed in programs such as ZDOCK and GRAMM [34, 35]. Yet, higher freedom dimensions as flexible ligand–protein docking mainly require stochastic methods. MSCA is an example of a stochastic method-based docking software [23, 36].

Following the conformational search, scoring functions evaluate the quality of the docked poses (obtained from search algorithm), in an aim to determine a suitable ligand conformation [31]. A scoring function is a tool that enables the differentiation between the active and inactive ligands after determining the best poses according to the lowest energy values along with anticipation of affinity constants and ranking according to potency [31]. Scoring functions may employ empirical methods, knowledge-based methods, or molecular mechanics-based methods (like AMBER force field). The latter is included in AutoDock scoring, and also has two additional parameters: the first one is based on atomic solvation and calculates the change in desolvation energy upon binding, while the second one is empirical, which calculates the loss of conformation entropy upon binding [37–40].

Docking results are evaluated based on the satisfaction of chemical complementarity between ligand and protein. This includes satisfaction of hydrogen bond donors and acceptors in the ligand, satisfaction of opposite charged groups ligand and protein, and the interaction of hydrophobic groups between both [23].

5.4 Ligand-Based Drug Design

Ligand-based drug design is a very useful tool used by medicinal chemists and computational chemists in the absence of the 3D structures of potential active targets for drug discovery and lead optimization in a time-efficient manner. If the target structure is unknown, some methods used in structure-based drug design, like docking and biochemical tests, cannot be applied. When the target is unknown, the best way to detect preliminary hits is to use a cell culture assay, although *in vivo* testing can be used as a final resort. Similar compounds can be tested if there are already known active chemicals. If there are only a few known active compounds, testing a diversified library of compounds is the best approach. To confirm that the assay functions correctly, there should be at least one known active compound [41]. In cases that there are known active compounds, a diverse library is usually screened. This is known as "lead hopping," and it is done in the hopes of discovering new classes of active chemicals [42].

Ligand-based drug design (LBDD) includes pharmacophore and 3D-QSAR models. Due to its potentially higher accuracy, 3D-QSAR is marginally more popular. It is more commonly used for testing computationally designed molecules. Pharmacophore models are more commonly employed to screen big databases of existing substances since pharmacophore searches are much faster [41]. LBDD, in contrast with SBDD, does not have the advantage of analyzing actual drug–target interactions and structural requirements in order to optimize ligand binding [28].

5.5 Pharmacophore

A pharmacophore is defined as the three-dimensional shape of a compound that makes it suitable for the active site of a target, including hydrogen bond donors and acceptors, bulky groups, etc. A pharmacophore model can be generated even in the absence of adequate information on the geometry of the active site of the protein. It is usually performed by, first, creating a training set, which is a database containing the 3D structures of a collection of predefined ligands that reflect the ligands' important interactions with the target. Each molecule in the training set undergoes conformational search in order to account for the ligands' conformational flexibility. This is followed by multiple ligand alignment and constructing the pharmacophore model based on the essential chemical features, found in common between all the aligned ligands [24].

Ligand-based pharmacophore modeling encounters numerous significant challenges. These are mainly related to modeling of ligand flexibility as well as properly selecting the compounds of the training set.

In order to deal with the ligand flexibility problem, two approaches have been developed. These are the pre-enumerating method and the on-the-fly method. In the first method, multiple conformations are computed for each molecule before the pharmacophore modeling process. In the second method, the conformational analysis of each molecule takes place during the pharmacophore modeling process. The advantage of pre-enumerating is that it requires lower computing cost in order to perform molecular alignment. However, a major disadvantage is that it requires a large storage capacity. The advantage of the on-the-fly method is that it does not require large storage. However, it needs higher CPU time in order to perform rigorous optimization of the pharmacophore model. The pre-enumerating method has been shown to surpass the on-the-fly computation method. The proper selection of the training set compounds poses a challenge, since the final pharmacophore model generated is mainly changed based on the chemical diversity of the dataset, its size and the type of ligands [24, 43].

Constructing pharmacophore models using the geometry of the active site is an alternative to generating them from active compounds, if protein structure is known. Errors owing to the limited set of training compounds are eliminated because it is feasible to observe the interaction of the pharmacophore model with the active site.

The structure-based pharmacophore modeling methodology entails a review of the active site's complimentary chemical properties and their spatial correlations, followed by the creation of a pharmacophore model employing selected features. The structure-based pharmacophore modeling methods can be performed through a macromolecule–ligand complex-based approach or a macromolecule-based approach. Finding the ligand-binding site and identifying the crucial points of interaction between the macromolecule and the ligand are possible using the first method. The requirement for the 3D structure of a macromolecule–ligand complex represents a major limitation of this approach. Therefore, it cannot be employed when the ligand targeting the macromolecule is unknown, even if the binding site of the target is defined. The macromolecule-based approach can overcome this problem, where pharmacophore models can be derived from apo structures. LigandScout [44] and MOE (Molecular Operating Environment) [45] can perform pharmacophore modeling based on the two approaches [46].

If both the target structure and all of its ligands are unknown, pharmacophore modeling becomes impossible. The only option for implementing the pharmacophore principle is to build a diverse library using a diversity measure based on pharmacophore fingerprints to ensure adequate variability of the library, which would consist of a wide range of molecules with unique pharmacophore feature composition [26, 46].

Pharmacophore models can be employed in virtual screening of huge databases. In docking simulations, pharmacophore models can be used to discard the docking results that fail to bind to the active site based on the pharmacophore query or compounds with good scores. Pharmacophore models can also be implemented in predicting ADMET by identifying the interactions between drugs and metabolic enzymes based on a known drug with a well-known ADMET profile [26].

5.6 QSAR

The QSAR method is a prominent approach for ligand-based drug discovery. QSAR is a computational method for determining the relationship between a sequence of chemical structures and a certain chemical or biological process. The QSAR technique is based on the notion that similar structural or physio-chemical features result in similar activity [28]. To begin, a set of chemical entities or lead molecules that exhibit the desired biological activity of interest are discovered. The physicochemical characteristics of active molecules and biological activity are linked in a quantifiable way. Descriptors are the key molecular attributes in a QSAR equation. A descriptor can be any numerical value that describes a molecular property. Descriptors can range in complexity from topological indices to something as simple as the molecular weight. In QSAR, descriptors quantifying a molecular feature by a single value are employed [47].

The active chemicals are then optimized using the created QSAR model to maximize the required biological activity. The anticipated compounds are then put to the test in the lab to see if they have the desired activity. As a result, the QSAR approach can be utilized as a guiding tool for finding chemical changes that improve activity. The process of developing a QSAR model is as follows [28]:

1. Identify a set of active compounds with experimentally measured biological activity with satisfying chemical diversity.
2. Identify suitable molecular descriptors associated with the studied molecules and their physio-chemical properties.
3. Build correlations between the selected molecular descriptors and the required biological activity aligned with the variability of the set molecules.
4. Test the predictability and validity of the QSAR model with another set of compounds.

Traditional QSAR is best for computing properties that are a function of nonspecific interactions between the molecule and its surroundings. Minor changes in molecule structure usually result in small changes in the property for these attributes. Traditional QSAR is the preferred approach for applications including typical boiling temperatures and passive intestinal absorption, in addition to blood–brain barrier permeability. 3D-QSAR, on the other hand, is better at calculating highly specific interactions, such as how strongly a molecule will bind to the active site of a single protein [47].

5.6.1 Advantages of the QSAR Method [47]

1. It enables the quantification of the relationship between structure and activity on the basis of physio-chemical characteristics.
2. It is feasible to anticipate the properties of intended compounds before the chemical synthesis of novel compounds.
3. It could explain how functional groups of designed compounds relate to the activity of the enzyme or protein that they are intended to affect.

5.6.2 Limitations of the QSAR Method [48–50]

1. Insufficient chemical representation in the training sets, which impairs the applicability domain of the models and, consequently, their suitability across various chemical classes.
2. Endpoints with mechanistically complex mechanisms have a low predictability.
3. The effect of the quantity and quality of the underlying training data.
4. The predictions from several QSAR models could result in contradiction.
5. The performance of QSAR techniques varies depending on the chemical set under investigation [51].
6. Inconsistent predictions create questions about interpretation, validation, and adequacy.
7. Reducing false positives and false negatives is crucial. For example:
   - A medicine that is known to cure depression can be allowed if it causes skin sensitization but not if it causes tumors.
   - A chemical that is wrongly predicted to be non-carcinogenic may get regulatory approval but will create cancer exposure risk.

8. In the absence of experimental data, selecting a suitable technique for evaluating hazardous consequences is difficult.

5.6.3 Examples of Applications of QSAR in Drug Discovery

1. During the drug design and discovery process, quantitative structure–activity relationships (QSAR) can be exploited. With regard to anticipated biological activity, QSAR models can be used to screen a large number of compounds or to eliminate test compounds that do not seem promising.
2. QSAR-based techniques can predict the toxicity endpoints of chemicals towards organisms [52]. Take, for example, the oral rat fatal dose of 50% (LD50) [53].
3. For the prediction of skin sensitization and activity towards (hERG) K^+ ion channels (human ether-à-go-go related

gene), QSAR models have been included into computational methods. Cardiotoxicity is linked to the latter [54, 55].

5.7 Virtual Screening

There is an increasing need for the development of new successful drugs for the market. The difficult challenge of selecting which molecules to synthesize has always plagued medicinal chemists. A chemist is required to select from thousands or millions of potential molecules at every stage of the drug design process. As a result, virtual screening (VS) has attracted much interest. Virtual screening is known as the use of a set of computational techniques for the detection of novel bioactive lead compounds from a huge virtual library.

The purpose of high-throughput screening (HTS) is roughly the same: to find new hits. HTS and VS, on the other hand, are built on opposing ideas. VS strives to rationalize compound selection to limit the number of possibilities for experimental evaluation as much as feasible, whereas HTS aims to test a growing number of compounds in the most efficient way possible [56].

VS approaches can be classified into ligand-based and structure-based approaches. Structure-based computational approaches can be employed when the target protein's structure is identified. Such approaches involve the molecular docking of every compound, in the database, into the target's binding site. The binding mode and the quality of fitting of each ligand are determined. Library compounds are ranked based on the generated information and a small subset is then biologically tested [57]. When the structure of the target protein is unknown, ligand-based VS is the preferred strategy. Compounds, proven to interact with the target, provide valuable information that is utilized to filter the compounds with similar properties in the database. Similarity searching, substructure searching, pharmacophore matching, and 3D shape matching are some of the strategies that can be used to accomplish this [58].

5.7.1 Structure-Based Virtual Screening

It is required to establish the database of compounds to be screened in the early stages of a structure-based VS project. The database's source is usually a database of chemicals available from vendors or corporate database of physically available compounds. Multiple physical and chemical filters (such as molecular weight, clogP, functional groups, etc.) are employed to minimize the size of the original database, resulting in a database of compounds having physical and chemical attributes that are similar to the majority of known active ligands. Lipinski's rule-of-five is a common filtering method for drug-likeness properties.

The active site is prepared by adding hydrogens, followed by energy minimization and assigning the appropriate protonation states for ionizable amino acids. In order to perform VS, tightly bound crystallographic water molecules may be required in some cases.

After the preparation of the compound database and the active site, each molecule is subjected to dock into the target binding site. It is possible that the conformation of a chemical bound to a target differs from the unbound state. Considering the molecules' conformational flexibility during the docking procedure is therefore critical. This can be accomplished by performing a conformational search to the filtered and prepared database, and then tightly docking each conformer into the binding site. Instead, most docking tools perform the conformational search as the docking progresses. Because proteins are not fixed by nature, flexibility of the protein target is a key aspect in the docking process. To account for protein flexibility, some algorithms allow the mobility and relaxation of the side chains in the binding site during the docking process of the ligand.

Current virtual screening scoring functions are usually insufficient in estimating a ligand's actual binding mode for a receptor. To reduce the number of false positives, a variety of post-analysis procedures can be implemented. Consensus scoring is one prevalent technique used for post-analysis. In this method, the obtained top-ranked ligand poses in the target receptor are rescored using several scoring functions. This method has been found to improve hit rates from VS [59]. Geometric analysis (based on the geometry of the binding pocket) can also be implemented in docking post-analysis. This employs calculations to determine ligand–protein complementarity, the percentage of exposed ligand surface area in the predicted complex, and the unexposed total surface area of the ligand–protein complex. Another approach of geometric post-analysis is based on protein–ligand interactions. Knowledge of the desirable key protein–ligand interactions can be applied to retain all the docked poses that possess those interactions, resulting in the minimized number of false-positive poses in the final selected list of compounds. Compounds are inspected visually, prior to choosing a candidate for experimental testing, to ensure that ligand–protein interactions as well as the ligand-bound conformation are reasonable [60].

5.7.2 Ligand-Based Virtual Screening

Ligand-based VS approaches identify and optimize leads by utilizing information contained in known active ligands as opposed to the target structure. New ligands are identified by evaluating the similarity (in terms of structure, pharmacophoric features, etc.) between potential compounds and the recognized active ligand. Therefore, an effective similarity measure and a consistent scoring method are two crucial components of a ligand-based computational technique [61]. Ligand-based VS employ descriptors that describe compounds by representing their molecular characteristics in numbers. Descriptors can be categorized based on a variety of criteria, including the type of the features they describe (e.g., geometrical, electronic, or topological descriptors). Another common classification is based on a molecule's geometrical structure, which can be classified as 1D, 2D, or 3D [58].

In order to construct similarity measurements, geometrical data of hypothetical, arbitrary items designated on the structures are used. The classification of an object varies depending on the approach used; however, it may be divided into three categories: pharmacophores, molecular shapes, and molecular fields. Pharmacophore-based approaches establish patterns of distances between predetermined molecular features such as hydrogen-bond donors, hydrogen-bond acceptors, and/or aromatic systems. Such patterns are then compared to calculate the similarity value among the compounds in the screened library. The goal of molecular shape techniques is to maximize shape overlap and calculate a similarity value based on that overlap. During the ranking phase, a scoring technique for ligand-based screening should effectively distinguish actives from inactives and be able to quickly choose a few bioactive molecules from a library with a high number of inert compounds [61].

5.8 Pharmacophore-Based VS

A pharmacophore model is employed to screen huge chemical libraries during a VS. Compounds that fit the model are stored in a hit list. Because these molecules meet the model's requirements, they are very likely to be biologically active. As a result, VS can be used to filter promising compounds from vast compound collections and highlight active hits in chemical databases to be experimentally evaluated. Individual studies reported that hit rates from prospective pharmacophore-based VS vary, usually ranging between 5% and 40%. However, randomly identifying active molecules from a sample of investigated chemicals has a hit probability of less than 1% [62].

5.9 Similarity-Based VS

Similarity-based VS involves screening a database to find molecules that are structurally similar to an active ligand. According to the similar property concept, compounds with similar structures are more likely to exhibit similar properties, making them good candidates for biological screening. A structural representation is utilized to characterize the molecules, and a similarity coefficient is used to quantify the degree of similarity regarding such representations between pairs in a molecular similarity measure. The Tanimoto coefficient and molecular "fingerprints" are used in most current methods for similarity-based VS and are found to be highly effective [63].

5.10 Homology Modeling and Protein Folding

The molecular details of a protein's 3D structure are crucial for interpreting its biological function and interaction properties. Nevertheless, known protein sequences far outnumber experimentally determined protein structures [64]. Computational techniques are utilized to predict 3D protein models due to the lack of experimental structures to shed light on the composition and functionality of these proteins. Homology modeling, also known as comparative modeling, is a computational technique that is used to determine the 3D model of a protein from its amino acid sequence. Ideally, the objective of homology modeling is to accurately predict and obtain the structure of a protein from its sequence similar to the experimentally determined structure. Users would be able to reliably use *in silico* protein models that have been developed rapidly. Homology modeling is based on two main observations. Firstly, the amino acid sequence of the protein determines its 3D structure. Secondly, the protein structure is more conserved during evolution than the primary sequence, and therefore identical structures can be adopted from similar primary sequences [65, 66].

The homology modeling process includes seven standard steps [65, 66]:

5.10.1 Template Selection

The first step is to identify other homologous proteins with known 3D structures to be used as templates. A protein Basic Local Alignment Search Tool

(BLAST) search [service provided by National Center for Biotechnology Information (NCBI) at https://blast.ncbi.nlm.nih.gov/] is performed with the query being the target protein sequence and the database being the worldwide Protein Data Bank (wwPDB).

5.10.2 Alignment Correction

The step described above is only an initial screening after which promising model templates are identified. Better alignment is achieved *via* more sophisticated methods. A common challenge is observed when there is low sequence identity between two sequences in a region, resulting in a challenging alignment. This will cause alignment issues brought on by additional gaps and mutations, which will result in false folding assignments. In that case, other sequences from homologous proteins can be used for this purpose. Multiple sequence alignment can improve the modeling process and is performed using alternative methods, such as position-specific iterated BLAST (psi-BLAST) and Hidden Markov Models (HMMs). These methods can reduce shifts and gaps between amino acid sequences.

5.10.3 Backbone Model Building

The information obtained from the second step is used to generate the 3D structure of the protein represented by the coordinates of each atom in the protein. This step is performed using one of four approaches, which are rigid-body assembly, segmented matching, spatial restraint, and combining alignments and structures. Rigid-body assembly involves breaking the structure of the parent-aligned proteins down into conserved core regions, loops, and side chains. Rigid bodies' obtained proteins are then fitted and assembled together. Tools like 3D-JIGSAW and COMPOSER utilize this method. The segment matching method constructs the protein 3D structure by using the positions of a subset of atoms, such as the C_α atoms, from template proteins as a guide. Segments matching these guiding positions are identified and assembled. The spatial restraint method identifies many possible constraints from protein structures aligned with the target sequence. These constraints are used as a guide to build the 3D structure of the target sequence. Stereochemical restrictions on bond lengths, bond angles, dihedral angles, and non-bonded atom–atom interactions, which are generated from a molecular mechanics force field, are usually employed to enhance these homology-derived restraints. All restrained violations are then minimized in order to create the model.

The spatial restraints approach is implemented in the computer software MODELLER. Combining alignments comprises an iteration of template selection and alignment, followed by model construction and evaluation, until no improvement is detected in the model, in order to improve the homology model [67].

5.10.3.1 Loop Modeling

Loops can play a role in fundamental protein functions. The accuracy of this step can have a big impact on the model's overall value. Loop prediction is accomplished using two methods: database search approaches and conformational search approaches (*ab initio*). Approaches to database searching rely on comparisons with all identified proteins in the Protein Data Bank. Conformational search approaches are more direct. They are based on scoring function optimization.

5.10.3.2 Adding Amino Acid Side Chains

Amino acid residues, conserved in proteins that are structurally similar, possess side-chain rotamers with similar torsion angles about the C_α–C_β bond. The process involves the trial of various rotamers, from a rotamer library, and using different energy functions for scoring. The rotamer library is built based on the conditions of backbone conformation (backbone-dependent rotamer library), where the rotamers are taken from proteins with high-resolution structures [65, 66]. Side-chain rotamers can be predicted using several programs, such as OPUS-Rota2 and FASPR, which are freely available for use [68, 69].

5.10.3.3 Model Optimization

Energy minimization using molecular mechanics force fields is normally the first step in model optimization. A few major errors can be minimized with each energy minimization, yet many more minor errors are incorporated and accumulated simultaneously. To reduce model defects upon optimization, atom positions are restrained, energy minimization is performed with a few hundred steps, and more accurate force fields are employed such as quantum force fields and self-parameterizing force fields. To further improve the model, techniques like Monte Carlo and MD can be employed [66].

5.10.3.4 *Model Evaluation and Validation*

Accuracy can greatly vary between different models. It can also vary within the same protein model among different regions. Variable loop regions in a model usually show less accuracy than highly conserved core regions [70]. It is therefore crucial to verify and validate the constructed homology model since its accuracy affects its application in the future. Several factors affect the accuracy of the protein model. These are environmental parameters, sequence similarity, and template quality. One crucial requirement is stereochemistry analysis. Checking bond length parameters, torsion angle parameters, and rotational angles is done in order to perform this analysis. In homology modeling, common tools for determining the model's stereochemistry include WHATIF, PROCHECK, and MOLPROBITY. The quality of a protein model can be determined using the Ramachandran plot, where residues with defective stereochemistry will be excluded from the favored regions of the Ramachandran plot. Other tools are available for determining the model spatial features such as VERIFY3D and PROSAII, which compare model environmental parameters with the expected environmental conditions [66].

5.11 *In Silico* Pharmacokinetics

To avoid failure, an ADMET profile (absorption, distribution, metabolism, excretion, and toxicity) that meets the necessary standards is just as critical as a drug's therapeutic effectiveness. *In vitro* high-throughput ADMET screening is available. However, computational ADMET prediction is extremely helpful. In the drug discovery process, using computational ADMET in conjunction with *in vivo* and *in vitro* predictions eliminates a lot of safety problems [71]. In order to build a good model for predicting ADMET properties, the correct molecular descriptors, mathematical equations, and sufficient experimental data are required [72].

5.11.1 Absorption

Single molecular features, such as log D or hydrogen-bonding, can be used to estimate the permeability of a molecule penetrating the intestinal membrane *via* passive diffusion. Nevertheless, many chemicals are impacted by biological processes, such as the influence of active transporters. Oral absorption predictions are frequently made *in vitro* using methods such as Caco-2. Computational models for predicting oral absorption have also been developed with considerable effort [73]. Examples of absorption–simulation programs include GastroPlus and Idea [74].

5.11.2 Bioavailability

A combination of two processes determines a drug's bioavailability. These are absorption of the drug from the gastrointestinal tract (GIT) as well as its metabolism. The solubility of a substance, its permeability, and its interactions with active transporters and metabolic enzymes in the gastro-intestinal wall all influence absorption in return. Molecular size, hydrogen bonding capacity, lipophilicity, shape, and flexibility appear to be important features for influencing the permeability of a compound. This makes the prediction of bioavailability a difficult challenge. However, recent studies have documented the first attempts to directly estimate bioavailability from molecular structure [72, 75].

5.11.3 Metabolism

Drug metabolism is crucial in the research and development of novel drugs. Metabolism significantly guides the pharmacokinetics and pharmacodynamics of any drug in the human body. In order to predict metabolism, *in vitro* metabolism studies may be substituted with QSAR and molecular modeling approaches. Computational strategies for predicting metabolism include pharmacophore and protein (involved in ADME) models, QSAR, and three-dimensional-QSAR studies, in addition to predictive databases. Metabolism predication of drugs can be performed using efficient computer-based techniques [72]. MetaSite is a metabolism predictor that uses a pharmacophore representation derived from protein structure interaction regions and a pharmacophoric pattern for the proposed substrate [76]. QSAR-based metabolism prediction tools perform the drug metabolism prediction process in four steps. Metabolism parameters should be first determined and then molecular descriptors are calculated and variables are selected. Training and test datasets are used to construct and evaluate the model. The last step involves the use of an external validation set to predict the metabolism of new molecules [77]. Some QSAR-based models are available for metabolism prediction such as isoCYP, a commercial software developed by Molecular Networks; SMARTCyp, a free tool offered by University of Copenhagen; and RS-Predictor, a free tool offered by the Rensselaer Polytechnic Institute [78–81]. Expert systems like MetabolExpert (Compudrug), Meteor (Lhasa), or META (MultiCASE) as well as the databases Metabolite (MDL) and Metabolite (MDL) are among the tools that employ databases to predict metabolism that are either in development or accessible [72]. Metabolism predictor software usually performs the prediction beginning with a parent molecule that is subjected to a pattern of recognition methods and machine learning algorithms in order to identify metabolism sites in the molecule and the chemical products of a biotransformation metabolic sites. The majority of computational metabolism prediction programs are based on specific types of reactions or metabolic pathways, whether only phase I or only phase II metabolism. SMARTCyp and isoCYP are used to predict phase I metabolism only. Meteor is employed for both phase I and phase II metabolism [78, 79, 82, 83].

5.11.4 Toxicity

Many potential drugs either fail to reach the market or are withdrawn from the market after approval due to toxicity issues. There are two types of commercially available computational algorithms for predicting toxicity concerns. The first method makes use of expert systems to construct models based on the expertise of toxicologists and literature. The second method relies on the statistical analysis of correlations between the calculated chemical descriptors and the toxicological endpoint. The accuracy of the data used to build the models is crucial to the success of any *in silico* technique. However, due to the shortage of publicly available toxicology data, commercially accessible methods can only predict a finite number of toxicological endpoints. Modern software packages are primarily concerned with carcinogenicity and mutagenicity, while some also include models and knowledge bases for teratogenicity, irritation, neurotoxicity, immunotoxicology, and sensitization [72].

Knowledge-based approaches include structural alerts (SAs) and read-across models are used for toxicity calculation. SAs are also known as toxicophores, which are molecular structures associated with toxicity properties. Compared to a single SA, a combination of SAs could lead to more toxicity. SAs can be implemented to determine the toxicity of drugs. The drawbacks of SAs include the use of only binary features, which only describe the presence or absence of chemical structures, and only qualitative endpoints (whether toxic or non-toxic). Read-across is a technique for predicting a chemical's unknown toxicity by comparing it to analogous molecules (also known as chemical analogues) from the same chemical class [84].

QSAR, representing the second approach, is a collection of models correlating molecular descriptor chemicals' toxic properties. Chemical structures with similar QSAR models are likely to have similar mechanisms of action. QSAR models that estimate toxicity are known as quantitative structure–toxicity relationship (QSTR) models. As mentioned earlier, a QSAR (also QSTR) model is developed by calculating molecular descriptors followed by creating models to represent the data. Compounds can be described by a variety of molecular descriptors such as functional groups (1D descriptor), topological indices (2D descriptor), and hydrophobicity (3D descriptor). OECD QSAR Toolbox, Derek Nexus, and METEOR are some examples of tools that offer pre-built QSTR models, as well as SA lists [84].

5.12 Conclusion

The effectiveness of Computer-Aided Drug Design (CADD) approaches is being utilized to overcome a variety of obstacles in drug discovery, including the optimization of ligand binding, selectivity, and physicochemical properties. We have emphasized, in this chapter, key *in silico* drug design methodologies, including Ligand-Based Drug Design (LBDD) and Structure-Based Drug Design (SBDD) techniques, in addition to homology modeling, as well as *in silico* pharmacokinetics. With the aid of a target molecule's three-dimensional structure, tiny molecules that bind to and influence the activity of the target are designed and created using SBDD techniques. SBDD employs methods such as molecular docking, molecular dynamics (MD), and Structure-Based Virtual Screening (SBVS). When the target structure is not known, a drug designer will opt for LBDD techniques or homology modeling. In LBDD, novel compounds that might be able to bind to the same target are found by analyzing the 3D structure of an existing ligand. LBDD includes pharmacophore modeling, 3D-QSAR modeling, and Ligand-Based Virtual Screening (LBVS). Homology modeling is utilized to determine the 3D structure of a target protein from its primary structure, which can be further used in SBDD. *In silico* pharmacokinetics utilizes QSAR models to simulate the absorption, distribution, metabolism, excretion, and toxicity of drugs in the body (ADMET). This can help minimize the time and cost associated with drug development by offering a more effective way to assess novel drugs before they are evaluated in clinical trials.

References

1. Hughes, J.P., Rees, S., Kalindjian, S.B., Philpott, K.L., Principles of early drug discovery. *Br. J. Pharmacol.*, 162, 1239, 2011.
2. Zhong, F., Xing, J., Li, X., Liu, X., Fu, Z., Xiong, Z., Lu, D., Wu, X., Zhao, J., Tan, X., Li, F., Luo, X., Li, Z., Chen, K., Zheng, M., Jiang, H., Artificial intelligence in drug design. *Sci. China Life Sci.*, 61, 1191, 2018.
3. Tagliapietra, S., Binello, A., Bucciol, F., Trukhan, V., Colia, M., Cravotto, G., Green Enabling Technologies for Competitive Synthesis of Pharmaceutical Lead Compounds. *Curr. Pharm. Des.*, 26, 5700, 2020.
4. Yu, W. and MacKerell, A.D., Jr., Computer-Aided Drug Design Methods. *Methods Mol. Biol.*, 1520, 85, 2017.
5. Anderson, A.C., The process of structure-based drug design. *Chem. Biol.*, 10, 787, 2003.
6. Lind, K.E., Du, Z., Fujinaga, K., Peterlin, B.M., James, T.L., Structure-based computational database screening, *in vitro* assay, and NMR assessment of compounds that target TAR RNA. *Chem. Biol.*, 9, 185, 2002.
7. Filikov, A.V., Mohan, V., Vickers, T.A., Griffey, R.H., Cook, P.D., Abagyan, R.A., James, T.L., Identification of ligands for RNA targets via structure-based virtual screening: HIV-1 TAR. *J. Comput.-Aided Mol. Des.*, 14, 593, 2000.
8. Gadek, T.R. and Nicholas, J.B., Small molecule antagonists of proteins. *Biochem. Pharmacol.*, 65, 1, 2003.
9. Tilley, J.W., Chen, L., Fry, D.C., Emerson, S.D., Powers, G.D., Biondi, D., Varnell, T., Trilles, R., Guthrie, R., Mennona, F., Identification of a small molecule inhibitor of the IL-2/IL-2Rα receptor interaction which binds to IL-2. *J. Am. Chem. Soc.*, 119, 7589, 1997.
10. Wiesmann, C., Christinger, H.W., Cochran, A.G., Cunningham, B.C., Fairbrother, W.J., Keenan, C.J., Meng, G., de Vos, A.M., Crystal structure of the complex between VEGF and a receptor-blocking peptide. *Biochemistry*, 37, 17765, 1998.
11. Anighoro, A., Deep Learning in Structure-Based Drug Design. *Methods Mol. Biol.*, 2390, 261, 2022.
12. Kalyaanamoorthy, S. and Chen, Y.P., Structure-based drug design to augment hit discovery. *Drug Discov. Today*, 16, 831, 2011.
13. Hospital, A., Goni, J.R., Orozco, M., Gelpi, J.L., Molecular dynamics simulations: advances and applications. *Adv. Appl. Bioinf. Chem.*, 8, 37, 2015.
14. Perilla, J.R., Goh, B.C., Cassidy, C.K., Liu, B., Bernardi, R.C., Rudack, T., Yu, H., Wu, Z., Schulten, K., Molecular dynamics simulations of large macromolecular complexes. *Curr. Opin. Struct. Biol.*, 31, 64, 2015.
15. Mazurek, A.H., Szeleszczuk, L., Gubica, T., Application of Molecular Dynamics Simulations in the Analysis of Cyclodextrin Complexes. *Int. J. Mol. Sci.*, 22, 9422, 2021.

16. Wei, G., Xi, W., Nussinov, R., Ma, B., Protein Ensembles: How Does Nature Harness Thermodynamic Fluctuations for Life? The Diverse Functional Roles of Conformational Ensembles in the Cell. *Chem. Rev.*, 116, 6516, 2016.
17. Liu, X., Shi, D., Zhou, S., Liu, H., Liu, H., Yao, X., Molecular dynamics simulations and novel drug discovery. *Expert Opin. Drug Discov.*, 13, 23, 2018.
18. Verlinde, C.L. and Hol, W.G., Structure-based drug design: progress, results and challenges. *Structure*, 2, 577, 1994.
19. Cruciani, G., Pastor, M., Guba, W., VolSurf: a new tool for the pharmacokinetic optimization of lead compounds. *Eur. J. Pharm. Sci.*, 11 Suppl 2, S29, 2000.
20. Klebe, G., *Optimization of Lead Structures*, pp. 153–172, Springer Berlin Heidelberg, Berlin, Heidelberg, 2013.
21. Norel, R., Sheinerman, F., Petrey, D., Honig, B., Electrostatic contributions to protein-protein interactions: fast energetic filters for docking and their physical basis. *Protein Sci.*, 10, 2147, 2001.
22. Palma, P.N., Krippahl, L., Wampler, J.E., Moura, J.J., BiGGER: a new (soft) docking algorithm for predicting protein interactions. *Proteins*, 39, 372, 2000.
23. Morris, G.M. and Lim-Wilby, M., Molecular docking, in: *Molecular Modeling of Proteins*, vol. 443, p. 365, 2008.
24. Yang, S.-Y., Pharmacophore modeling and applications in drug discovery: challenges and recent advances. *Drug Discov. Today*, 15, 444, 2010.
25. Kontoyianni, M., Docking and virtual screening in drug discovery, in: *Proteomics for drug discovery: Methods and Protocols*, I.M. Lazar, M. Kontoyianni, A.C. Lazar (Eds.), pp. 255–266, Springer New York, New York, NY, 2017.
26. Qing, X., Lee, X.Y., De Raeymaecker, J., Tame, J.R., Zhang, K.Y., De Maeyer, M., Voet, A., Pharmacophore modeling: advances, limitations, and current utility in drug discovery. *J. Recept. Ligand Channel Res.*, 7, 81, 2014.
27. Carlson, H.A., Masukawa, K.M., McCammon, J.A., Method for including the dynamic fluctuations of a protein in computer-aided drug design. *J. Phys. Chem. A*, 103, 10213, 1999.
28. Acharya, C., Coop, A., E Polli, J., D MacKerell, A., Recent advances in ligand-based drug design: relevance and utility of the conformationally sampled pharmacophore approach. *Curr. Comput.-Aided Drug Des.*, 7, 10, 2011.
29. Mashiach, E., Nussinov, R., Wolfson, H.J., FiberDock: Flexible induced-fit backbone refinement in molecular docking. *Proteins*, 78, 1503, 2010.
30. Boehr, D.D., Nussinov, R., Wright, P.E., The role of dynamic conformational ensembles in biomolecular recognition. *Nat. Chem. Biol.*, 5, 789, 2009.
31. Guedes, I.A., de Magalhães, C.S., Dardenne, L.E., Receptor–ligand molecular docking. *Biophys. Rev.*, 6, 75, 2014.
32. Beusen, D.D., Berkley Shands, E.F., Karasek, S.F., Marshall, G.R., Dammkoehler, R.A., Systematic search in conformational analysis. *J. Mol. Struct.: THEOCHEM*, 370, 157, 1996.

33. Sousa, S.F., Fernandes, P.A., Ramos, M.J., Protein-ligand docking: current status and future challenges. *Proteins*, 65, 15, 2006.
34. Chen, R. and Weng, Z., Docking unbound proteins using shape complementarity, desolvation, and electrostatics. *Proteins Struct. Funct. Bioinf.*, 47, 281, 2002.
35. Tovchigrechko, A. and Vakser, I.A., GRAMM-X public web server for protein–protein docking. *Nucleic Acids Res.*, 34, 310, 2006.
36. Goodsell, D.S. and Olson, A.J., Automated docking of substrates to proteins by simulated annealing. *Proteins Struct. Funct. Bioinf.*, 8, 195, 1990.
37. Cornell, W.D., Cieplak, P., Bayly, C., II, Gould, I.R., Merz, K.M., Ferguson, D.M., Spellmeyer, D.C., Fox, T., Caldwell, J.W., Kollman, P.A., A second generation force field for the simulation of proteins, nucleic acids, and organic molecules. *J. Am. Chem. Soc.*, 117, 5179, 1995.
38. Taylor, R.D., Jewsbury, P.J., Essex, J.W., A review of protein-small molecule docking methods. *J. Comput.-Aided Mol. Des.*, 16, 151, 2002.
39. Huey, R., Morris, G.M., Olson, A.J., Goodsell, D.S., A semiempirical free energy force field with charge-based desolvation. *J. Comput. Chem.*, 28, 1145, 2007.
40. Morris, G.M., Goodsell, D.S., Halliday, R.S., Huey, R., Hart, W.E., Belew, R.K., Olson, A.J., Automated docking using a Lamarckian genetic algorithm and an empirical binding free energy function. *J. Comput. Chem.*, 19, 1639, 1998.
41. Lee, C.-H., Huang, H.-C., Juan, H.-F., Reviewing ligand-based rational drug design: the search for an ATP synthase inhibitor. *Int. J. Mol. Sci.*, 12, 5304, 2011.
42. Martin, Y.C., Muchmore, S., Beyond, Q.S.A.R., Lead Hopping to Different Structures. *QSAR Comb. Sci.*, 28, 797, 2009.
43. Wolber, G., Seidel, T., Bendix, F., Langer, T., Molecule-pharmacophore superpositioning and pattern matching in computational drug design. *Drug Discov. Today*, 13, 23, 2008.
44. Wolber, G. and Langer, T., LigandScout: 3-D pharmacophores derived from protein-bound ligands and their use as virtual screening filters. *J. Chem. Inf. Model.*, 45, 160, 2005.
45. Fayed E.A., Ebrahim M.A., Fathy U., Elawady A.M., Khalaf W.S., Ramsis, T.M., Pyrano-coumarin hybrids as potential antimicrobial agents against MRSA strains: Design, synthesis, ADMET, molecular docking studies, as DNA gyrase inhibitors. *J. Mol. Struct.*, 1295, 136663, 2024.
46. Schaller, D., Šribar, D., Noonan, T., Deng, L., Nguyen, T.N., Pach, S., Machalz, D., Bermudez, M., Wolber, G., Next generation 3D pharmacophore modeling. *Wiley Interdiscip. Rev.: Comput. Mol. Sci.*, 10, e1468, 2020.
47. Young, D.C., *Computational Drug Design*, pp. 171–179, John Wiley & Sons Inc. Hoboken, New Jersey, United States of America, 2009.
48. Bhattacharjee, S. and Brayden, D.J., Development of nanotoxicology: implications for drug delivery and medical devices. *Nanomedicine*, 10, 2289, 2015.

49. Benz, R.D., Toxicological and clinical computational analysis and the US FDA/CDER. *Expert Opin. Drug Metab. Toxicol.*, 3, 109, 2007.
50. Yang, C., Valerio Jr., L.G., Arvidson, K.B., Computational toxicology approaches at the US Food and Drug Administration. *Altern. Lab. Anim.*, 37, 523, 2009.
51. Pradeep, P., Povinelli, R.J., White, S., Merrill, S.J., An ensemble model of QSAR tools for regulatory risk assessment. *J. Cheminf.*, 8, 1, 2016.
52. Toropova, A., Toropov, A., Martyanov, S., Benfenati, E., Gini, G., Leszczynska, D., Leszczynski, J., CORAL: Monte Carlo method as a tool for the prediction of the bioconcentration factor of industrial pollutants. *Mol. Inf.*, 32, 145, 2013.
53. Gonella Diaza, R., Manganelli, S., Esposito, A., Roncaglioni, A., Manganaro, A., Benfenati, E., Comparison of *in silico* tools for evaluating rat oral acute toxicity. *SAR QSAR Environ. Res.*, 26, 1, 2015.
54. Braga, R.C., Alves, V.M., Muratov, E.N., Strickland, J., Kleinstreuer, N., Trospsha, A., Andrade, C.H., Pred-skin: a fast and reliable web application to assess skin sensitization effect of chemicals. *J. Chem. Inf. Model.*, 57, 1013, 2017.
55. Braga, R.C., Alves, V.M., Silva, M.F., Muratov, E., Fourches, D., Lião, L.M., Tropsha, A., Andrade, C.H., Pred-hERG: A novel web-accessible computational tool for predicting cardiac toxicity. *Mol. Inf.*, 34, 698, 2015.
56. Ripphausen, P., Nisius, B., Bajorath, J., State-of-the-art in ligand-based virtual screening. *Drug Discov. Today*, 16, 372, 2011.
57. Cheng, T., Li, Q., Zhou, Z., Wang, Y., Bryant, S.H., Structure-based virtual screening for drug discovery: a problem-centric review. *AAPS J.*, 14, 133, 2012.
58. Banegas-Luna, A.-J., Cerón-Carrasco, J.P., Pérez-Sánchez, H., A review of ligand-based virtual screening web tools and screening algorithms in large molecular databases in the age of big data. *Future Med. Chem.*, 10, 2641, 2018.
59. Charifson, P.S., Corkery, J.J., Murcko, M.A., Walters, W.P., Consensus scoring: A method for obtaining improved hit rates from docking databases of three-dimensional structures into proteins. *J. Med. Chem.*, 42, 5100, 1999.
60. Stahl, M. and Böhm, H.J., Development of filter functions for protein-ligand docking. *J. Mol. Graphics Modell.*, 16, 121, 1998.
61. Hamza, A., Wei, N.-N., Zhan, C.-G., Ligand-based virtual screening approach using a new scoring function. *J. Chem. Inf. Model.*, 52, 963, 2012.
62. Kaserer, T., Beck, K.R., Akram, M., Odermatt, A., Schuster, D., Pharmacophore models and pharmacophore-based virtual screening: concepts and applications exemplified on hydroxysteroid dehydrogenases. *Molecules*, 20, 22799, 2015.
63. Willett, P., Similarity-based approaches to virtual screening. *Biochem. Soc. Trans.*, 31, 603, 2003.

64. Bordoli, L., Kiefer, F., Arnold, K., Benkert, P., Battey, J., Schwede, T., Protein structure homology modeling using SWISS-MODEL workspace. *Nat. Protoc.*, 4, 1, 2009.
65. Krieger, E., Nabuurs, S.B., Vriend, G., Homology modeling. *Methods Biochem. Anal.*, 44, 509, 2003.
66. Muhammed, M.T. and Aki-Yalcin, E., Homology modeling in drug discovery: Overview, current applications, and future perspectives. *Chem. Biol. Drug Des.*, 93, 12, 2019.
67. Fiser, A., Template-based protein structure modeling, in: *Computational Biology*, D. Fenyö (Ed.), pp. 73–94, Humana Press, Totowa, NJ, 2010.
68. Xu, G., Ma, T., Du, J., Wang, Q., Ma, J., OPUS-Rota2: an improved fast and accurate side-chain modeling method. *J. Chem. Theory Comput.*, 15, 5154, 2019.
69. Huang, X., Pearce, R., Zhang, Y., FASPR: an open-source tool for fast and accurate protein side-chain packing. *Bioinformatics*, 36, 3758, 2020.
70. Schwede, T., Kopp, J.r., Guex, N., Peitsch, M.C., SWISS-MODEL: an automated protein homology-modeling server. *Nucleic Acids Res.*, 31, 3381, 2003.
71. Moroy, G., Martiny, V.Y., Vayer, P., Villoutreix, B.O., Miteva, M.A., Toward *in silico* structure-based ADMET prediction in drug discovery. *Drug Discov. Today*, 17, 44, 2012.
72. van de Waterbeemd, H. and Gifford, E., ADMET *in silico* modelling: towards prediction paradise? *Nat. Rev. Drug Discov.*, 2, 192, 2003.
73. Norinder, U. and Österberg, T., Theoretical calculation and prediction of drug transport processes using simple parameters and partial least squares projections to latent structures (PLS) statistics. The use of electrotopological state indices. *J. Pharm. Sci.*, 90, 1076, 2001.
74. Parrott, N. and Lavé, T., Prediction of intestinal absorption: comparative assessment of gastroplus™ and idea™. *Eur. J. Pharm. Sci.*, 17, 51, 2002.
75. Yoshida, F. and Topliss, J.G., QSAR model for drug human oral bioavailability. *J. Med. Chem.*, 43, 2575, 2000.
76. Cruciani, G., Carosati, E., De Boeck, B., Ethirajulu, K., Mackie, C., Howe, T., Vianello, R., MetaSite: Understanding Metabolism in Human Cytochromes from the Perspective of the Chemist. *J. Med. Chem.*, 48, 6970, 2005.
77. Li, H., Sun, J., Fan, X., Sui, X., Zhang, L., Wang, Y., He, Z., Considerations and recent advances in QSAR models for cytochrome P450-mediated drug metabolism prediction. *J. Comput.-Aided Mol. Des.*, 22, 843, 2008.
78. Terfloth, L., Bienfait, B., Gasteiger, J., Ligand-Based Models for the Isoform Specificity of Cytochrome P450 3A4, 2D6, and 2C9 Substrates. *J. Chem. Inf. Model.*, 47, 1688, 2007.
79. Liu, R., Liu, J., Tawa, G., Wallqvist, A., 2D SMARTCyp reactivity-based site of metabolism prediction for major drug-metabolizing cytochrome P450 enzymes. *J. Chem. Inf. Model.*, 52, 1698, 2012.
80. Zaretzki, J., Rydberg, P., Bergeron, C., Bennett, K.P., Olsen, L., Breneman, C.M., RS-Predictor models augmented with SMARTCyp reactivities: robust

metabolic regioselectivity predictions for nine CYP isozymes. *J. Chem. Inf. Model.*, 52, 1637, 2012.

81. Peach, M.L., Zakharov, A.V., Liu, R., Pugliese, A., Tawa, G., Wallqvist, A., Nicklaus, M.C., Computational tools and resources for metabolism-related property predictions. 1. Overview of publicly available (free and commercial) databases and software. *Future Med. Chem.*, 4, 1907, 2012.
82. Djoumbou-Feunang, Y., Fiamoncini, J., Gil-de-la-Fuente, A., Greiner, R., Manach, C., Wishart, D.S., BioTransformer: a comprehensive computational tool for small molecule metabolism prediction and metabolite identification. *J. Cheminf.*, 11, 2, 2019.
83. Marchant, C.A., Briggs, K.A., Long, A., *In Silico* Tools for Sharing Data and Knowledge on Toxicity and Metabolism: Derek for Windows, Meteor, and Vitic. *Toxicol. Mech. Methods*, 18, 177, 2008.
84. Raies, A.B. and Bajic, V.B., *In silico* toxicology: computational methods for the prediction of chemical toxicity. *Wiley Interdiscip. Rev.: Comput. Mol. Sci.*, 6, 147, 2016.

6

Comparative Study on Tannase Sequence and Structure of *Lactiplantibacillus*: An *In Silico* Protein Variability Analysis and Its Impact on Microbial Speciation

Ishita Biswas[1,2], Debanjan Mitra[1] and Pradeep K. Das Mohapatra[1]*

[1]Department of Microbiology, Raiganj University, Raiganj Uttar Dinajpur, West Bengal, India
[2]Department of Botany, S.B.S. Government College, Hili Dakshin Dinajpur, West Bengal, India

Abstract

Tannins are plant polyphenols with diversified activities. Tannins can be converted into more useful components with the presence of an inducible enzyme tannase. Tannase has several commercial applications such as production of gallic acid, acron wine, coffee-flavored soft drinks, clarification of fruit juices, etc. Application of probiotic tannase producers is more suitable for the production of tannin-derived biomolecules. The probiotic members of the genus *Lactobacillus* now designated as *Lactiplantibacillus* are promising members in terms of tannase production. Sequence and structure comparison of the tannase protein derived from six *Lactiplantibacillus* have been performed. Homology modeling was performed to model the 3D structure of tannase isolated from five *Lactiplantibacillus* sp. by considering the available 3D structure of the *Lactiplantibacillus plantarum* template. Amino acid distributions in Ramachandran plot revealed the accuracy of the model structures. Variations in the amino acid composition was observed among the tannase proteins. Charged polar amino acid residues were higher in number in *L. garii* followed by *L. paraplantarum* while a polar one is present in *L. mudanjiangensis* followed by *L. plantarum*. Secondary structure assessment of the proteins was done and psi loop formation was observed in *L. mudanjiangensis* and *L. garii*. Intra-protein interactions revealed the higher number of isolated salt

**Corresponding author*: pkdmvu@gmail.com

Hrudayanath Thatoi, Sonali Mohapatra, Swagat Kumar Das and Sukanta Kumar Pradhan (eds.) Applied Biotechnology and Bioinformatics: Agriculture, Pharmaceutical Research and Environment, (139–156) © 2025 Scrivener Publishing LLC

bridges in *L. paraplantarum* followed by *L. plantarum* and network salt bridges in *L. mudanjiangensis* followed by *L. garii*. Network aromatic–sulfur interaction was found only in *L. mudanjiangensis*. Due to the lower deviation and the lowest surface area, the tannase of *L. paraplantarum* appeared as the most stable among them.

***Keywords*:** Tannin, probiotics, homology modeling, Ramachandran plot

6.1 Introduction

Tannins are a high-molecular-weight polyphenolic class of plant secondary metabolites present in different parts of plants. Tannins are reported to have antioxidant, antibacterial, antiviral, antiparasitic, anti-inflammatory, and antidiarrheal activities [1]. Tannase is an intracellular or extracellular inducible enzyme that catalyzes the hydrolysis of tannin's ester and depside bonds [2]. Tannase is mainly used in the production of gallate, the hydrolyzed product of tannin, which, in turn, is involved in the production of antimalarial drug trimethoprim and the antioxidant propyl gallate. Tannase is also used as a clarifier in the making of fruit juices, coffee-flavored drinks, wines, and also found to reduce the bitterness of the fruit juice [3]. Using tannase-producing probiotic bacteria with satisfactory metabolizing capacity is a promising way to increase the release of bioactive tannin metabolites. Among the probiotic species, *Lactobacillus* gains more acceptability as a tannase producer [4]. Sabu *et al.* (2006) reported on tannase production by *Lactobacillus* sp. [5]. Iwamoto *et al.* (2008) isolated and cloned the tannase encoding gene tanLpl from *Lactobacillus plantarum* [6]. *Lactobacillus plantarum* is now being allocated into the new genus *Lactiplantibacillus* and renamed as *Lactiplantibacillus plantarum. L. plantarum* is a bacteria of interest due of its extensive array of tannases and other polyphenol related enzymes [7]. Biswas *et al.* (2020) first report on the influence of salt bridges and its microenvironment components in tannase proteins of bacteria [8]. A new idea in structural biology is the finding of the microenvironment of the salt bridge. Sequence and structure analysis of the intragenic protein sheds light on the diversity function and evolutionary relationships among the species.

Although there are several available reports present in the online database regarding the tannase protein sequences of *Lactiplantibacillus* species, no such reports on sequence and structure analysis have yet been published. In the current study, this is the first report on comparative sequence and structure analysis of tannase proteins of six species of *Lactiplantibacillus*.

6.2 Materials and Methods

6.2.1 Dataset

Tannase protein sequences from six *Lactiplantibacillus* species were retrieved from the largest protein sequence database UniProt [9]. A total of six different species were identified from the genus *Lactiplantibacillus* that produces tannase enzymes. Only the tannase 3D structure of *Lactiplantibacillus plantarum* from this group of species was available in the RCSB PDB database [10, 11].

6.2.2 Analysis of Protein Sequences

All the selected tannase sequences were subjected to refinement and further analyzed by the ProtParam server to check their physicochemical properties [12, 13].

6.2.3 Homology Modeling

The 3D structures of tannase from *Lactiplantibacillus* sp. (except *Lactiplantibacillus plantarum*) that are not available in RCSB-PDB were predicted using the homology modeling method-based tool Modeller V.10.4 [14] and has been presented in Figure 6.1. The suited computational approaches for the prediction of the 3D structure of a protein from its amino acid sequence using its template 3D structure is homology modeling. It is considered as computational structure prediction approach having the best accuracy level. Here, the tannase of *Lactiplantibacillus plantarum* (3WA7) was taken as a template structure for performing homology modeling. The sequence similarity of the template tannase was compared with tannase of other *Lactiplantibacillus* sp. and showed a range of identity from 64.58% to 97.64% (Table 6.1). Validation of the model was done using the PROCHECK Server [15].

6.2.4 Analysis of Crystal Structure

All PDB 3D structures of tannase available in PDB were minimized in 1000 steps through UCSF Chimera 1.15rc with an Amber force field [16]. The secondary structure of the tannase enzyme from all the *Lactiplantibacillus* sp. was investigated using the PDBsum server [17]. Intra-protein interactions like salt bridges, aromatic–aromatic interactions, aromatic–sulfur, cation–pi, etc. were identified through the PIC server [18, 19].

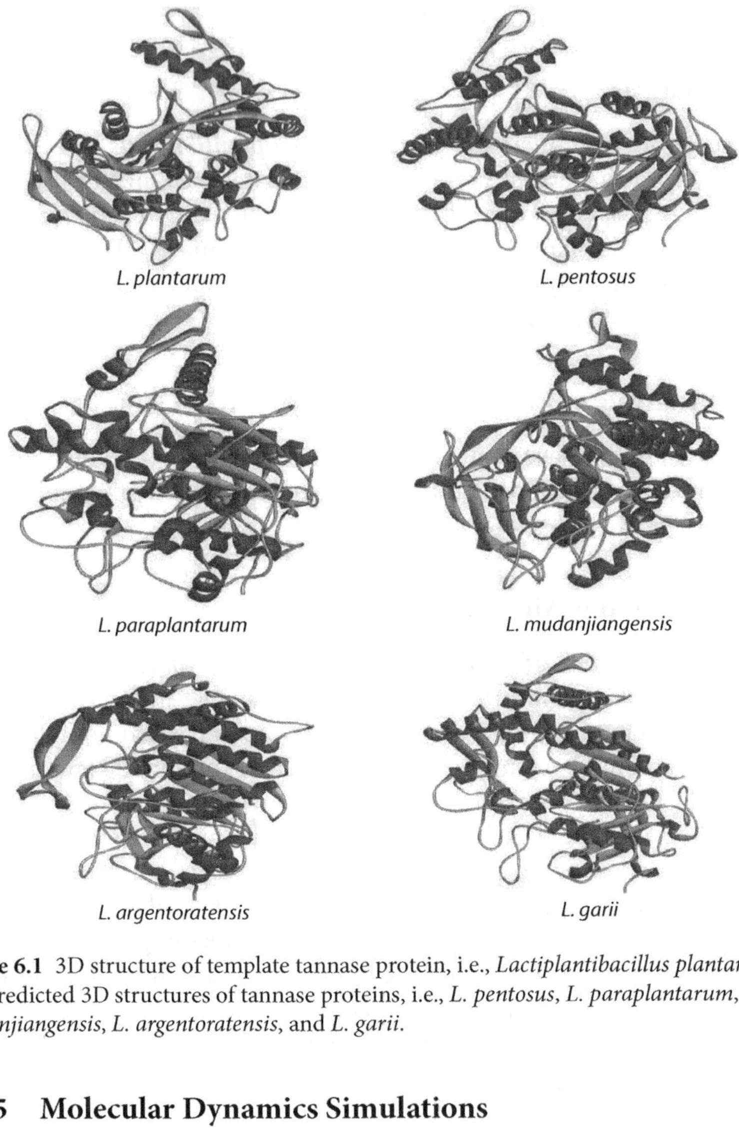

Figure 6.1 3D structure of template tannase protein, i.e., *Lactiplantibacillus plantarum*, and predicted 3D structures of tannase proteins, i.e., *L. pentosus*, *L. paraplantarum*, *L. mudanjiangensis*, *L. argentoratensis*, and *L. garii*.

6.2.5 Molecular Dynamics Simulations

Molecular dynamics (MD) is a computer-based simulation method for analyzing the physical movements of atoms and molecules of a protein. The atoms and molecules present in a protein are allowed to interact for a fixed period of time, giving a view of the dynamic "evolution" of the system. The experimental protein 3D structure available in the Protein Data Bank (PDB) is used as the basis for MD simulations of proteins and/or enzymes [20]. GROMACS [21, 22] and the GROMOS96 43a1 forcefield

Table 6.1 Name of target organisms with their UniProt ID and similarity of tannase with template sequence.

Name	Target ID	Identity with template sequence (%)
Lactiplantibacillus pentosus	A0A2P1JP46	72.53
Lactiplantibacillus paraplantarum	A0A2I9CLW7	82.58
Lactiplantibacillus mudanjiangensis	A0A660EAS4	64.58
Lactiplantibacillus argentoratensis	A0A8D4N8G2	97.64
Lactiplantibacillus garii	A0A3R8J7D4	65.52

were used to perform the molecular dynamics simulations. After equilibration, 5000 steps of the steepest descent approach were employed to minimize the energy on the solvated systems. The final production run consisted of 50-ns molecular dynamic simulations at 300 K temperatures. Molecular dynamics simulations were used to determine the radius of gyration (Rg), solvent accessible surface area (SASA), root mean square deviation (RMSD), and root mean square fluctuation (RMSF), as well as hydrogen bonds [23].

6.3 Results and Discussion

6.3.1 Physicochemical Properties of Tannase

The physicochemical properties of a protein are generally determined by the amino acid composition and function. Sequence analysis revealed variations in amino acid compositions among the tannase proteins of the six species. The abundance of charged polar amino acid residues has been shown in *L. garii* followed by *L. paraplantarum* (Figure 6.2), while the uncharged polar amino acid residues have a higher abundance in *L. mudanjiangensis* followed by *L. plantarum. L. mudanjiangensis* was found to have a higher number of some specific amino acids like C, K, M, S, T, and Y. Amino acids like Q, I, and L were present in higher quantities in *L. argentoratensis,* whereas the specific abundance of A, R, P, and V was observed in *L. garii.*

The presence of specific amino acids in abundant number is related to the specific function of a protein. For example, cysteine and methionine

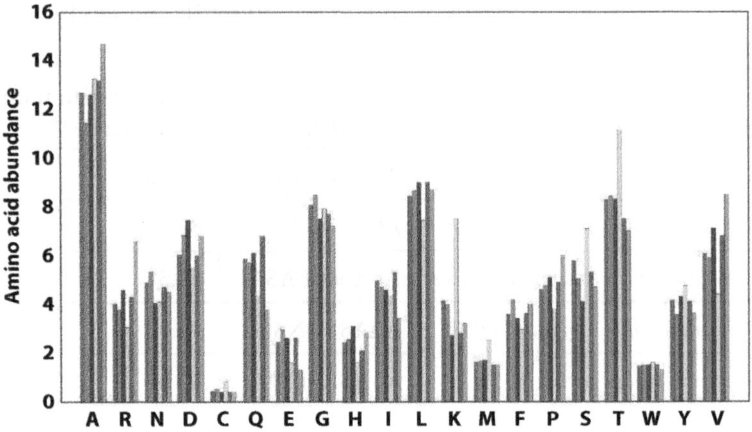

Figure 6.2 Amino acid abundance in tannase of *Lactiplantibacillus plantarum* (red), *L. pentosus* (green), *L. paraplantarum* (blue), *L. mudanjiangensis* (yellow), *L. argentoratensis* (pink), and *L. garii* (cyan).

residues in proteins are vulnerable to reversible oxidation and reduction, mediated either enzymatically or non-enzymatically, and is one of their key characteristics [24]. Proline interferes with the backbone's ability to adopt an alpha-helix or beta-sheet shape, disrupting the protein secondary structure. Another interpretation is that proline imposes a unique type of secondary structure with a constrained phi angle that supersedes all other secondary structures [25].

6.3.2 Validation of Modeled Protein Structures

Model validation is an important procedure by which the quality of the protein structures is evaluated. The evaluation was performed mainly based on the Ramachandran plot by which the amino acid distribution was observed and has been presented in Table 6.2. The template structure, i.e., tannase, of *L. plantarum* showed 87.9% residues in its allowed regions and 11.4% in its additional allowed regions and 0.2% of residues fall under disallowed regions. The target structure of *L. pentosus* showed almost similar numbers in the most favored regions but it had 0.5% residues in its disallowed regions. *L. paraplantarum* and *L. mudanjiangensis* showed a lower abundance of amino acid residues in their most favored regions than the template structure. The number of amino acid residues, i.e., 0.7% in each, respectively, in disallowed regions was slightly higher than the template structure. *L. argentoratensis* and *L. garii* showed a higher number of

Table 6.2 Amino acid distribution of tannase from *Lactiplantibacillus* sp. in the respective area of the Ramachandran plot.

Name	Residues present in most favored regions	Residues present in additional allowed regions	Residues present in generously allowed regions	Residues present in disallowed regions
Lactiplantibacillus plantarum	87.9	11.4	0.5	0.2
Lactiplantibacillus pentosus	87.3	11.4	0.7	0.5
Lactiplantibacillus paraplantarum	85.2	13.6	0.5	0.7
Lactiplantibacillus mudanjiangensis	86.7	11.9	0.7	0.7
Lactiplantibacillus argentoratensis	88.6	10.6	0.2	0.5
Lactiplantibacillus garii	88	10.7	1	0.2

residues in their most favored regions, which indicates that these structures were quite better than the template structure. They had 10.6% and 10.7% amino acid in additional allowed regions, respectively. The Ramachandran plot of these tannase proteins of the studied *Lactiplantibacillus* species is presented in Figures 6.3 and 6.4.

6.3.3 Secondary Structure Assessment

The sequence of amino acid residues of a polypeptide chain when remained arranged in a particular pattern is known as the secondary structure. It is formed by the hydrogen bonds between the amide hydrogens and carbonyl oxygens of the peptide chain. α-helices and β-structures are the two main secondary structure formations of a protein. *L. plantarum* is constituted by 4 sheets, 3 beta–alpha–beta units, 5 beta hairpins, 3 beta bulges, 15 strands, 19 helices, 23 helix–helix interactions, 34 beta turns, and 7 gamma turns. Numerous proteins with a single transmembrane-helix can oligomerize by helix–helix interactions. The interactions are quite particular and involve

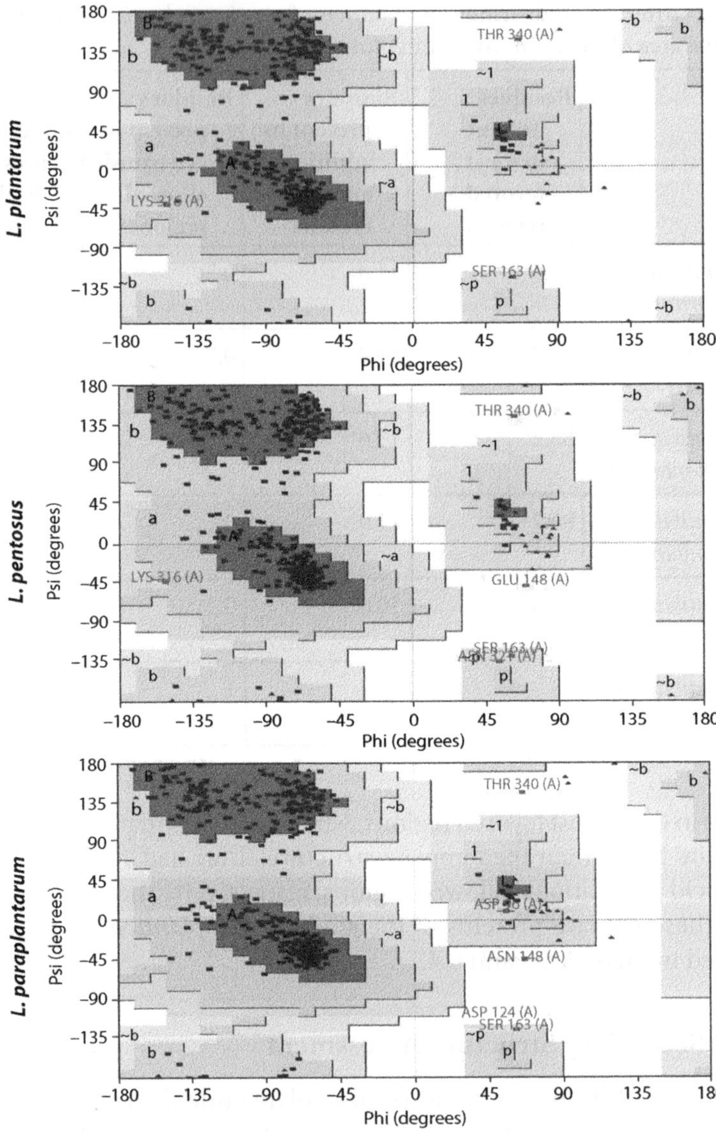

Figure 6.3 Ramachandran plot showing amino acid distribution of tannase belonging to *Lactiplantibacillus plantarum*, *L. pentosus*, and *L. paraplantarum*.

hydrogen bonds, interhelical salt bridges, or exact packing interactions [26]. *L. pentosus* showed almost the same arrangement but just differ in some specific regions like 5 sheets, 17 strands, 25 helix–helix interactions, and 8 gamma turns. *L. plantarum* just differs from *L. pentosus* in the number

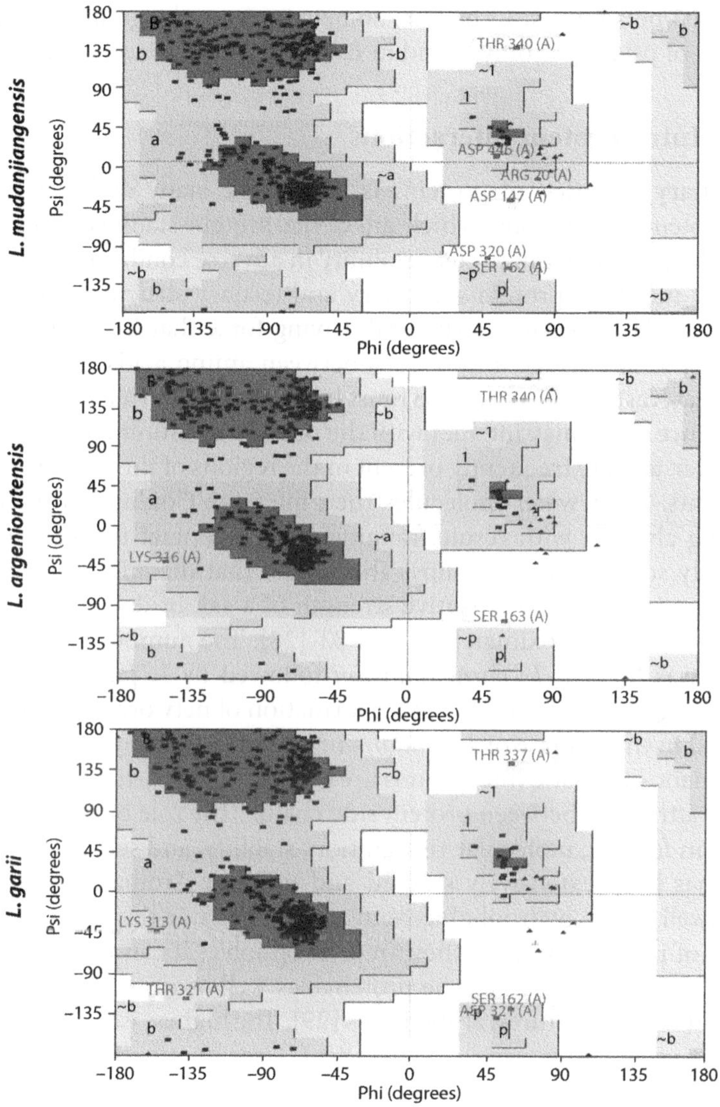

Figure 6.4 Ramachandran plot showing amino acid distribution of tannase belonging to *Lactiplantibacillus mudanjiangensis*, *L. argentoratensis*, and *L. garii*.

of 18 helices. *L. mudanjiangensis* consisted of 3 sheets, 3 beta-alpha-beta units, 4 beta hairpins, 1 psi loop, 2 beta bulges, 13 strands, 19 helices, 24 helix-helix interactions, 34 beta turns, and 11 gamma turns. *L. argentoratensis* differs from the template structure of *L. plantarum* in 33 beta turns. *L. garii* showed an almost similar structure to *L. mudanjiangensis* but just differs in 3 beta bulges, 17 helices, 36 beta turns, and 9 gamma turns. The presence of psi loop is only found in *L. mudanjiangensis* and *L. garii.*

6.3.4 Intra-Protein Interactions

The tertiary structure of proteins is formed by many different types of intra-protein interactions, which affect the protein stability. These interactions help to maintain protein stability in adverse conditions [27]. Being the basic motifs of protein secondary and tertiary structure, salt bridges are frequently linked to a structural driving force that improves stability. To form the salt bridge interactions between amino acids with opposing charges, at least two heavy atoms must be located within a hydrogen bonding distance. They may interact with the outside environment, particularly with water, and are frequently present in the regions of the protein exposed to solvents. Polar water molecules then filter the Coulomb attraction of opposing charges, with results ranging from a contact ion pair to independently solvated ionic groups. This shows that the environment has a significant impact on the relative strength of a salt bridge, which is often estimated to be approximately a few kBT [28]. The number of isolated salt bridges was high in *L. paraplantarum* followed by *L. plantarum* and *L. mudanjiangensis* (Table 6.3). Higher formation of network salt bridges was observed in the tannase of *L. mudanjiangensis* followed by *L. garii.*

Aromatic–aromatic interactions are one of the important non-covalent force of attraction between protein side chains. The role of aromatic interactions in folding, molecular recognition, stability, and self-assembly processes has been extensively studied and utilized. Aromatic amino acids often dwell in the hydrophobic center of proteins or the transmembrane regions of proteins because they are hydrophobic. It can play a variety of roles in soluble and membrane proteins, as well as in the stabilization of helix and sheet structures, at such sites [29]. The highest isolated aromatic–aromatic interactions were observed in *L. argentoratensis* followed by *L. mudanjiangensis* and *L. pentosus.* The highest network aromatic–aromatic interactions were found in *L. plantarum* and *L. paraplantarum* (Table 6.3).

The bulk of protein structures include aromatic–sulfur interactions, yet little is understood about how these interactions operate in ion channels. In comparison to only hydrophobic interactions, the aromatic–sulfur interactions offer extra stabilization [30, 31]. *L. mudanjiangensis* has the highest isolated aromatic–sulfur interactions. Most importantly, only *L. mudanjiangensis* has a network aromatic–sulfur interaction.

A cation's stabilizing electrostatic contact to the polarizable pi electron cloud is known as the cation–pi interaction. In biological systems, cation

Table 6.3 Intra-protein interaction in tannase from species of *Lactiplantibacillus*.

	Salt bridge		Aromatic–aromatic		Aromatic–sulfur		Cation–pi	
Species name	**Isolated**	**Network**	**Isolated**	**Network**	**Isolated**	**Network**	**Isolated**	**Network**
Lactiplantibacillus plantarum	14	2	5	6	1	0	10	1
Lactiplantibacillus pentosus	11	3	7	5	1	0	7	1
Lactiplantibacillus paraplantarum	15	2	5	6	1	0	11	1
Lactiplantibacillus mudanjiangensis	14	5	7	4	3	1	5	2
Lactiplantibacillus argentoratensis	12	2	8	4	1	0	10	1
Lactiplantibacillus garii	11	4	5	5	1	0	13	0

interactions are frequent, and several structural analyses have identified the aromatic box as a typical motif [32]. *L. garii* has the highest isolated cation–pi interactions followed by *L. paraplantarum* (Table 6.3), while the network cation–pi interaction is highest in *L. mudanjiangensis* (Table 6.3).

6.3.5 Stability in Simulation Studies

In recent years, the significance of the molecular dynamics (MD) simulations study has been increasing in the field of molecular biology and drug development. These simulations accurately depict the behavior of proteins and other biomolecules at the atomic level and at very fine temporal resolutions. The understanding of the workings of proteins and other biomolecules, the discovery of the structural underpinnings of illness, and the design and optimization of small molecules, peptides, and proteins have all benefited from simulations [33]. A common similarity metric in the research of macromolecular structures and dynamics is the root mean square deviation (RMSD). *L. plantarum* and *L. mudanjiangensis* showed almost similar deviations and follow almost equal trajectories throughout the path. On the other hand, *L. pentosus, L. paraplantarum, L. argentoratensis*, and *L. garii* initially showed deviation at the start but represent a similar trajectory after 20 ns to the end. *L. mudanjiangensis* showed the highest deviations among all organisms at 40 ns to 50 ns (Figure 6.5a).

Root mean square fluctuation (RMSF) showed that *L. argentoratensis* has multiple fluctuation sites throughout its path. The RMSF of tannase of other organisms was lower than that of *L. argentoratensis*. Some specific common fluctuation sites like residue numbers 55, 122, 244, and 383 were observed (Figure 6.5b).

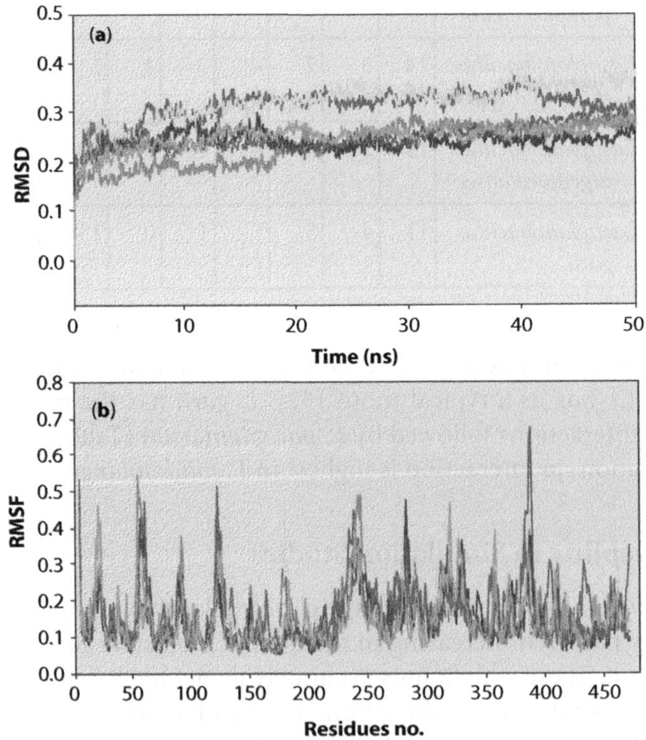

Figure 6.5 (a) RMSD and (b) RMSF plots of tannase of *Lactiplantibacillus plantarum* (red), *L. pentosus* (green), *L. paraplantarum* (blue), *L. mudanjiangensis* (yellow), *L. argentoratensis* (pink), and *L. garii* (cyan).

The radius of gyration serves as a measurement of compactness of protein structures. How regular secondary structures may be tightly packed into a protein's three-dimensional structure is of concern. In the protein size range under consideration, proteins exhibit the maximum radius of gyration, indicating a less tightly packed state compared to β- and (α + β)-proteins. The smallest gyroscopic radius and, hence, the most tightly packing are properties of α/β-proteins [34]. *L. argentoratensis* showed the lowest radius; hence, it has the highest tight packing. *L. pentosus* showed the highest radius, i.e., lower packing, among all organisms (Figure 6.6a).

Proteins' solvent-accessible surface area (SASA) has long been regarded as a key variable in research on protein folding and stability. It is described as the surface characterized by a hypothetical solvent sphere's center and the molecule's Van der Waals contact surface all around a protein [35, 36]. *L. paraplantarum* showed the lowest surface area, which means it has a

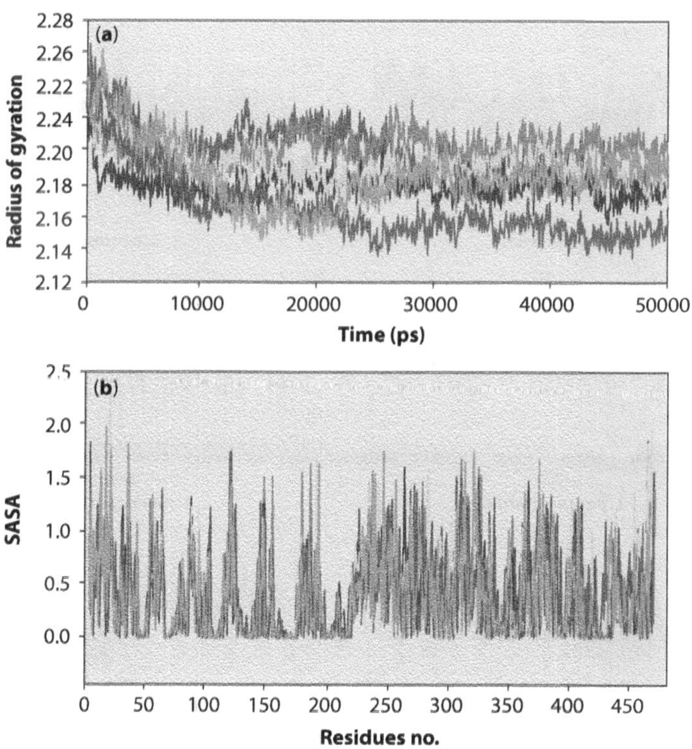

Figure 6.6 (a) Rg and (b) SASA of tannase of *Lactiplantibacillus plantarum* (red), *L. pentosus* (green), *L. paraplantarum* (blue), *L. mudanjiangensis* (yellow), *L. argentoratensis* (pink), and *L. garii* (cyan).

higher protein folding. The high surface area was observed in *L. mudanjiangensis*, which means the folding of the tannase of this organism was weak (Figure 6.6b).

6.3.6 Hydrogen Bonds

Protein structures are formed and stabilized in part by hydrogen bonds or H-bonds. When a protein deforms, such as when going from an inactive to an active state, they form and break. Since it enables the evaluation of sizable databases of organic compounds in virtual screening projects targeted at a protein of interest, quick and accurate assessment of the hydrogen bond potential energy has a substantial influence on drug design and development [37]. *Lactiplantibacillus mudanjiangensis* showed the highest numbers of hydrogen bond followed by *L. paraplantarum*. However,

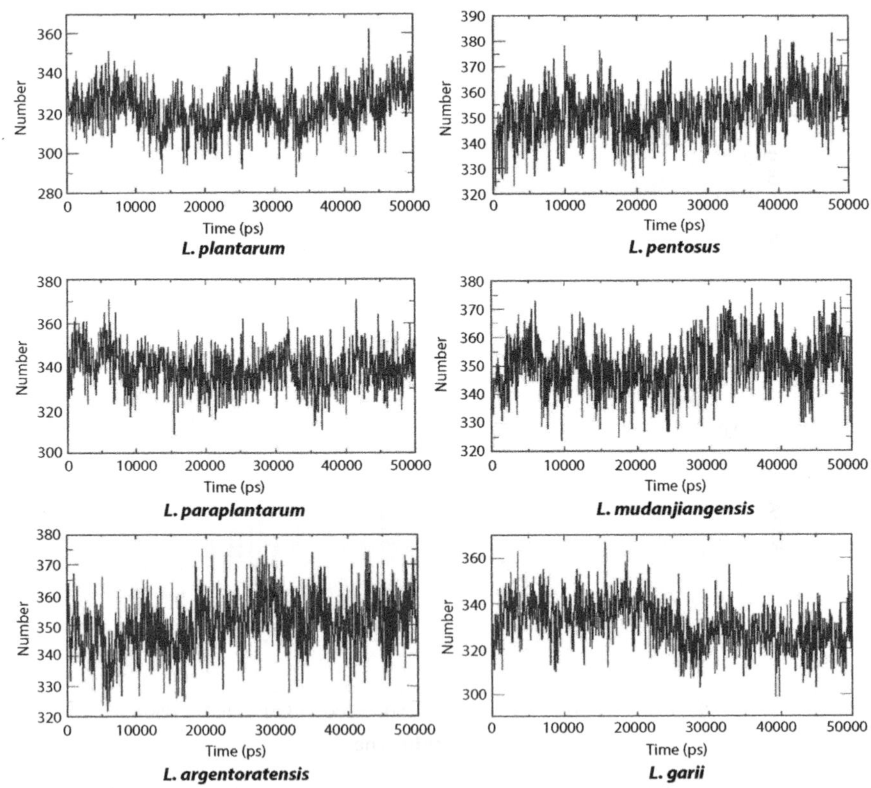

Figure 6.7 Hydrogen bonds in tannase of *Lactiplantibacillus plantarum, L. pentosus, L. paraplantarum, L. mudanjiangensis, L. argentoratensis,* and *L. garii.*

L. argentoratensis formed a higher number of hydrogen bonds than others at 50 ns (Figure 6.7).

6.4 Conclusion

The present study reveals the similarity and differences among the tannase of *L. plantarum*, *L. pentosus*, *L. paraplantarum*, *L. mudanjiangensis*, *L. argentoratensis*, and *L. garii*. This is the first report where homology modeling was applied exhaustively to model the unavailable structures of tannase from different species of *Lactiplantibacillus*. A higher abundance of charged polar amino acid residues was observed in *L. garii* and *L. paraplantarum*, whereas the uncharged polar amino acid residues have a higher presence in *L. mudanjiangensis* followed by *L. plantarum*. The secondary structures of all those organisms showed their conserved nature. However, *L. mudanjiangensis* and *L. garii* have an extra psi loop formation in their secondary structure. *L. paraplantarum* and *L. mudanjiangensis* increase their intra-protein interactions rather than other *Lactiplantibacillus* sp. Molecular dynamics simulations revealed their stability, folding pattern, and tight packing characteristics. This study will help to understand the structural formation of tannase and will help in the protein engineering of tannase for better industrial application.

References

1. Tong, Z., He, W., Fan, X., Guo, A., Biological function of plant tannin and its application in animal health. *Front. Vet. Sci.*, 8, 803657, 2022.
2. Biswas, I. and Mitra, D., and Mohapatra, P.K.D., Structural and catalytic advancement of fungal tannase: A proteomic contribution in industrial applicability. *Bioresour. Technol. Rep.*, 19, 101103, 2022.
3. Belur, P.D., Goud, R., Goudar, D.C., Optimization of culture medium for novel cell-associated tannase production from Bacillus massiliensis using response surface methodology. *J. Microbiol. Biotechnol.*, 22, 2, 199–206, 2012.
4. Biswas, I., Soren, J.P., Mohapatra, P.K.D., Preparation of metabiotic curd and optimization of tannase and gallic acid production by probiotic *Lactiplantibacillus plantarum* PKI15. *Biocatal. Agric. Biotechnol.*, 103041, 2024.
5. Sabu, A., Augur, C., Swati, C., Pandey, A., Tannase production by Lactobacillus sp. ASR-S1 under solid-state fermentation. *Process Biochem.*, 41, 3, 575–580, 2006.

6. Iwamoto, K., Tsuruta, H., Nishitaini, Y., Osawa, R., Identification and cloning of a gene encoding tannase (tannin acylhydrolase) from Lactobacillus plantarum ATCC 14917T. *Syst. Appl. Microbiol.*, 31, 4, 269–277, 2008.
7. Pulido-Mateos, E.C., Lessard-Lord, J., Guyonnet, D., Desjardins, Y., Roy, D., Comprehensive analysis of the metabolic and genomic features of tannin-transforming Lactiplantibacillus plantarum strains. *Sci. Rep.*, 12, 1, 22406, 2022.
8. Biswas, I., Mitra, D., Bandyopadhyay, A.K., Mohapatra, P.K.D., Contributions of protein microenvironment in tannase industrial applicability: An in-silico comparative study of pathogenic and non-pathogenic bacterial tannase. *Heliyon*, 6, 11, e05359, 2020.
9. UniProt Consortium, UniProt: a hub for protein information. *Nucleic Acids Res.*, 43, D1, D204–D212, 2015.
10. Berman, H.M., Westbrook, J., Feng, Z., Gilliland, G., Bhat, T.N., Weissig, H., Shindyalov, I.N., Bourne, P.E., The protein data bank. *Nucleic Acids Res.*, 28, 235–242, 2000.
11. Kouranov, A., Xie, L., de la Cruz, J., Chen, L., Westbrook, J., Bourne, P.E., Berman, H.M., The RCSB PDB information portal for structural genomics. *Nucleic Acids Res.*, 34, suppl_1, D302–D305, 2006.
12. Gasteiger, E., Hoogland, C., Gattiker, A., Duvaud, S.E., Wilkins, M.R., Appel, R.D., Bairoch, A., *Protein identification and analysis tools on the ExPASy server*, pp. 571–607, Humana Press, Switzerland, 2005.
13. Mitra, D., Pal, A.K., Das Mohapatra, P.K., Intra-protein interactions of SARS-CoV-2 and SARS: a bioinformatic analysis for plausible explanation regarding stability, divergency, and severity. *Syst. Microbiol. Biomanuf.*, 2, 4, 653–664, 2022a.
14. Webb, B. and Sali, A., Comparative protein structure modeling using MODELLER. *Curr. Protoc. Bioinf.*, 54, 1, 5–6, 2016.
15. Laskowski, R.A., MacArthur, M.W., Moss, D.S., Thornton, J.M., PROCHECK: a program to check the stereochemical quality of protein structures. *J. Appl. Cryst.*, 26, 283–291, 1993.
16. Pettersen, E.F., Goddard, T.D., Huang, C.C., Couch, G.S., Greenblatt, D.M., Meng, E.C., Ferrin, T.E., UCSF Chimera—a visualization system for exploratory research and analysis. *J. Comput. Chem.*, 25, 13, 1605–1612, 2004.
17. Laskowski, R.A., PDBsum: summaries and analyses of PDB structures. *Nucleic Acids Res.*, 29, 1, 221–2, 2001.
18. Tina, K.G., Bhadra, R., Srinivasan, N., PIC: protein interactions calculator. *Nucleic Acids Res.*, 35, suppl_2, W473–W476, 2007.
19. Mitra, D. and Das Mohapatra, P.K., *In silico* comparative structural and compositional analysis of glycoproteins of RSV to study the nature of stability and transmissibility of RSV A. *Syst. Microbiol. Biomanuf.*, 3, 2, 312–327, 2023.

20. Salo-Ahen, O.M., Alanko, I., Bhadane, R., Bonvin, A.M., Honorato, R.V., Hossain, S., Juffer, A.H., Kabedev, A., Lahtela-Kakkonen, M., Larsen, A.S., Lescrinier, E., Molecular dynamics simulations in drug discovery and pharmaceutical development. *Processes*, 9, 1, 71, 2020.
21. Berendsen, H.J., van der Spoel, D., van Drunen, R., GROMACS: A message-passing parallel molecular dynamics implementation. *Comput. Phys. Commun.*, 91, 1-3, 43–56, 1995.
22. Abraham, M.J., Murtola, T., Schulz, R., Páll, S., Smith, J.C., Hess, B., Lindahl, E., GROMACS: High performance molecular simulations through multi-level parallelism from laptops to supercomputers. *SoftwareX*, 1, 19–25, 2015.
23. Mitra, D., Paul, M., Thatoi, H., Mohapatra, P.K.D., Study of potentiality of dexamethasone and its derivatives against Covid-19. *J. Biomol. Struct. Dyn.*, 40, 20, 10239–10249, 2022b.
24. Lim, J.M., Kim, G., Levine, R.L., Methionine in proteins: it's not just for protein initiation anymore. *Neurochem. Res.*, 44, 247–257, 2019.
25. Agnihotry, S., Pathak, R.K., Singh, D.B., Tiwari, A., Hussain, I., Protein structure prediction, in: *Bioinformatics*, pp. 177–188, Academic Press, USA, 2022.
26. Lemmon, M.A. and Engelman, D.M., Helix-helix interactions inside lipid bilayers. *Curr. Opin. Struct. Biol.*, 2, 4, 511–518, 1992.
27. Mitra, D. and Das Mohapatra, P.K., Cold adaptation strategy of psychrophilic bacteria: an *in-silico* analysis of isocitrate dehydrogenase. *Syst. Microbiol. Biomanuf.*, 1, 4, 483–493, 2021a.
28. Pylaeva, S., Brehm, M., Sebastiani, D., Salt bridge in aqueous solution: Strong structural motifs but weak enthalpic effect. *Sci. Rep.*, 8, 1, 13626, 2018.
29. Makwana, K.M. and Mahalakshmi, R., Implications of aromatic–aromatic interactions: From protein structures to peptide models. *Protein Sci.*, 24, 12, 1920–1933, 2015.
30. Yeung, P.S., Ing, C.E., Yamashita, M., Pomès, R., Prakriya, M., A sulfur-aromatic gate latch is essential for opening of the Orai1 channel pore. *Elife*, 9, e60751, 2020.
31. Mitra, D. and Das Mohapatra, P.K., Discovery of novel cyclic salt bridge in thermophilic bacterial protease and study of its sequence and structure. *Appl. Biochem. Biotechnol.*, 193, 6, 1688–1700, 2021b.
32. Davis, M.R. and Dougherty, D.A., Cation–π interactions: computational analyses of the aromatic box motif and the fluorination strategy for experimental evaluation. *Phys. Chem. Chem. Phys.*, 17, 43, 29262–70, 2015.
33. Hollingsworth, S.A. and Dror, R.O., Molecular dynamics simulation for all. *Neuron*, 99, 1129–1143, 2018.
34. Lobanov, M.Y., Bogatyreva, N.S., Galzitskaya, O.V., Radius of gyration as an indicator of protein structure compactness. *Mol. Biol.*, 42, 623–628, 2008.
35. Ausaf Ali, S., Hassan, I., Islam, A., Ahmad, F., A review of methods available to estimate solvent-accessible surface areas of soluble proteins in the folded and unfolded states. *Curr. Protein Pept. Sci.*, 15, 5, 456–476, 2014.

36. Mitra, D., Afreen, S., Das Mohapatra, P.K., Abdalla, M., Threat of respiratory syncytial virus infection knocking the door: a proposed potential drug candidate through molecular dynamics simulations, a future alternative. *J. Mol. Model.*, 29, 4, 91, 2023.
37. Williams, M.A. and Ladbury, J.E., Hydrogen bonds in protein-ligand complexes, in: *Protein-ligand interactions: from molecular recognition to drug design*, pp. 137–61, 2003.

7

Probiotics: A Novel Natural Therapy for Oral Health

Preeti Pallavi[1], Vikas Kumar[1], Sangeeta Prakash[2] and Sangeeta Raut[1]*

[1]Centre for Biotechnology, School of Pharmaceutical Sciences, Siksha 'O' Anusandhan (Deemed to be University), Bhubaneswar, Odisha, India
[2]School of Agriculture and Food Sustainability, The University of Queensland, Brisbane, Queensland, Australia

Abstract

The oral environment constitutes an integral part of the human microbiome. Among the prevalent non-communicable oral diseases, dental caries and periodontal disorders stand out, resulting from imbalances in the microbial populations within susceptible hosts' mouths. It is widely recognized that oral biofilms play a pivotal role in initiating and progressing various oral disorders. Notably, treating these disorders ranks as the fourth most expensive condition in numerous developed nations. Antibacterial treatment is thought to be a crucial strategy for treating oral diseases. While antibacterial treatment represents a crucial approach to tackling oral diseases, the limited local drug concentrations and the potential for drug resistance emergence constrain conventional antibiotics' effectiveness. Hence, addressing the pathogenicity of oral biofilms has become a primary focus in preventing infectious oral diseases through the utilization of probiotics. Commonly accepted knowledge suggests that probiotics have shown promise in effectively integrating into oral biofilms, mitigating prevalent oral infections, and influencing local immune responses through the release of bioactive substances. Using probiotics proves to be an efficacious method for managing oral diseases. The chapter comprehensively elucidates the current understanding of etiology, drawing from existing research on applying probiotic species to address issues such as oral candidiasis, periodontal disease, and caries prevention, thereby contributing to improved oral health management.

**Corresponding author*: research.sangeeta@gmail.com

Hrudayanath Thatoi, Sonali Mohapatra, Swagat Kumar Das and Sukanta Kumar Pradhan (eds.) Applied Biotechnology and Bioinformatics: Agriculture, Pharmaceutical Research and Environment, (157–180) © 2025 Scrivener Publishing LLC

***Keywords*:** Biofilm, dental caries, oral health, periodontitis, probiotics

7.1 Introduction

Worldwide, oral health is intricately linked to both quality of life and overall health. Deteriorating oral health not only disrupts daily life but also elevates the risk of developing systemic diseases [1, 2]. The oral microbiota constitutes a complex and diverse microbial community, and its disruption can profoundly affect oral health, triggering the onset of oral diseases [3, 4]. These oral diseases are chronic and progressively escalating. A staggering 70% of the world's population confronts various oral diseases stemming from diverse risk factors like sugar consumption, host genetic status, reduced saliva flow, tobacco usage, alcohol consumption, and poor hygiene practices. Inadequate dietary habits disrupt the balanced state of resident microbiomes, consequently fostering the emergence of oral ailments such as dental caries, periodontitis, gingivitis, and endodontic and orthodontic infections.

In recent years, emerging evidence has demonstrated the close association between oral diseases, particularly those induced by bacteria, and several systemic diseases [6, 7]. Research has demonstrated that a history of caries in primary teeth is a robust and reliable predictor of future dental caries development [5]. This arises due to the production of acid, which erodes tooth enamel. Pathogenic bacteria break down sugar to generate acid, leading to enamel demineralization through calcium and phosphate loss, a phenomenon termed demineralization. Dental caries materializes with repeated demineralization over an extended period of time [8–10]. The oral commensals strive to maintain a neutral pH when the microbiome is in symbiosis. Some bacteria in the biofilm use refined carbohydrates as an energy source, yielding organic acid as by-products. Extended exposure to these organic acids within the biofilm environment can induce a substantial decrease in pH. This low pH impacts both the composition of the biofilm and the tooth surface. The initial phase in the formation of the bacterial biofilm linked to dental caries involves the reversible attachment of bacteria to the enamel surface, followed by an irreversible attachment that leads to the synthesis of an extracellular polymeric substance (EPS) matrix by bacterial microcolonies, culminating in the creation of a three-dimensional biofilm [11].

A noteworthy pathogen in this context is *Streptococcus mutans*, a Gram-positive bacterium that depends on sucrose for its growth, found within the human oral cavity. Conditions favoring sucrose fermentation by-products lower the pH in the oral environment, upsetting the balance between demineralization and remineralization. This process contributes to enamel erosion and facilitates the formation of dental biofilms. Addressing dental caries continues to be an elaborate and costly process, with predominant strategies focused on inhibiting the growth of dental biofilms through mechanical means like brushing, scaling, and flossing, often complemented by the use of chemical dentifrices [12].

Furthermore, one of the most prevalent diseases is periodontal disease, encompassing various pathological conditions of the periodontium, triggered by bacterial infections residing in dental plaques on the tooth surface [13]. The advancement of periodontal disease is linked to the accumulation of subgingival plaque, prompting a transition in the microflora composition from a healthy to an unhealthy state. This condition has been identified by a polymicrobial inflammatory state of the periodontium. Gingivitis signifies the initial and least severe stage of periodontal disease, characterized by inflammation of the gums. If left unaddressed, gingivitis progresses to periodontitis, a more severe condition marked by the loss of gingiva, bone, and ligaments. Periodontitis, being a bacterial infection, is a gradual and damaging ailment that can ultimately result in tooth loss [14]. Severe periodontitis can lead to significant loss of supportive structures and teeth, affecting around 10%–15% of the global population.

The composition of dental biofilms dictates the initiation and persistence of gingivitis and periodontitis. With the progression of gingivitis, dental plaque gains more organization, forming multi-layered biofilms containing bacteria stratified by metabolism and aerotolerance. Though gingivitis typically precedes periodontitis, the shift from gingivitis to periodontitis depends on host susceptibility, pathogenic bacteria, and protective bacteria. It only occurs universally across all patients or sites. Periodontal disease is not attributed to a single pathogen, but to dysbiosis within the biofilm. When disrupted, as in activities like flossing, microorganisms within the biofilm begin to exhibit pathogenic characteristics that aggravate and stimulate the gingival inflammation. The microbiota associated with periodontal health includes early or initial colonizers such as *Streptococcus sanguinis, Gemella* spp., *Streptococcus mitis, Fusobacterium nucleatum, Atopobium* spp., and *Capnocytophaga* [14], while the etiology of periodontal disease is associated with biofilm microorganisms [15]. Numerous risk factors have been identified for the disease, encompassing (1) smoking, (2) diabetes mellitus, (3) socioeconomic and demographic variables, (4) psychosocial factors such as stress, (5) genetic predispositions, (6) hormonal characteristics like pregnancy, and (7) compromised immunity.

7.2 Background

7.2.1 Methods for Treatment of Oral Diseases

Dentists have long considered tooth restoration a treatment for dental caries, dating back to the start of the 20th century. During a period when there was little understanding of caries prevention, the emphasis on tooth repair and retention represented a step forward from the prior treatment strategy of tooth extraction. There are additional therapies available, like root canals, dental implants, and orthodontics, which are sometimes known as non-restorative or endodontic treatments since they aim to treat the cause of caries rather than returning the tooth to its initial state.

Likewise, fluoride compounds have recently demonstrated substantial success in halting and delaying the growth of tooth caries. Fluoride's anti-cariogenic and antibacterial qualities help to reduce early dental caries in a number of ways. The use of fluoride can lead to a decrease in pH levels. Consequently, bacteria are compelled to spend more energy to maintain their pH homeostasis; as a result, they will not have much energy to continue their growth and produce acid. According to the findings of one microbiological investigation, the impact may be caused by fluoride inhibiting the development of *S. mutans* biofilms. Nevertheless, excessive fluoride usage might have negative consequences, such as dental fluorosis, a condition in which the bone is radiologically dense yet brittle [16]. Children under the age of six experience more than 80% of fluoride poisoning due to ingesting fluoridated toothpaste or mouthwash [17]. In extreme circumstances, this might lead to renal and cardiac failure, unconsciousness, and death [16]. There is a chance of fractures and calcification of ligaments and tendons, which might restrict joint motion. Fluoridation has recently been the target of mounting criticism on a global scale, with the possibility and significant risk of toxicity being highlighted. Hence, neglecting oral health and maintaining inadequate cleanliness are still seen as the major causes of tooth deterioration [18].

In addition to fluoride, additional substances, including chlorhexidine, triclosan, and sodium dodecyl sulfate (SDS), are used to treat dental cavities. Type I and Type IV hypersensitivity responses, followed by severe anaphylaxis, are the most dangerous adverse drug reactions (ADRs) associated with chlorhexidine (CHX) oral use. Antimicrobial resistance (AMR), a major negative side effect, is a developing problem with the usage of CHX. Apart from the development of calculus and external staining on teeth over prolonged use, numerous self-reported adverse drug reactions (ADRs) comprised alterations in taste, mouth and tongue numbness, mouth and tongue discomfort, dry mouth (xerostomia), and subjective discoloration. Concern over consumer safety and the search for natural substitutes are a result of their negative impacts. Preventive measures against dental caries mainly aim to lower the risk of tooth demineralization and stop it before teeth are permanently harmed [19]. Consequently, looking for innovative approaches to prevent tooth decay can produce worthwhile outcomes. The ancient strategy known as bacteriotherapy involves employing benign bacteria to attack dangerous germs and has significantly increased in recent years. Probiotics are among the advantageous bacteria that are commonly used for medicinal purposes.

Probiotics are healthy bacteria that are safe for ingestion by humans and that, when consumed in appropriate amounts, positively impact a person's health. Probiotics are primarily used to maintain or restore normal microbiota and prevent the growth of pathogenic bacteria [20]. Their properties include the production of bioactive or regulatory metabolites, inactivating toxins and their receptors, inhibiting pathogenic bacteria, modifying the environment, and competing for nutrients and adhesion sites, preventing colonization by other bacteria. They can also produce bacteriocins, which inhibit pathogenic bacteria. They exert their effects in various ways, including preventing the growth of dangerous bacteria by secreting antibacterial peptides, lowering pH levels by producing acids, fighting for nutrients, trying for adhesion sites on the mucosa, and finally activating the host immune system [21]. Inhibiting infections and affecting the host's adaptive immune response are two effects of probiotics [19, 22]. The oral cavity harbors a varied microbiota, encompassing bacteria associated with conditions like dental caries, gingivitis, periodontitis, candidiasis, and halitosis, making this approach potentially beneficial [10]. Additionally, probiotics could influence the immune factors present in saliva [20]. The use of probiotics for preventive and therapeutic objectives has been employed due to the inherent variability in saliva quality and an individual's oral environment, influenced by their dietary habits [22–25]. The probiotics from the genera *Lactobacillus* and *Bifidobacterium* are often utilized for oral diseases. Discussing the topic is crucial, given the disagreements about probiotics' impact on dental health. In enhancing oral health management, this chapter elaborates on the existing etiological insights into the application of probiotics for the treatment of oral candidiasis, management of periodontal diseases, and prevention of dental caries.

7.2.2 Probiotics in Prevention of Caries

Researchers have studied and created new methods to enhance oral health that seek to control inflammation and lessen the quantity of biofilm.

Probiotics, especially strains like *Lactobacillus* and *Bifidobacterium*, are one of the cutting-edge methods for treating dental biofilm disorders because they cling to and colonize a variety of surfaces in the oral cavity [26]. In order to effectively minimize or prevent tooth caries, probiotic needs to be able to stick to dental surfaces, blend in with the bacterial populations inside the dental biofilm, and outcompete them [27]. Moreover, when some probiotics coexist with the microbial community of the biofilm, they can alter *S. mutans* cariogenicity. Bacterial co-aggregation, a process involving the interaction among microorganisms either within the same species or between different species, plays a significant role in promoting the adhesion of these microbes to tooth surfaces. This phenomenon could serve as an alternative approach for the prevention and management of dental caries. *S. mutans* had a stronger capacity to co-aggregate with beneficial strains than pathogens during an *in vitro* investigation where several probiotic strains examined co-aggregation with oral pathogens. The best co-aggregation was shown by *Lactobacillus acidophilus* among probiotic bacteria [28]. Encouraging findings from *in vitro* research demonstrate that beneficial strains inhibit and reduce the development of cariogenic bacteria [29]. Its anti-cariogenic action is linked to the probiotics' ability to alter the salivary pellicle's protein composition, both of which are particularly hinder the adhesion of *S. mutans* [30]. Probiotics are therefore given to patients to preserve or restore their natural microflora against pathogen growth linked to the emergence of serious oral illnesses. Several experimental investigations have assessed the impact of probiotics on dental caries and their associated risk factors [31].

Probiotics for dental plaque have been identified as strains of the bacteria *Streptococcus salivarius. S. salivarius* competes with cariogenic bacteria due to its capacity to colonize soft tissues. *S. salivarius* strain M18 supplementation resulted in considerably decreased plaque scores and *S. mutans* counts in children. Increased *S. salivarius* M18 colonization was associated with lower plaque scores and fewer cariogenic bacteria [32]. The residents of dental plaque are inhibited by *S. salivarius* M18 bacteriocin. *S. salivarius* M18 produces the enzymes urease and dextranase, which reduce plaque formation and neutralize salivary acidity, respectively. Due to the presence of bacteriocins and enzymes, *S. salivarius* JH is recognized as a potent beneficial strain for combating plaque [33]. Additionally, supplementation with *L. paracasei* SD1 has been shown to elevate levels of human neutrophil peptide 1–3 (HNP1–3) and reduce the counts of *S. mutans*. These findings suggest that the observed decline in new pit and fissure caries may be attributed to the competitive interactions among microbial populations [34]. The existing body of literature strongly indicates that the mechanisms underlying the probiotic actions against caries and plaque are linked to several factors.

7.2.3 Probiotics Against Periodontal Disease

Periodontitis is a complex disease with far-reaching implications, causing tooth loss and elevating the risk of broader health issues and even certain cancers [35]. A layered network of bacterial cells is at the heart of the intrinsic pathogenesis of biofilm-related periodontitis. Among these, *P. gingivalis, Treponema denticola*, and *Tannerella forsythia*—three anaerobic oral bacteria—are widely recognized as the primary culprits, owing to their virulence and strong association with disease-related sites [36].

While antibiotics and antiseptics temporarily relieve periodontal disease, they fall short of providing lasting solutions. As a result, attention has turned towards diminishing the presence of pathogenic bacteria, which play a central role in driving periodontal inflammation. Probiotics have emerged as a potential strategy to suppress these periodontitis-causing microorganisms. Their mechanisms encompass the production of antimicrobial substances, competition with pathogens, and modulation of immune responses [37]. The incorporation of beneficial microorganisms through products holds promise in preventing periodontal diseases, encompassing both gingivitis and periodontitis [37]. Probiotic microorganisms wield a range of effects that extend beyond microbiota modulation. They impact immunological factors, epithelial permeability, bacterial translocation, and the generation of bioactive or regulatory metabolites [38].

These mechanisms hold substantial importance for periodontal health, as the host's response plays a pivotal role in the degradation of the periodontium, with bacterial challenges serving as key contributors to this process. Singular probiotic microorganisms, such as *L. reuteri*, have demonstrated their potential to enhance various periodontal clinical parameters, including plaque index, bleeding on probing, and pocket depth [39]. *L. reuteri*, used alongside for periodontal treatment, has shown promising reductions in pocket depth and moderate improvement. Additionally, the consumption of *L. salivarius* probiotics has led to improvements in periodontal clinical parameters [40], accompanied by a significant reduction in the count of periodontopathic bacteria within the subgingival plaque. This suggests that the introduction of probiotics containing beneficial microorganisms could hold potential as a treatment avenue, promoting a balanced bacterial environment and reducing pathogenic bacteria.

In addition, there is evidence that consuming *L. casei* Shirota significantly lowers the plaque index and gingival irritation [41]. In a clinical study involving probiotic milk consumption, immunomodulatory effects were observed in the gingival region. Pathogens commonly associated with periodontitis, like *P. gingivalis* and *A. actinomycetemcomitans*, prompted the release of factors influencing collagen degradation. These factors could serve as discriminative biomarkers for site-specific periodontitis diagnosis.

A systematic review revealed that probiotics did not induce significant changes in the levels of microorganisms associated with periodontitis (*A. actinomycetemcomitans*, *P. gingivalis*, and *P. intermedia*); they did have a positive impact on bleeding, gingival index, and pocket probing depth. However, it was noted that this analysis highlighted high heterogeneity and potential publication bias across all evaluated outcomes [42]. Similarly, two other reviews focusing on the effects of probiotics on periodontal disease reported substantial heterogeneity among the included randomized clinical trials. This variability likely stems from differences in probiotic strains, concentrations, delivery methods, and the clinical characteristics of the studied populations [43, 44]. Probiotics show promise in countering periodontitis by targeting pathogenic microorganisms and influencing host responses, although challenges in standardization and study heterogeneity need to be addressed.

7.2.4 Probiotics Against Halitosis

Halitosis, commonly called bad breath, emanates from the mouth, nose, face, and pharyngeal sinuses, producing an unpleasant odor. Historical accounts of halitosis date back to ancient times, with Hippocrates suggesting mouth rinsing with wine, anise, and dill seeds for maintaining fresh breath [45]. Halitosis can stem from various factors such as reduced salivary flow, certain food consumption, metabolic irregularities, and respiratory tract infections, but its primary cause often involves an imbalance in the oral commensal microbiota [46]. Among the main contributors to halitosis is the accumulation of bacteria and food particles in the grooves and posterior areas of the tongue. This condition is specifically initiated by volatile sulfur compounds, which encompass dimethyl sulfide, hydrogen sulfide, and methyl mercaptan. Additionally, the bacterial decomposition of food remnants, saliva proteins, and dental biofilm results in the release of other pungent chemicals such as putrescine, indole, skatole, and cadaverine. Dentists treat halitosis, oral pathological halitosis, and pseudo-halitosis, whereas patients are concerned with halitosis prevention. Eliminating or decreasing all bacteria and food residues appears to be essential for effectively managing halitosis. To lessen the amount of these germs and detritus, mechanical treatments, including tooth brushing, flossing, and tongue cleaning, are used [47].

To achieve a sustained reduction in halitosis, it is essential to promptly reintroduce beneficial bacteria through targeted microbiota restocking following mechanical or chemical therapy. This proactive colonization of the oral cavity with non-virulent, commensal microbes appears to be a viable strategy for preventing the resurgence of odor-causing organisms. The identification of a probiotic strain that effectively colonizes oral surfaces without producing malodorous metabolic by-products would be of considerable utility. According to Burton *et al.* [48], the following species might be probiotics:

- By competing with species that produce volatile sulfides in colonization sites, this bacterium reduces the impact of volatile sulfides [48]. This species shares a close relationship with *Streptococcus thermophilus*, a bacterium commonly employed in the dairy food industry. However, it exhibits a limited ability to produce volatile sulfur compounds and is not likely to make a significant contribution to oral odor (formerly known as *S. salivarius* spp. *thermophilus*).
- In an *in vitro* and placebo-controlled trial involving the consumption of yogurt containing *Lactobacillus rhamnosus* L8020, fermented with bovine milk, a notable reduction in the oral carriage of *Streptococcus mutans* and four periodontal pathogens *Porphyromonas gingivalis*, *Prevotella intermedia*, *Tannerella forsythia*, and *Fusobacterium* spp. observed. This effect was not evident in the placebo yogurt group, suggesting the potential of the *L. rhamnosus* L8020-containing yogurt to decrease the risk of periodontal disease, dental cavities, and the presence of bacteria associated with bad breath [49, 50].

The effectiveness of probiotics in treating halitosis has been demonstrated in various studies involving probiotic lozenges, tablets, and gums [48]. However, there is a dearth of supporting evidence regarding the usage of probiotics in dairy products, particularly beverages, for addressing malodor, and this area remains underexplored.

7.3 Mechanism in Oral Diseases Prevention by Probiotics

7.3.1 Caries Prevention by Probiotics

To effectively impede the occurrence of dental caries, probiotics must demonstrate competence in nutrient competition and adhesion rivalry with pathogens. They need to integrate within biofilms and establish a robust physical barrier to hinder the adhesion of pathogenic microorganisms. These factors hold significant scientific importance for preventing and mitigating dental caries [1, 5]. Consequently, these actions restrain pathogen proliferation and establish a thriving oral colonization that contributes to oral health.

Moreover, specific probiotics have the capacity to modify the cariogenic nature of *Streptococcus mutans*, enhancing their effectiveness in preventing dental caries [51]. This ability adds another dimension to the mechanisms through which probiotics exert their effects. Among these mechanisms, the antagonistic relationship they establish with infections stands out. This multifaceted anti-pathogenic activity involves synthesizing antimicrobial compounds and implementing competitive exclusion. Probiotic strains with strong adhesion capabilities are pivotal in thwarting pathogen attachment, engaging in competitive interactions for binding sites on host cells. Illustrating this concept further, Tytgat and colleagues (2016) elucidated a competitive exclusion mechanism wherein *Lactobacillus rhamnosus* GG (LGG) outcompetes *Enterococcus faecium*. The researchers demonstrated that LGG's mucus-binding pili, particularly the SpaCBA pili, and SpaC proteins or antibodies targeting SpaC, effectively hindered *E. faecium*'s adhesion. This study gains significance as *E. faecium* possesses similar pili structures, underscoring the finding's importance.

The positive impact of probiotics on dental caries and plaque is governed by a range of strain-specific mechanisms, arising from a synergy of competitive actions. Key criteria for effective probiotic products, outlined by Collins and Gibson [52], encompass aspects such as pathogenicity absence, toxicity avoidance, viability during storage, host detectability, and elicitation of beneficial effects. These mechanisms of probiotic action can be broadly classified into three categories, as proposed by Oelschlaeger [53]: (i) the antimicrobial effect arises from microbial metabolism; (ii) competitive exclusion involves competition for nutrients and colonization sites; and (iii) immunomodulation influences the host's immune systems. Pathogenic microorganisms can disrupt the harmonious connection between the host and its normal flora.

However, in order to colonize these pathogens, they must first locate free receptor sites on the host cell wall for adhesion, which are typically polysaccharide molecules [54]. This impedes pathogen colonization and prevents further disruption to the host's natural flora [28]. Furthermore, probiotics can secrete a mucosal coating that eliminates binding sites and prevents pathogen adhesion to the host [29]. These mechanisms are unique to each probiotic strain and vital to maintaining a healthy oral environment.

Probiotics influence the immune system of the host through two distinct pathways: immunomodulation and immunoregulation [55]. Products resulting from probiotic metabolism can influence the immune system by modulating cytokine and chemokine production, as well as immune cell participation in these processes [55]. This immunomodulatory effect can contribute favorably to the host's innate and adaptive immune responses, protecting against potential pathogens. It is important to note that the specific immunomodulatory mechanisms can vary based on the employed probiotic strain. Demonstrating promise in combating cancer and infections, probiotics stimulate the production of interleukin-12, activating natural killer cells and fostering T-helper 1 cell development. Additionally, probiotics exhibit potential as anti-virulence agents against infectious biofilms. Compared to conventional antibiotics, they apply reduced selective pressure to resistant isolates and exhibit lower cytotoxicity than QS-suppressing agents, rendering them an appealing option for tackling biofilms [55].

Probiotics can interfere with the activity and adhesion of pathogenic bacteria using various mechanisms. It is crucial to emphasize that adherence capacity involves the binding of bacterial strains to the host and their adherence to other bacterial strains of different species (co-aggregation) or the same species (auto-aggregation) [56]. In this context, numerous studies have emphasized the efficient protective function of probiotics by facilitating the attachment of pathogens within co-aggregates, thus hindering the biofilm formation process, a common factor in infections. A recent *in vitro* study provides a concrete example wherein two strains of *Lactobacillus reuteri* exhibited the ability to co-aggregate with pathogens in the process of biofilm development. This collaborative interaction established an antagonistic environment that effectively inhibited the proliferation of yeast. Additionally, these strains contributed to lowering environmental pH, producing indole (a signaling molecule in quorum sensing), and reducing both biofilm biomass, ultimately thwarting the creation of harmful biofilms and promoting their elimination. These multi-faceted actions underscore the intricate and strain-specific ways probiotics contribute to oral health. Figure 7.1 shows the mechanisitic action of probiotics on oral health, highlighting the reduction in harmful bacteria and improvement in gum health.

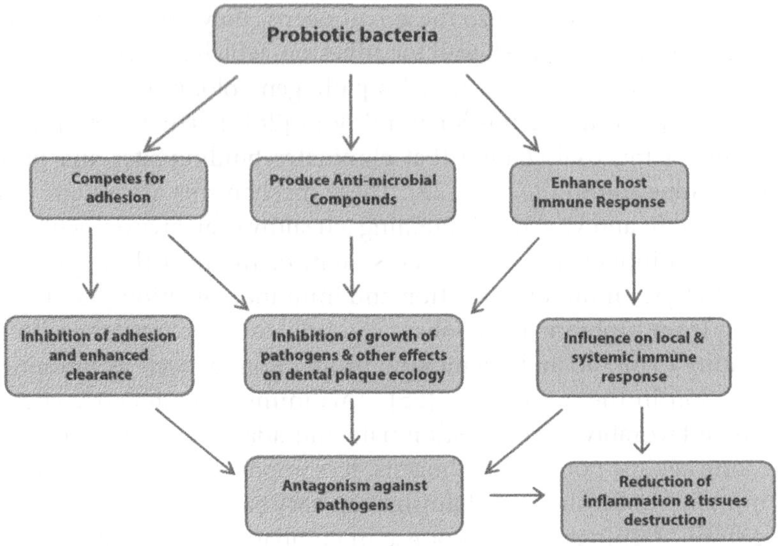

Figure 7.1 Mechanistic actions of probiotics against dental caries.

7.3.2 Periodontics Prevention by Probiotics

Probiotics offer a multifaceted mechanism of action against periodontitis. These beneficial microorganisms are pivotal in suppressing periodontitis-associated pathogens through various means. Firstly, probiotics produce antimicrobial substances that directly inhibit the growth and activity of pathogenic bacteria, particularly the notorious trio—*P. gingivalis*, *Treponema denticola*, and *Tannerella forsythia*. Secondly, probiotics engage in competitive exclusion, effectively outcompeting pathogens for essential nutrients and adhesion sites within the oral environment. This interference disrupts the ability of pathogenic bacteria to establish themselves and thrive, thereby aiding in maintaining a more balanced microbial community [44]. Additionally, probiotics exert immune-modulatory effects by interacting with the host's immune system. They can modulate immunological parameters, resulting in a more controlled and balanced immune response to the bacterial challenge, thus reducing excessive inflammation contributing to periodontal tissue destruction. Moreover, probiotics have been shown to influence epithelial permeability, bacterial translocation, and the production of bioactive metabolites, all collectively contributing to a healthier periodontal environment [43, 44]. Through these multifaceted actions, probiotics hold potential as an adjunctive therapeutic approach to mitigate periodontitis, offering a promising avenue to address this complex disease.

7.3.3 Halitosis Prevention by Probiotics

The mechanism underlying halitosis prevention by probiotics involves targeted colonization of the oral cavity with beneficial bacteria following mechanical or chemical therapy. This colonization strategy aims to counteract the resurgence of odor-causing microorganisms. Certain probiotic species are engaged in co-aggregation with pathogen, thereby reducing the production of volatile sulfide compounds. Notably, *Lactobacillus* spp. display inhibitory effects on cariogenic and periodontal pathogens associated with halitosis. Additionally, dairy-fermented products have demonstrated a reduction in oral carriage of periodontal pathogens, potentially mitigating the risk of both periodontal disease and bad breath. The documented effectiveness of probiotics in forms such as lozenges and tablets further support their role in halitosis treatment. Nevertheless, the potential of probiotics in dairy products, particularly beverages, for addressing malodor remains an area warranting exploration. This mechanistic approach underscores the pivotal role of probiotics in restoring a balanced oral microbiota, thus contributing to the prevention and management of halitosis.

7.4 Probiotic Formulation

The initial stage in probiotic product development involves identifying a master strain of probiotic bacteria and ensuring its proper preservation. To safeguard the therapeutic potential of the strain, meticulous monitoring of storage and handling conditions is imperative to prevent mutations or alterations in physiological and biochemical characteristics [57]. Strain identification is a pivotal step, achieved through microbiological, biochemical, and immunological techniques.

Subsequently, the propagation stage becomes critical, necessitating the optimization of physical factors such as temperature, agitation rate, and oxygen content. Flow velocity and exchange rate during tangential flow must also be precisely controlled to maximize production efficiency. Following the harvest phase, employing techniques like centrifugation and filtration, the harvested material is evaluated for culture purity, expressing a viable cell count per unit volume.

During the formulation phase, concentrated live organisms are packaged into specified doses. The uniformity of these doses is ensured through careful monitoring of weights and volumes during the filling of vials, capsules, sachets, or tablets. In the probiotic product development process, the "specification" stage entails establishing acceptance criteria for the product. Ensuring control of the harvested bio-product requires conducting specific tests to verify the identification, purity, and viability of the microorganisms.

Transitioning to the "control of dosage form" phase, the product's potency, safety, and identity are rigorously tested once packaged into final form. The cultivation and maintenance of probiotic bacteria entail a sequence of procedures, encompassing amplification, scaling up in fermentation, freeze-drying, and subsequent downstream processing. These steps are orchestrated to produce the probiotic product while upholding the quality and consistency of the master strain [58].

Probiotic-rich foods encompass a range of items such as dairy, fruits, and vegetable products. Additionally, supplements are accessible in diverse formats, including tablets, powders, capsules, pills, caplets, liquids, and gel caps [59]. These dietary options support gastrointestinal and immunological health, extending to oral well-being. Specifically for oral health, these beneficial components are delivered through methods like pills, lozenges, mouthwashes/oral rinses, and toothpaste, facilitating the extended presence of beneficial bacteria within the oral cavity and thereby contributing to the management of oral conditions (Table 7.1).

Table 7.1 Various oral probiotic supplements have medicinal benefits. Sourced from Mishra *et al.*, 2020 [59].

Name of the product	Product type	Microbial content	Properties
Oral Biotic	Lozenges	*S. salivarius* BLIS K12	Enhances the overall well-being of the oral cavity
Great Oral Health, children oral probiotics	Chewable tablets	*S. salivarius* BLIS K12 and BLIS M18	Promotes oral health
Probiora Plus	Tablets	*Streptococcus uberis* KJ2, *Streptococcus oralis* KJ3, and *Streptococcus rattus* JH145	Brightens tooth enamel, addresses bad breath, and contributes to the well-being of gums

(Continued)

Table 7.1 Various oral probiotic supplements have medicinal benefits. Sourced from Mishra *et al.*, 2020 [59]. (*Continued*)

Name of the product	Product type	Microbial content	Properties
Cultured Care, oral probiotic gum	Chewing gum	*S. salivarius* BLIS K12	Combats halitosis and reinstates a balanced oral microflora
Life Extension, florassist oral hygiene	Lozenges	*S. salivarius* and *Bacillus coagulans* BLIS M18	Promotes the well-being of teeth and gums while regulating microbial balance
Oral Care Probiotic	Chewable tablets	*S. salivarius*	Reestablishes and sustains optimal oral well-being
Oral Complete	Chewable tablets	*L. acidophilus*, *Bacillus subtilis*, *B. coagulans*, *L. rhamnosus*, and *Bifidobacterium bifidum*	Revitalizes breath, promotes gum health, and reinstates a balanced oral microflora
Solaray Probiotic, mouth and throat formula	Lozenges	*S. salivarius* BLIS K12, M18, and *B. coagulans*	Preserves equilibrium in oral flora while refreshing breath
Advanced Orthomolecular Research, Lava Rox	Flavored Popin tablets	*S. salivarius* K12	Promotes oral health
Flora udo's Choice Super 5	Lozenges	*L. acidophilus*, *L. casei*, *L. rhamnosus*, *L. salivarius*, and *B. bifidum*	Nurtures gum health and encourages fresh breath
E.N.T. Biotic	Lozenges	*S. salivarius* BLIS K12	Fosters the well-being of beneficial oral bacteria while imparting a refreshing breath

(*Continued*)

Table 7.1 Various oral probiotic supplements have medicinal benefits. Sourced from Mishra *et al.*, 2020 [59]. (*Continued*)

Name of the product	Product type	Microbial content	Properties
Probiotic Toothpaste, fennel-F-free and sodium lauryl sulfate-free probiotic xylitol toothpaste	Toothpaste	*L. paracasei*	Sustains the well-being of oral health
CuchuPe Mouthwash, fresh mint stick L-8020 10 mL	Mouthwash	*Lactobacillus* species	Acts as a deterrent to cavity formation and combats the occurrence of bad breath
ProbioRinz Concentrate	Mouthwash/ rinse	*Lactobacillus* species	Minimizes the development of biofilms, enhances oral freshness, and fosters a more optimal environment within the mouth

7.5 Prevention and Oral Health Management

Advancements in research have expanded our understanding of dental caries beyond genetic factors, now encompassing the influence of epigenetic and environmental elements. In particular, epigenetic modifications induced by inflammation, bacteria, smoking, and diet have been recognized as potent drivers of oral health, impacting genes associated with immune responses in the oral mucosa. The intricate interplay of these factors underscores the multifaceted nature of dental caries onset and progression.

Among the prominent environmental influences, dietary habits, especially excessive carbohydrate consumption, significantly contribute to dental caries. The breakdown of carbohydrates fosters acid production in dental plaque, triggering demineralization of tooth structures. This deterioration, which originates within the bacterial biofilm–dental plaque adhering to tooth surfaces, is perpetuated by continuous acid production that overwhelms saliva's protective effects. The ensuing frequent pH drops alter the composition of the oral microbiota, favoring acid-tolerant microorganisms [60]. The dynamic equilibrium between demineralization and remineralization ultimately dictates the course of dental caries, with both processes occurring regularly throughout the day [61]. Re-mineralization, facilitated by decreased calcium and phosphate levels coupled with trace amounts of fluoride, is pivotal in counteracting demineralization's impact. Remarkably, fluoride, even in minute concentrations (<1 ppm), acts as a catalyst, influencing reactions that dissolve and transform various calcium phosphate phases within tooth structures and adjacent plaque [62].

The role of bacteria within the oral biofilm further amplifies the disease process. Microbes like *Lactobacillus* spp., *Streptococcus mutans*, and *Streptococcus sobrinus* release weak organic acids during carbohydrate fermentation, compounding the acid-mediated degradation of dental tissues. Dietary habits also factor into the equation, where the prolonged exposure of teeth to sugary substances, as seen with soft drinks and hard candies, intensifies the risk of caries formation. Holding sugary beverages in the mouth or consuming them continuously fosters an environment conducive to demineralization [63].

Recent research underscores the significance of vitamin D in calcium metabolism and mineralization of tooth structures. Insufficient vitamin D levels have been linked to enamel hypoplasia, elevating dental caries risk. Beyond its role in mineralization, vitamin D emerges as a key player in the immune system. It stimulates specific genes controlling the production of antimicrobial peptides like cathelicidin, which combat various pathogens. These peptides and their immune regulatory functions impact oral microbiome homeostasis, inflammation resolution, and other vital immune processes [64].

Dental caries represents a complex interplay between genetics, epigenetics, and environmental factors. In conjunction with dietary choices and microbial activity, epigenetic modifications triggered by inflammation, bacteria, smoking, and diet significantly influence the initiation and progression of dental caries. Understanding these intricate relationships enhances our approach to preventive strategies and underscores the broader impact of oral health on overall well-being.

7.6 Concluding Remarks

The importance of oral health to an individual's overall well-being and quality of life cannot be overstated. Oral diseases can significantly impact a person's ability to eat, speak, and smile, leading to substantial personal and social distress. In recent years, both *in vitro* and *in vivo* research have highlighted probiotic strains' crucial role in preventing oral health issues, including caries, halitosis, and periodontal diseases.

This chapter concludes by asserting that the judicious use of potent probiotics can effectively avert dental caries and curtail its risk across individuals from childhood to adulthood. Most studies have focused on examining *Lactobacillus* and *Streptococcus* species, revealing their potent anti-cariogenic potential. Probiotics have shown promise as a bio-therapeutic avenue and have also gained recognition as functional foods, nanoparticles, dietary supplements, and anticancer agents. By enhancing normal physiological processes and bolstering immunity, these probiotic agents have demonstrated the ability to prevent infections, oxidative stress, and inflammatory conditions.

Notably, various probiotic products have been meticulously crafted to safeguard oral health. Some have demonstrated the ability to reduce *S. mutans* populations, enhance gingival well-being, and mitigate dental plaque formation. While numerous probiotic formulations are available in the Indian pharmaceutical market, issues with sourcing and marketing have somewhat limited their application. Nevertheless, the potential of probiotics remain promising, envisioning a future where they emerge as a strategy to prevent enamel demineralization, promote periodontal health, and eradicate halitosis.

However, amidst this promise, an essential avenue beckons further exploration. Identifying the most fitting probiotic candidate for optimizing oral health necessitates sustained and meticulous investigation.

7.7 Future Aspects

The prospects for probiotics in oral health appear promising and extensive. As scientific understanding deepens, exploring the dynamic interplay between probiotics and oral microbiota could unlock innovative strategies for preventing and treating dental caries, plaque formation, and other oral conditions. Further investigations into strain-specific adhesion mechanisms and their impact on biofilm dynamics could pave the way for tailored probiotic interventions. Additionally, harnessing the immuno-modulatory potential of probiotics offers exciting avenues for enhancing host defense mechanisms and fortifying oral health resilience.

Certainly, here are some future perspectives on the utilization of probiotics in oral health:

- Innovative Delivery Systems: In innovative delivery, harnessing nanoparticles to encapsulate probiotics offers precise control over their release within the oral cavity. This advancement can potentially revolutionize probiotic-infused oral care products like toothpaste, mouthwash, and chewing gum. This approach can achieve prolonged and targeted probiotic delivery, maximizing their therapeutic efficacy for optimal oral health benefits.
- Biofilm Disruption: Exploring probiotic strains with enhanced biofilm disruption capabilities to prevent the formation of pathogenic biofilms, a significant contributor to oral infections.
- Combination Therapies: Investigating synergistic effects of probiotics when combined with other oral health interventions, such as fluoride treatment or mechanical plaque removal methods.
- Pediatric Oral Health: Investigating the potential of early-life probiotic interventions to shape the oral microbiome and prevent the development of dental caries and other oral conditions in children.
- Mechanistic Insights: Deepening our understanding of the molecular mechanisms behind probiotic actions in the oral cavity, shedding light on their impact on host–microbe interactions and immune responses.
- Microbiome Modulation: Investigating the long-term effects of probiotics on the overall oral microbiome composition and diversity, assessing their potential for reshaping microbial communities in a beneficial manner.
- Educational Initiatives: Promoting public awareness and healthcare professional education about the benefits of probiotics for oral health, fostering informed decision-making, and adoption of probiotic interventions.
- Long-Term Health Implications: Studying the potential systemic effects of oral probiotics, including their impact on systemic inflammation, cardiovascular health, and the gut–brain axis.

In conclusion, integrating nanoparticle formulations into probiotic-infused oral care products represents an innovative approach with substantial potential to enhance the efficacy of probiotics in maintaining oral health. By harnessing the advantages of nanoparticles, researchers and industry professionals can revolutionize oral care routines and elevate the impact of probiotics on promoting a healthy oral environment.

References

1. Gao, L., Xu, T., Huang, G., Jiang, S., Gu, Y., Chen, F., Oral microbiomes: more and more importance in oral cavity and whole body, protein. *Cell*, 9, 488–500, 2018.
2. Peres, M.A., Macpherson, L.M.D., Weyant, R.J., Daly, B., Venturelli, R., Mathur, M.R., Listl, S., Celeste, R.K., Guarnizo-Herreño, C.C., Kearns, C., Benzian, H., Allison, P., Watt, R.G., Oral diseases: a global public health challenge. *Lancet*, 394, 249–260, 2019.
3. El Tantawi, M., Folayan, M.O., Mehaina, M., Vukovic, A., Castillo, J.L., Gaffar, B.O., Arheiam, A., Al-Batayneh, O.B., Kemoli, A.M., Schroth, R.J., Lee., G., Prevalence and data availability of early childhood caries in 193 United Nations countries, 2007–2017. *Am. J. Public Health*, 108, 1066–1072, 2018.
4. Cheng, A.C., Yeh, S.P., Hu, S.Y., Lin, H.L., Liu, C.H., Intestinal microbiota of white shrimp, Litopenaeusvannamei, fed diets containing Bacillus subtilis E20-fermented soybean meal (FSBM) or an antimicrobial peptide derived from B. subtilis E20-FSBM. *Aquacult. Res.*, 51, 41–50, 2020.
5. Markowiak, P. and Śliżewska, K., Effects of probiotics, prebiotics, and synbiotics on human health. *Nutrients*, 9, 1021, 2017, https://doi.org/10.3390/nu9091021.
6. Zhao, M., Chen, X., Yang, Z., Yang, X., Peng, Q., Bacteria and tumor: understanding the roles of bacteria in tumor genesis and immunology. *Microbiol. Res.*, 261, 127082, 2022.
7. Kitamoto, S., Nagao-Kitamoto, H., Hein, R., Schmidt, T.M., Kamada, N., The bacterial connection between the oral cavity and the gut diseases. *J. Dent. Res.*, 99, 1021–1029, 2020.
8. Lemos, J.A. and Burne, R.A., A model of efficiency: stress tolerance by Streptococcus mutans. *Microbiology*, 154, 3247–3255, 2008.
9. Jamal, M., Tasneem, U., Hussain, T., Andleeb, S., Bacterial biofilm: its composition, formation and role in human infections. *Res. Rev: J. Microbiol. Biotechnol.*, 4, 1–14, 2015.
10. Azizi, A., Shohrati, P., Goudarzi, M., Lawaf, S., Rahimi, A., Comparison of the effect of Photodynamic Therapy with curcumin and Methylene Blue on Streptococcus mutans bacterial colonies Photodiagnosisphotodyn. *Ther.*, 27, 203–209, 2019, 10.1016/j.pdpdt.2019.06.002.
11. Forssten, S.D., Bjorklund, M., Ouwehand, A.C., Streptococcus mutans, caries and simulation models. *Nutrients*, 2, 290–298, 2010.

12. Marsh, P.D., Are dental diseases examples of ecological catastrophes? *Microbiology*, *149*, 2, 279–294, 2003.
13. Hu, D., Gong, J., He, B., Chen, Z., Li, M., Surface properties and Streptococcus mutans - Streptococcus sanguinis adhesion of fluorotic enamel. *Arch. Oral. Biol.*, 121, 1–8, 2021, https://doi.org/10.1016/j.archoralbio.2020.10497011.
14. Houle, M.A., Grenier, D., Plamondon, P., Nakayama, K., The collagenase activity of Porphyromonasgingivalis is due to Arg-gingipain. *FEMS Microbiol. Lett.*, *221*, 2, 181–185, 2003.
15. Kinane, D.F., Stathopoulou, P.G., Papapanou, P.N., Periodontal diseases. *Nat. Rev. Dis. Primers*, *3*, 1, 1–14, 2017.
16. Qiu, W., Zhou, Y., Li, Z., Huang, T., Xiao, Y., Cheng, L., Peng, X., Zhang, L., Ren, B., Application of antibiotics/antimicrobial agents on dental caries. *BioMed. Res. Int.*, *2020*, p. 5658212 (1–11), 2020.
17. Aoun, A., Darwiche, F., Al Hayek, S., Doumit, J., The fluoride debate: the pros and cons of fluoridation. *Prev. Nutr. Food Sci.*, *23*, 3, 171, 2018.
18. Wimalawansa, S.J., The role of ions, heavy metals, fluoride, and agrochemicals: critical evaluation of potential aetiological factors of chronic kidney disease of multifactorial origin (CKDmfo/CKDu) and recommendations for its eradication. *Environ. Geochem. Health*, *38*, 3, 639–678, 2016.
19. Pitts, N.B., Zero, D.T., Marsh, P.D., Ekstrand, K., Weintraub, J.A., Ramos-Gomez, F., Tagami, J., Twetman, S., Tsakos, G., Ismail, A., Dental caries. *Nat. Rev. Dis. Primers*, *3*, 1, 1–16, 2017.
20. Yamamoto, Y., Fujino, K., Saruta, J., Takahashi, T., To, M., Fuchida, S., Shimizu, T., Kamata, Y., Misawa, K., Tsukinoki, K., Effects of yogurt fermented with Lactobacillus delbrueckii ssp. bulgaricus OLL1073R-1 on the IgA flow rate of saliva in elderly persons residing in a nursing home: A before-after non-randomised intervention study. *Gerodontology*, 34, 479–485, 2017.
21. taab, B., Eick, S., Knofler, G., Jentsch, H., The influence of a probiotic milk drink on the development of gingivitis: A pilot study. *J. Clin. Periodontol.*, 36, 850–856, 2009.
22. Parra, D. and Martínez, J.A., Amino acid uptake from a probiotic milk in lactose intolerant subjects. *Br. J. Nutr.*, 98, 101–104, 2007.
23. deVrese, M., Kristen, H., Rautenberg, P., Laue, C., Schrezenmeir, J., Probiotic lactobacilli and bifidobacteria in a fermented milk product with added fruit preparation reduce antibiotic associated diarrhea and Helicobacter pylori activity. *J. Dairy. Res.*, 78, 396–403, 2011.
24. Tongtawee, T., Dechsukhum, C., Leeanansaksiri, W., Kaewpitoon, S., Kaewpitoon, N., Loyd, R.A., Matrakool, L., Panpimanmas, S., Effect of Pretreatment with Lactobacillus delbrueckii and Streptococcus thermophillus on Tailored Triple Therapy for Helicobacter pylori Eradication: A Prospective Randomized Controlled Clinical Trial. *Asian Pac. J. Cancer. Prev.*, 16, 4885–4890, 2015.
25. Cagetti, M.G. *et al.*, The use of probiotic strains in caries prevention: a systematic review. *Nutrients*, *5*, 7, 2530–2550, 2013.

26. Stamatova, I. and Meurman, J.H., Probiotics: health benefits in the mouth. *Am. J. Dent.*, 22, 6, 329, 2009.
27. Bonifait, L., Chandad, F., Grenier, D., Probiotics for oral health: myth or reality? *J. Can. Dent. Assoc.*, *75*, 8, 585–590, 2009.
28. Hedberg, M., Hasslöf, P., Sjöström, I., Twetman, S., Stecksén-Blicks, C., Sugar fermentation in probiotic bacteria–an *in vitro* study. *Oral. Microbiol. Immunol.*, *23*, 6, 482–485, 2008.
29. Söderling, E.M., Marttinen, A.M., Haukioja, A.L., Probiotic lactobacilli interfere with Streptococcus mutans biofilm formation *in vitro*. *Curr. Microbiol.*, *62*, 618–622, 2011.
30. Bizzini, B., Pizzo, G., Scapagnini, G., Nuzzo, D., Vasto, S., Probiotics and oral health. *Curr. Pharm. Des.*, *18*, 34, 5522–5531, 2012.
31. Laleman, I., Teughels, W., Laleman, I., Probiotics in the dental practice. *Quintessence Int. Oral. Med.*, *46*, 255–264, 2015.
32. Di Pierro, F., Zanvit, A., Nobili, P., Risso, P., Fornaini, C., Cariogram outcome after 90 days of oral treatment with Streptococcus salivarius M18 in children at high risk for dental caries: results of a randomized, controlled study. *Clin. Cosmet. Investig. Dent.*, *7*, 107–113, 2015.
33. Sivamaruthi, B.S., Kesika, P., Chaiyasut, C., A review of the role of probiotic supplementation in dental caries. *Probiotics Antimicrob. Proteins*, *12*, 1300–1309, 2020.
34. da Cruz, M.F., Magno, M.B., Jural, L.A., Pimentel, T.C., Ferreira, D.M.T.P., Esmerino, E.A., Ramos, G.L.P.A., Gomila, J.V., Silva, M.C., da Cruz, A.G., da Silva Fidalgo, T.K., Probiotics and dairy products in dentistry: A bibliometric and critical review of randomized clinical trials. *Food Res. Int.*, *157*, 111228, 2022.
35. Isola, G., Santonocito, S., Lupi, S.M., Polizzi, A., Sclafani, R., Patini, R., Marchetti, E., Periodontal Health and Disease in the Context of Systemic Diseases. *Mediators Inflamm.*, *2023*, 9720947(1–19), 2023.
36. Seneviratne, C.J., Suriyanarayanan, T., Widyarman, A.S., Lee, L.S., Lau, M., Ching, J., Delaney, C., Ramage, G., Multi-omics tools for studying microbial biofilms: current perspectives and future directions. *Crit. Rev. Microbiol.*, *46*, 6, 759–778, 2020.
37. Nguyen, T., Brody, H., Radaic, A., Kapila, Y., Probiotics for periodontal health—Current molecular findings. *Periodontol. 2000*, *87*, 1, 254–267, 2021.
38. Pandey, K.R., Naik, S.R., Vakil, B.V., Probiotics, prebiotics and synbiotics-a review. *J. Food Sci. Technol.*, *52*, 7577–7587, 2015.
39. Muniz, F.W.M.G., Zanatta, F.B., Muñoz, M.D.S., Aguiar, L.M., Silva, F.H., Montagner, A.F., Antiplaque and antigingivitis efficacy of medicated and non-medicated sugar-free chewing gum as adjuncts to toothbrushing: systematic review and network meta-analysis. *Clin. Oral. Investig.*, *26*, 2, 1155–1172, 2022.
40. Laleman, I., Teughels, W., Laleman, I., Probiotics in the dental practice. *Quintessence Int. Oral. Med.*, *46*, 255–264, 2015.

41. Kohar, M., Emmanuel, V., Astuti, L., Comparison between probiotic lozenges and drinks towards periodontal status improvement of orthodontic patients. *Dent. J. (MajalahKedokteran Gigi)*, *48*, 126, 2015.
42. Matsubara, V.H., Bandara, H.M.H.N., Ishikawa, K.H., Mayer, M.P.A., Samaranayake, L.P., The role of probiotic bacteria in managing periodontal disease: a systematic review. *Expert. Rev. Anti. - Infect. Ther.*, *14*, 7, 643–655, 2016.
43. Nguyen, T., Brody, H., Lin, G.H., Rangé, H., Kuraji, R., Ye, C., Kamarajan, P., Radaic, A., Gao, L., Kapila, Y., Probiotics, including nisin-based probiotics, improve clinical and microbial outcomes relevant to oral and systemic diseases. *Periodontol. 2000*, *82*, 1, 173–185, 2020.
44. Matsubara, V.H., Fakhruddin, K.S., Ngo, H., Samaranayake, L.P., Probiotic Bifidobacteria in Managing Periodontal Disease: A Systematic Review *Int. Dent. J.*, *73*, 11–20, 2022.
45. Achmada, H., Rieuwpassab, I.E., Djaisc, A.I., Riyantid, E., Saptarini, R., Primartid, A.I.I., Gambier Extract (UncariagambierRoxb.) as herbal treatment for the oral cavity: A systematic review. *Infection*, *9*, 11, 12, 2021.
46. Nipanikar, S., Agnihotri, S., Panicker, S., Shelke, N., Evaluation of Anti-Microbial (*In-Vitro*) Activity of Amarantha Herbal Mouthwash. *Int. J. Sci. Res.*, 9, 1624–1627, 2020.
47. Haraszthy, V.I. and Sreenivasan, P.K., Microbiological and clinical effects of an oral hygiene regimen. *Contemp. Clin. Trials Commun.*, *8*, 85–89, 2017.
48. Burton, J.P., Chilcott, C.N., Tagg, J.R., The rationale and potential for the reduction of oral malodour using Streptococcus salivarius probiotics. *Oral. Dis.*, *11*, 29–31, 2005.
49. Romani Vestman, N., Timby, N., Holgerson, P.L., Kressirer, C.A., Claesson, R., Domellöf, M., Öhman, C., Tanner, A.C., Hernell, O., Johansson, I., Characterization and *in vitro* properties of oral lactobacilli in breastfed infants. *BMC Microbiol.*, *13*, 1–12, 2013.
50. Endo, K., Mine, Y., Shuto, T., Taji, T., Murayama, T., Nikawa, H., Comprehensive analysis of transcriptional profiles in oral epithelial-like cells stimulated with oral probiotic Lactobacillus spp. *Arch. Oral. Biol.*, *118*, 104832, 2020.
51. Lin, X., Chen, X., Tu, Y., Wang, S., Chen, H., Effect of probiotic lactobacilli on the growth of streptococcus mutans and multispecies biofilms isolated from children with active caries. *Med. Sci. Monit.*, 23, 4175–4181, 2017, https://doi.org/10.12659/msm.
52. Collins, M.D. and Gibson, G.R., Probiotics, prebiotics, and synbiotics: approaches for modulating the microbial ecology of the gut. *Am. J. Clin. Nutr.*, *69*, 5, 1052s–1057s, 1999.
53. Nobbs, A.H., Lamont, R.J., Jenkinson, H.F., Streptococcus adherence and colonization. *Microbiol. Mol. Biol. Rev.*, *73*, 3, 407–450, 2009.
54. Rajoka, M.S.R., Wu, Y., Mehwish, H.M., Bansal, M., Zhao, L., Lactobacillus exopolysaccharides: New perspectives on engineering strategies,

physiochemical functions, and immunomodulatory effects on host health. *Trends Food Sci. Technol.*, *103*, 36–48, 2020.

55. Collado, M.C., Meriluoto, J., Salminen, S., Adhesion and aggregation properties of probiotic and pathogen strains. *Eur. Food Res. Technol.*, *226*, 1065–1073, 2008.
56. Matsubara, V.H., Wang, Y., Bandara, H.M.H.N., Mayer, M.P.A., Samaranayake, L.P., Probiotic lactobacilli inhibit early stages of Candida albicans biofilm development by reducing their growth, cell adhesion, and filamentation. *Appl. Microbiol. Biotechnol.*, *100*, 6415–6426, 2016.
57. Arora, M. and Baldi, A., Good manufacturing practice regulations for probiotic based pharmaceuticals: Current scenario and suggestive recommendations. *Appl. Clin. Res. Clin. Trials Regul. Aff.*, *2*, 3, 165–175, 2015.
58. Wang, W. and Singh, M. (Eds.), *Biological drug products: development and strategies*, John Wiley & Sons, Hoboken, NJ, 2013.
59. Mishra, S., Rath, S., Mohanty, N., Probiotics—A complete oral healthcare package. *J. Integr. Med.*, *18*, 6, 462–469, 2020.
60. Abebe, G.M., Oral biofilm and its impact on oral health, psychological and social interaction. *Int. J. Oral. Dent. Health*, *7*, 127, 2021.
61. Pitts, N.B., Zero, D.T., Marsh, P.D., Ekstrand, K., Weintraub, J.A., Ramos-Gomez, F., Tagami, J., Twetman, S., Tsakos, G., Ismail, A., Dental caries. *Nat. Rev. Dis. Primers*, *3*, 1, 1–16, 2017.
62. Ramos, C.T., Enamel decalcification and fluoride release among Fuji ortho LC and composite resins: An *in vitro* and *in vivo* study, West Virginia University, Morgantown, WV, 2000.
63. Damle, S., Goyal, P., Sakhare, D., Role of Diet in Dental Caries, in: *llustrated Pediatric Dentistry (Part I)*, p. 133, I, Haryana, India, 2022.
64. Foster, B.L. and Hujoel, P.P., Vitamin D in dentoalveolar and oral health, in: *Vitamin D*, pp. 497–519, 2018.

8

The Preventative and Curative Functions of Probiotics: A Paradigm of Food as Drug Revolution

Mohammad Zaki Shamim[1], Jibanjyoti Panda[2,3], Gargee Mohanty[3], Bhaswati Gogoi[2], Kaustuvmani Patowary[2], Bishwambhar Mishra[4] and Yugal Kishore Mohanta[2*]

[1]Faculty of Science, Assam down town University, Assam, India
[2]Nano-Biotechnology and Translational Knowledge Laboratory, Department of Applied Biology, School of Biological Sciences, University of Science and Technology Meghalaya (USTM), 9th Mile, Techno City, Baridua, Ri-Bhoi, Meghalaya, India
[3]Department of Zoology, Maharaja Sriram Chandra Bhanja Deo University, Baripada, Odisha, India
[4]Department of Biotechnology, Chaitanya Bharathi Institute of Technology (CBIT), Gandipet, Hyderabad, Telangana, India

Abstract

The body's normal microflora is disrupted by antibiotics, which do not distinguish between good and harmful bacteria. This results in vitamin deficiencies. They undermine the host's defenses by reducing the population of helpful bacteria in the gastrointestinal system and vaginal area. Due to the beneficial effects they have on human wellness, including the immune system and metabolism, probiotics are considered to be an "organ" of the metabolic system. From a therapeutic standpoint, these substances are employed for the purpose of both prevention and treatment across a diverse range of medical conditions. Diarrhea, colorectal cancer, diabetes, hypertension, *Helicobacter pylori*-related acute pancreatitis, ventilator-associated pneumonia, migraine headaches, and autistic spectrum disorder are all examples of such medical issues. Probiotics can influence immunological activity in a number of ways. These include triggering adaptive or innate immune responses, changing gut bacterial composition, increasing gut barrier

**Corresponding author*: ykmohanta@gmail.com; yugalkmohanta@ustm.ac.in

Hrudayanath Thatoi, Sonali Mohapatra, Swagat Kumar Das and Sukanta Kumar Pradhan (eds.) Applied Biotechnology and Bioinformatics: Agriculture, Pharmaceutical Research and Environment, (181–216) © 2025 Scrivener Publishing LLC

integrity, attaching competitively to mucosa and epithelium, and generating antimicrobial chemicals. This finding emphasizes the need for more probiotic research to cure and prevent illnesses. Searches of research databases employing probiotics, microbiota, prophylactics, and therapeutic purposes were extensive. This brief narrative chapter covers probiotics' background, selection, effects, newest preventive and therapeutic findings, and potential uses.

***Keywords*:** Probiotic effects, treatment, diseases, potential, therapeutic uses

8.1 Introduction

Antibiotics are commonly employed in the medical field for the dual purpose of prophylaxis and therapeutic intervention against a diverse range of potentially life-threatening bacterial infections [1]. The importance of antibiotics has resulted in widespread exposure to them among people everywhere. However, antibiotic therapy is associated with two problems: severe side effects and the disruption of natural microbiota, which can lead to a deficiency in vitamins B, K, and so on [2]. Widespread use of antibiotics can weaken a host's immune system by killing off good bacteria that normally live in the digestive tract and genital tract. Scientists everywhere are scrambling to find a replacement source because of these problems. Several gastrointestinal infections can be prevented or treated by adjusting the commensal microbiota [1]. This has been demonstrated by multiple groups of researchers. The term "human microbiota" is used to describe the wide variety of bacteria that live in and on the human body, particularly in the gastrointestinal, the respiratory system, cutaneous, buccal cavity, and genitourinary systems. The microbiota probably weighs anywhere between 1 and 2 kg [3]. The surface area of the GIT is second only to that of the respiratory system. The microbiota of the digestive tract is established during gestation and afterwards acquired from the mother and the surrounding environment through vertical transmission (through the birth canal). Researchers came to the conclusion that a newborn's GIT is essentially sterile, in stark contrast to the microbiota of an adult's GIT. The postnatal expansion of microflora is influenced by a number of factors, including the mother's microflora, inherited factors, and the use of medications, as well as the mode of delivery, the state of cleanliness of the childbirth environment, and the technique of feeding [4, 5]. According to a recent study using molecular techniques, the human gut microbiota has over 1000 phylotypes. Ninety-seven percent of all bacterial species are anaerobes [6]. Intestinal flora is dominated by Bifidobacterium and Bactericides, but the gut microbiome also includes Bacteroidetes (Porphyromonas, Prevotella, and Bacteroides), Firmicutes (Ruminococcus, Clostridium, Lactobacillus, and Eubacteria), and Actinobacteria (Bifidobacterium) [7].

Benefits to the host's health can be attributed to phenomena known as "microbial antagonism," in which beneficial microbes are actively opposed by the regular flora. By driving pathogenic bacteria out of the digestive tract, microbial competition has been shown to boost immune function by increasing levels of beneficial nutrients like short-chain fatty acids (SCFAs), vitamins, amino acids such as arginine, cysteine, and glutamine; growth factors; antioxidants; and more. Dietary supplements containing competitive probiotic bacteria are needed to develop effective epithelial barrier-protective therapeutics for gastrointestinal distress and other factors including mucosa-associated *Escherichia coli* [8]. The microbe used to control microorganisms must not be harmful to the host and must not alter the normal microbial flora of the human gastrointestinal tract [9]. While probiotics do increase specific species of beneficial gut bacteria, they do not increase the bacterial population in the gut as a whole. Protein-based components (bacteriocins) and organic acids, alcohol, hydrogen peroxide, and ethanol have all been linked to the antibacterial action of probiotics [10]. Once antibiotics are no longer effective due to antibiotic resistance, probiotics are the next best immune defense system, according to the WHO. This method of treatment is known as microbial interference therapy. In the field of nutraceutical science, "probiotics" (ingestible live microorganisms) are highly regarded due to their ability to offer consumers numerous health benefits [11].

The mammalian GIT has incredibly intricate defense systems to prevent bacterial invasion. The mucus layer in the intestines might decrease or vanish due to inflammation brought on by an imbalance between harmful and commensal bacteria. This allows bacteria and their metabolites to invade intestinal epithelial cells. Most mammals have two types of bacteria living in their guts: the "native microflora" and the "invasive pathogenic germs." A balanced gut microbiota is essential for maintaining intestinal immunity and stability. Pathophysiological effects may be altered if this balance is disturbed [12].

8.2 Criteria for Choosing Probiotics and the Bare Minimum Needed

Probiotic strains used in food applications must meet certain criteria in order to be considered safe for human consumption. These criteria include being of human origin, being able to survive in acid and bile, adhering to the lining of the digestive tract, competing with pathogenic bacteria, and being available in a dosage that is acceptable for human intake. Probiotics exhibit beneficial properties as they contribute to the well-being of the host by adhering to the membrane of intestinal epithelial cells, producing antibiotic compounds to combat infections, and maintaining the stability of the intestinal microbiota [13]. Fewer probiotics are needed as follows [14]:

1. There must be a sufficient number of viable microorganisms supplied.
2. The strains must be identified genetically, given labels (such as letters, numbers, or names), and placed in appropriate taxonomies.
3. In order to ascertain the suitability of probiotic strains for use on a host organism, such as humans, cows, or companion animals, it is imperative to employ meticulously crafted and adequately scaled testing methodologies to accurately identify these strains as probiotics.
4. Even though one strain has been proved to be effective against a particular disease, that does not mean it will be effective against another.
5. It is imperative to ensure that probiotic strains cultivated in human subjects and subsequently employed in animal studies are appropriately identified and labeled as such.

Research on probiotics is sparse, despite its usefulness for treating a wide range of disorders. This research aims to show the need of studying probiotics further for preventative and therapeutic uses against a wide range of disorders for which there may or may not be alternative treatments. Within this chapter, we shall establish a comprehensive definition of probiotics as a collective of advantageous microorganisms that possess the ability to readily acclimate to and endure the gastrointestinal environment, thereby bestowing health advantages. Additionally, we will provide a concise summary of their historical background, elucidate their mechanism of action, and furnish a detailed account of their preventive and remedial implications in relation to specific domains of human illness.

8.3 Action Mechanism of Probiotics

Probiotics are beneficial for a variety of reasons, and one of the ways they operate is by shifting the "microbial homeostasis" of the organism being treated or the surrounding environment. However, the precise molecular mechanisms by which they achieve this are rarely studied in depth because there are so many potential explanations. When taken orally, they prevent dangerous microbes from entering the body through the mucous membranes and stomach. Acid-producing probiotic bacteria, such as Bifidobacterium and Lactobacillus, reduce pH and prevent the growth of pathogenic bacteria. These acids include lactic acid, propionic acid, and acetic acid. It has also been proposed that they exert their effects via immunomodulation [15]. As of March 2020, there are 261 distinct species of Lactobacillus, each with their own unique phenotype, ecological niche, and genetic make-up. Because the genus Lactobacillus has been split into 25 distinct genera, the article notes that the names of several important probiotics have changed [16]. Probiotics' natural antimicrobial chemicals eliminate pathogens, restrict adhesion sites, reduce food availability, inhibit toxin receptors, and alter immune responses. This is a brief summary of the proposed probiotic mode of actions: suppressing the ability of pathogenic microorganisms to attach to the intestinal epithelium; the growth of pathogenic bacteria is stifled when they bind to Gram-negative bacteria; keeping your levels of SCFAs in check is the third goal; proliferation of colonocytes to restore intestinal barrier function; better intestinal electrolyte uptake; enhancing the immune system's response in the intestines; controlling lipid metabolism and pro-inflammatory cytokines in the intestines are dampened [17].

8.4 Probiotics in the Clinical Practice: A Growing Trend

Several acute and chronic infectious diseases, as well as gastrointestinal issues, may benefit from the use of probiotics, as depicted in Figure 8.1. Diseases such as inflammation, virus infection, cancer, and degeneration can develop when the body's defenses are compromised [18]. The normal composition of stomach flora can be altered by therapies such as antibiotics, irradiation, and immunosuppressive therapy. Functional food products, such as probiotics, are designed to boost cellular health and speed up the activation of the body's own defense systems. Several investigations investigating the impact of probiotics on human health, illness, disposition, behavior, and efficiency have been conducted. Many researches recommend incorporating pharmacological treatment, dietary modification, and complementary therapies like probiotics into the overall treatment

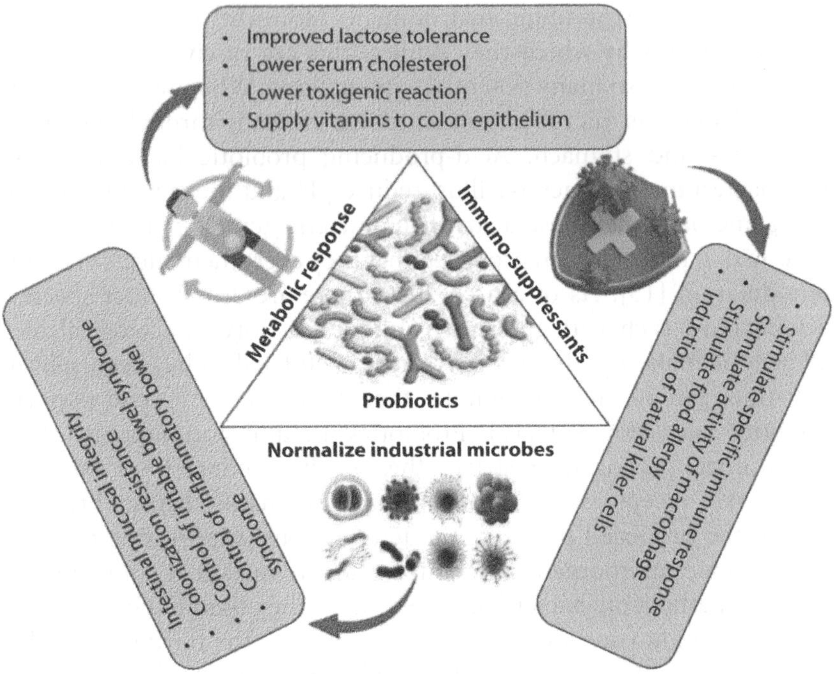

Figure 8.1 Schematic depiction of the numerous functions of probiotics.

strategy. Since probiotics can influence the host immune system, they show promise as a treatment and preventative measure for a wide range of diseases, including inflammatory conditions. Powders, pills, and beverages containing probiotic bacteria are all ways that people can replenish the good bacteria in their stomach and reap the health benefits of a stronger immune system [19].

8.5 Potential Preventative Roles of Probiotics

8.5.1 Inhibiting Carcinogenesis with Probiotics as a Preventative Measure

Cancer, a disease characterized by the uncontrolled growth and spread of abnormal cells, poses a significant global health burden. In the year 2012, it was responsible for roughly 14.1 million new cases and resulted in approximately 8.2 million deaths, making it the second most common cause of mortality worldwide. The aforementioned numerical value is anticipated to experience an upward trajectory, reaching a total of 21 million newly reported instances and 13.2 million fatalities by the year 2030. The accumulation of mutations in a cell's DNA is the root cause of cancer. Malignant tumors are characterized by uncontrolled cell proliferation, insensitivity to growth stimuli, and the ability to infect adjacent tissues, which are observed in the majority of cancer patients. Genetic abnormalities may account for 5%–10% of cancer occurrences, and environmental variables likely have a role in the development of 90%–95% of cancers [20]. One-third of all cancer deaths are attributable to factors including being overweight, not getting enough fruits and vegetables, not getting enough exercise, smoking, and drinking too much alcohol, as stated in the World Cancer Report, 2014.

Probiotic supplementation has long been linked to a variety of health benefits, including a lower risk of developing cancer and a slower rate at which the disease spreads. Cancers of the colon, stomach, breast, cervix, and myeloid leukemia have all been demonstrated to respond negatively to probiotics due to their antiproliferative or proapoptotic effects (Table 8.1) [21]. Probiotics *Lactobacillus rhamnosus* strain GG (LGG) and *Bifidobacterium adolescentis* SPM0212 were found to significantly inhibit human gastric cancer cells and three colonic cancer cell lines (HT-29, SW 480, and Caco-2) [22]. Kefir, which contains the probiotic bacterium *Lactobacillus kefiri*, has been shown to have an apoptotic effect on myeloid leukemia cell lines [23]. The potential of *Enterococcus lactis* IW5, which was taken from the human gut, as a probiotic for cancer patients was demonstrated by its ability to strongly suppress the growth of other pathogenic bacteria and reduce the survival of several cancerous cells (including HeLa, MCF-7, AGS, HT-29, and Caco2) [24].

Patients with malignancies of the colon, liver, breast, bladder, colon, and cervix have found that probiotic therapies are helpful in avoiding cancer and slowing its progression. These are the reasons why anticancer chemotherapy typically incorporates probiotics-based regimens as an adjuvant. However, more research is required to define the possible roles of probiotics in cancer prophylaxis, treatment, and progression. Probiotics have been shown to work against and treat cancer in a variety of different ways. These include (i) influencing the composition of gut microbes; (ii) improving intestinal barrier function; (iii) neutralizing possible carcinogens and (iv) shielding intestinal epithelial DNA from harm.

Colonic microbiota of persons with colorectal cancer showed an overrepresentation of Bacteroides and Prevotella, while Dorea spp. and feces-associated bacteria spp. were underrepresented in the microbiome of those with colorectal adenomas. These species can produce bile acids

Table 8.1 Potential anti-cancer effects of probiotics.

Probiotic strain	Subjects	Dose and duration of study	Effects ($P < 0.05$)	Reference
L. plantarum CGMCC, *L. acidophilus*-11 and *B. longum*-88	(i) Colorectal cancer patients (ii) 150 patients (1:1 ratio of probiotic and placebo group)	(i) *L. plantarum* CGMCC no. 1258; 1011(CFU)/g (ii) *L. acidophilus*; 1011CFU/g (iii) *B. longum*-881010 (CFU/g) (iv) Probiotics were given to the patients 6 days before surgery and again 10 days afterward	Inhibiting the p38 mitogen-activated protein kinase signaling pathway and lowering blood zonulin concentration, postoperative fever duration, antibiotic treatment duration, and postoperative infection complication rate were all benefits of taking probiotics.	[25]
L. rhamnosus LC705 and *P. freudenreichii subsp. Shermaniistrains*	(i) Aflatoxin induced liver cancer (ii) 90 male students with high aflatoxin level in urine	5 weeks, (1:1, wt:wt) at a dose of 2–5×10^{10} colony-forming units/day	Aflatoxin B1-N7-guanine (AFB N7-guanine) urine excretion is decreased by 61.5% as a biomarker for liver cancer.	[26]
L. casei Shirota (LcS)	(i) Breast cancer (ii) 968 breast cancer patients (306 probiotic group; 662 controls) aged 40 to 55	Soy product isoflavones and the probiotic Lactobacillus casei Shirota in Yakult were consumed regularly throughout a 2-year period.	Japanese women who regularly consumed LcS and isoflavones beginning in youth had a lower risk of developing breast cancer.	[27]
L. acidophilus L1	(i) Bladder cancer (ii) A total of 180 cases (mean age: 67 years, SD 10) and 445 population-based controls	200 g of yoghurt containing L. acidophilus L1 for 10 weeks	Regular consumption of lactic acid bacteria lowers the probability of developing bladder cancer.	[28]
L. casei Shirota (LcS)	(i) Cervical cancer (ii) 54 women with an HPV-positive intra-epithelial lesion	Daily administration of Yakult containing LcS for 6 months.	Infection with the human papillomavirus, or HP and cervical cancer risk factors decreased by 60%.	[29]

and heterocyclic amines, both of which are carcinogens and tumor promoters [30]. Some probiotic strains also exhibit anticancer characteristics in animals and humans, and they are effective at warding off infections. Future research should focus on the potential anticancer effects of probiotic microorganisms and their role in immunomodulation. SCFAs are essential for controlling cell death and proliferation, and they can be produced by specific bacterial species. These fatty acids also serve as immunological signaling molecules. Also, these bacteria have the potential to aid in the detection and eradication of cancer-causing substances. Górska *et al.* (2019) found that combining heat-killed probiotic bacteria with radiation improved the immune system's ability to recognize cancer cells [31]. Anticancer reactions have been found to be mediated by beneficial lactic acid bacteria (LAB) such Lactobacillus and Bifidobacterium in animal studies. For example, alpha-glucuronidase, azoreductase, and nitroreductase are all fecal enzymes that have been linked to colon cancer development, and their activity can be altered by lactic acid cultures [21]. Following a 12-week intervention utilizing a symbiotic combination comprising Bifidobacterium lactis Bb12, Lactobacillus rhamnosus GG, and oligofructose-enriched inulin, a notable improvement was observed in the gut microbiome of individuals diagnosed with colorectal cancer. Specifically, an increased abundance of lactobacilli and bifidobacteria, along with a reduced presence of *Clostridium perfringens* was observed. The anti-cancer effects of T cells, macrophages, natural killer cells, and T cells have been shown to be improved by the addition of *Lacticase bacilli casei* in multiple investigations. Radiotherapy and chemotherapy for colorectal cancer have been shown to have fewer negative side effects with this method [32].

The augmentation of cytokine levels, specifically IL-2 and IL-12, has been observed to contribute to the mitigation of carcinogenic substances and the reinforcement of the intestinal barrier's integrity and functionality. This phenomenon holds potential for the prophylaxis of colon and other malignancies. Additionally, the upregulation of mucin, defensin, and immunoglobulin A (IgA) synthesis, along with the modification of proinflammatory cytokine and chemokine responses, further aids in these protective mechanisms. Academics are interested in the conversion of linoleic acid to conjugated linoleic acid (CLA) because of the anti-inflammatory polyunsaturated fatty acid's (PUFA) possible therapeutic effects in cancer treatment [33–35], regulating apoptosis and differentiation of cells, which has antiproliferative effects [36, 37]. In response to Lactobacillus, the levels of certain biological factors increased, including cytokines as well as interleukins (IL-2 and IL-12), antioxidant substances (SOD, CAT, and GSH), indigenous microbes, interferons (IFN), and immune cells (TH cells and NK cells), while levels of other factors decreased, including damaging DNA, pathogens, inflammation, ulcers, tumor size, cancer-specific proteins, polyamine contents, and pro-carcinogens (Figure 8.2) [38].

Preventing the development of colon cancer in rats by administering probiotics and celecoxib one week beforehand was found to significantly reduce K-ras expression and increase tumor suppressor p53 transcription. These findings are consistent with findings that were discovered earlier. This finding could be attributable to the additive effect of the probiotics and celecoxib, as seen in Figure 8.3. This impact may have slowed cell growth by keeping cells active and in the cell cycle by reactivating tumor-suppressor genes. This type of combination therapy has been shown to help protect gut flora and improve immunological response, both of which are important in preventing future DNA damage. As a result of its ability to reduce inflammation and increase p53 expression, both of which can directly induce Bax-mediated apoptosis, celecoxib was found to be effective in reducing the incidence of colonic cancers [39]. The study bolstered prior findings by showing that the combination of beneficial bacteria (probiotic) and celecoxib reduced the levels of expression of the proto-oncogene K-ras and increased the expression level of the tumor suppressor p53 when given to rats for a week prior to the induction of colorectal cancer. By activating tumor-suppressor genes, the combined effects of probiotics and celecoxib may account for this phenomenon without interfering with cell activity or

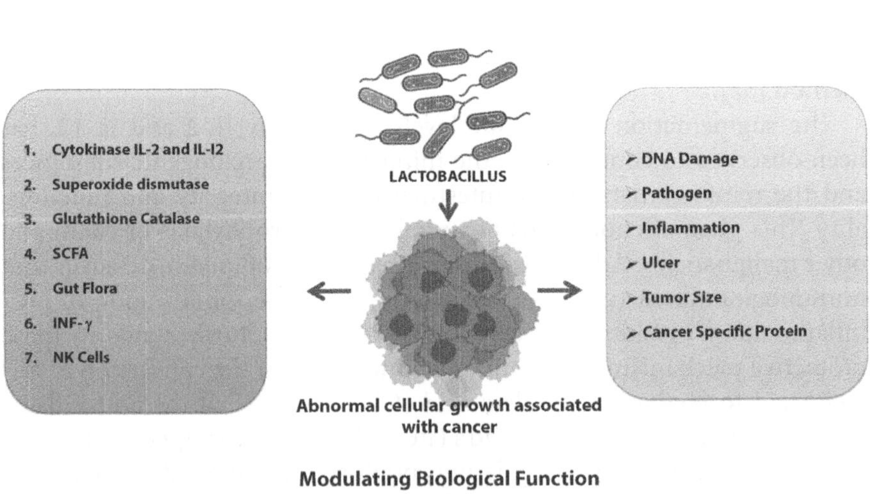

Figure 8.2 Changes in the biology of cancer cells were triggered by *Lactobacillus*.

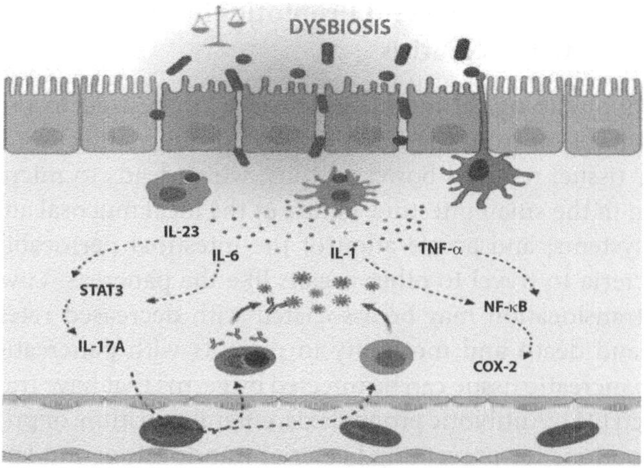

Figure 8.3 Intestinal inflammation can be triggered by bacteria in the gut and is the initial step in colorectal cancer development. The translocation of microbiota and metabolites from the gut to tissues and the systemic circulation, as well as immunological disorders, can be caused by dysbiotic bacteria that thrive in a leaking gut. This might lead to inflammation and colorectal cancer. Accordingly, IL-17A expression and CRC formation are caused by STAT3 activation by IL-6 and IL-23. The activity of COX-2 is induced by TNF-α and IL-1, which promotes inflammation and tumor formation, as well as angiogenesis and the inhibition of apoptosis in colorectal cancer.

the cell cycle (as shown in Figure 8.3). This type of treatment regimen may have helped preserve gut health and boost immunity, hence decreasing DNA damage. By increasing p53 and decreasing K-ras, two genes that can directly trigger Bax-mediated apoptosis, celecoxib has been demonstrated to diminish the incidence of colonic cancers [40].

a) DMH (Differential methylation hybridization) produces DNA methylation, which speeds up the Wnt pathway by increasing pro-inflammatory marker expression. The onset of cancer can be traced back to several factors, including exposure to noxious microbes, the creation of cancer-causing chemicals, and a decrease in proapoptotic markers.
b) By improving the gut microbiome and modulating the Wnt signaling pathway, the combination of probiotics and celecoxib may prevent carcinogenesis [39].

8.5.2 Enteral Nutrition and Probiotics in the Treatment of Acute Pancreatitis

Three pathophysiological mechanisms are hypothesized to precede bacterial translocation from the intestines and produce infection of necrotic pancreatic tissue: reduced bowel motility, which leads to microbial overpopulation in the small intestine; failure of the local mucosal and systemic immune systems; and acceleration of the intestinal permeability, which allows bacteria to travel to other organs like the pancreas. Lower rates of bacterial translocation may be associated with decreased rates of future infection and death and morbidity in patients with pancreatic necrosis. Necrotic pancreatic tissue can be infected by germs that have traveled from the stomach [41]. Antibiotic prophylaxis, enteral nutrition, or probiotics are all viable options for preventing bacterial translocation and late infection in patients at risk for pancreatic necrosis. The administration of probiotics subsequent to the induction of pancreatitis did not yield any significant enhancement in the functionality of the intestinal barrier. However, in a mouse model of acute pancreatitis, the functioning of the intestinal barrier improved after pretreatment with probiotics. Therefore, the timing of probiotic supplementation is critical to their efficacy. Although studies in animals have only shown a possible association between probiotic treatment timing and the development of pancreatitis, similar results may be seen in people [41, 42].

8.5.3 Infection with *Clostridium difficile* in Individuals with Colitis

Opportunistic infections, such *Clostridium difficile* colitis, occur when the natural gut flora is disturbed due to antibiotic use. Clindamycin, fluoroquinolones, broad-spectrum penicillin, and cephalosporin are the antibiotics most frequently used in conjunction with this condition. Infection with *Clostridium difficile* can range from mild to life-threatening, with the latter requiring hospitalization in a special facility. To prevent the harmful effects of *Clostridium difficile* toxin, probiotics (live, non-pathogenic bacteria) infiltrate the gut and secrete a lytic peptide. A protease produced by *Saccharomyces boulardii* blocks the action of *Clostridium difficile* toxin [43].

8.5.4 Probiotics Prevent Postoperative Infections

Manipulating the microbiota may enhance patient outcomes and provide host health benefits due to the growing body of evidence suggesting its widespread impact on both pathologic and normal processes. In order to combat *Clostridium difficile* (CD), prophylactic probiotics aim to maintain colonization by "good" bacteria. Some probiotics can prevent the harmful effects of the A and B toxins produced by *Clostridium difficile*. The results of *in vitro* and preclinical studies show that probiotics have protective effects [44]. Post-operative microbial dysbiosis, decreased immune suppression, and gastrointestinal mucosal injury leading to compromised gut barrier integrity and enhanced intestinal permeability are the prevailing factors contributing to bacterial translocation. Patients undergoing abdominal surgery for conditions such as biliary cancer, pancreaticoduodenectomy, and liver transplantation are at risk for a wide variety of infections, including UTIs, respiratory infection (pneumonia), infections of the wound, intra-abdominal abscesses, and cholangitis. Probiotics administered to people with non-surgical disorders diminish the prevalence of Clostridium difficile colitis that is resistant to antibiotics. This kind of infection control may help lessen the spread of antibiotic-resistant germs, shorten hospital stays, and alleviate pain for patients [18].

8.5.5 Newborns with Extremely Low Birth Weights, Late-Onset Sepsis, and Necrotizing Enterocolitis

Infants with extremely low birth weight or very low birth weight (VLBW) are particularly susceptible to nosocomial infection and necrotizing enterocolitis, two primary causes of mortality among VLBW infants. Both disorders are thought to have a complex etiology that includes endogenous host variables including gestational age and immaturity of the immune response. Enteral nutrition and exposures to an endemic healthcare facility play a key effect in aberrant gut colonization and the transfer of entero-pathogenic microorganisms through the compromised intestinal mucosa [45]. Surgery for necrotizing enterocolitis, abdominal surgery, and death from any cause were all associated with lower risks when probiotics containing *Lactobacillus acidophilus* and *Bifidobacterium infantis* were administered.

While it is true that Gram-negative bacteria are less commonly associated with causing *Listeria monocytogenes* infections compared to Gram-positive bacteria, their presence is often associated with a more severe clinical manifestation, higher mortality rates, and an elevated risk of morbidity in newborns. Being born preterm, with a low birth weight, or requiring critical care all increase the risk of developing sepsis. In the neonatal intensive care unit, premature infants are exposed to colonization bacteria.

Hospitalization for an extended period of time and the use of antibiotics after birth both reduce the diversity of microorganisms in the microbiome [42, 46]. Therapy with oral probiotics may improve colonization with commensals in this LOS-prone population. The strain, dosage, and clinical application all have a role in how effective probiotics are. In a recent randomized control experiment utilizing a probiotics combination consisting of *B. infantis, S. thermophiles*, and *B. lactis* , the length of stay (LOS) was significantly reduced in the preterm subgroup consisting of women who gave birth between 28 and 31 weeks of gestation [47].

Among infants born with a birth weight of less than 1,500 grams (1.5 kilograms), necrotizing enterocolitis is a leading cause of death and disability in the first month of life (20%–30%). Low and delayed colonization of the stomach by probiotics species like bifidobacteria and lactobacilli is common in premature newborns; this may be due to cesarean birth, a delay in breastfeeding, or postnatal antibiotic therapy [48]. Probiotics given by mouth reduce the risk of necrotizing enterocolitis and death in premature newborns [49]. Necrotizing enterocolitis and its accompanying morbidity may be prevented with the help of prophylactic enteral probiotics (live microbial supplements) by delaying microbial movement across the mucosa, competitively rejecting dangerous bacteria, and increasing the host's immune responses. Surgery for necrotizing enterocolitis, abdominal surgery, and death from any cause have all been linked to probiotics containing the microorganisms *Lactobacillus acidophilus* and *Bifidobacterium infantis* [47]. It is theorized that probiotics aid in the colonization of the gut by beneficial microorganisms, protect against pathogen colonization, promote gut mucosal barrier maturation and function, and influence the immune system. It is common practice to start giving probiotics to healthy newborns on day one, or after antibiotic treatment has been discontinued, for a total of two weeks [50].

8.5.6 Probiotics in Ventilator-Associated Pneumonia (VAP)

VAP is widely recognized as the predominant infection encountered within the intensive care unit (ICU) setting. This particular infection is associated with a considerable burden of both mortality and morbidity. Recently, probiotics have become an alternative strategy for combating VAP. By enhancing the function of the intestinal barrier, boosting host cell antimicrobial peptides, changing the composition of the microorganisms in the intestines, and decreasing pathogenic overgrowth of microorganisms and bacterial translocation, probiotic bacteria may be able to prevent the development of VAP. Multiple studies have shown that probiotics can help ICU patients who are at risk for or experiencing ventilator-associated pneumonia (VAP) [51].

8.6 Therapeutic Use of Probiotics

Probiotics have their most significant therapeutic effects when they operate on the gastrointestinal tract (GIT), either directly or indirectly. The reason behind this phenomenon can be attributed to the capacity of probiotic microflora to form symbiotic associations with the microflora residing on the mucous membranes of host cells, thereby manifesting their beneficial impacts upon internal consumption of the supplement. Because the human gut flora does not often contain all probiotic microorganisms, it is possible that the advantages associated with a particular species may not be transferable to those associated with another [52].

8.6.1 Role of Probiotics in Hypertension Treatment

Globally, cardiovascular, cerebrovascular, and renal sickness now has hypertension (HTN) as a prominent risk factor. Certain strains of gut microbiota have the potential to exert either a beneficial or detrimental influence on the development of hypertension. The development of hypertension (HTN) has been linked in recent studies to dysbiosis of the gut microbiota, in both animal models and humans [53]. The primary mechanisms by which probiotics exert their effects are as follows: lowering the luminal pH; secreting antibacterial peptides; hindering microbial spread; hindering microbial attachment to epithelial cells; increasing mucous secretion; triggering cytokine cascades and immune modulation; decreasing myeloperoxidase, the tumor necrosis factor alpha, and nuclear factor kappa B; and improving mucous secretion. By increasing the expression of growth factors and receptor sites and by strengthening the mucosal barrier, probiotics aid in maintaining the GI barrier's structural integrity. While increasing the production of peroxide from hydrogen and benzoic acid, probiotics promote the proliferation of acid-resistant commensal bacteria and suppress the growth of dangerous, acid-sensitive bacteria. The ability of SCFAs to reduce inflammation in colony epithelium cells and immune cells was demonstrated. The findings presented in Figure 8.4 provide compelling evidence that the consumption of fermented milk can effectively contribute to the reduction of blood pressure levels [11]. Recent research has linked the fermentation process, wherein bioactive peptides like ACE-inhibitory peptides are produced, to the hypotensive action of

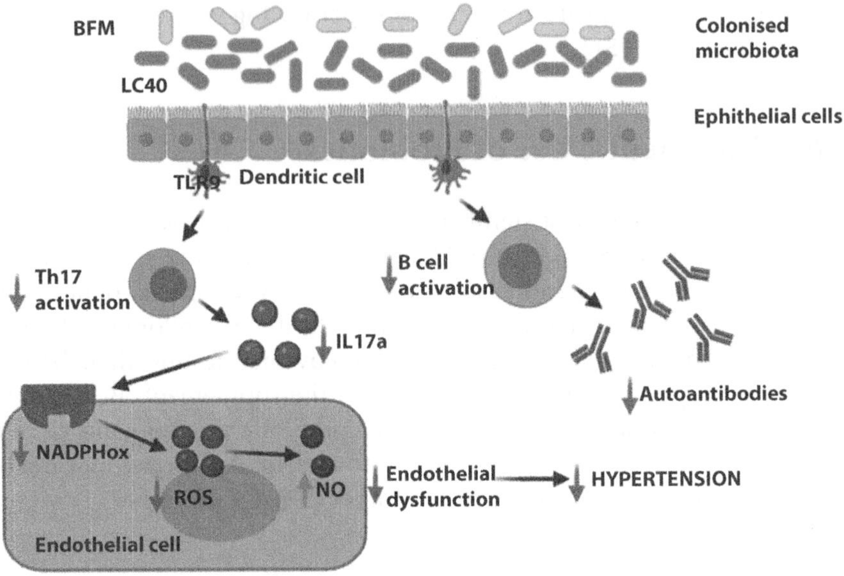

Figure 8.4 A schematic diagram exhibiting role of probiotics for hypertension treatment.

probiotics. ACE-inhibitor medications are used to reduce blood pressure. Two ACE-inhibiting tri-peptides are produced in lactic acid bacteria (*L. helveticus*) and yeast (*Saccharomyces cerevisiae*)-fermented sour milk [11]. Both valyl-prolyl-proline (Val-Pro-Pro) and isoleucyl-prolyl-proline (Ile-Pro-Pro) are essential amino acids. A considerable reduction in the risk of developing hypertension (HTN) has been linked to the intake of fermented milk items with *L. casei* strain Shirota (LcS) on at least a weekly basis. Specifically, the SG1-polysaccharide-glycopeptide complex is responsible for *L. casei*'s antihypertensive effects. Prostaglandin I2 secretion is stimulated and peripheral vascular resistance is lowered because of this complex [3].

8.6.2 Role of Probiotics Against Intervention for Inflammatory and Autoimmune Diseases

Molecular mimicry, self-antigen modification, bystander activation, and regulation of the immune response are four pathways that have been proposed as possible contributors to autoimmune disease. Probiotics are used to treat a variety of autoimmune diseases, including systemic lupus erythematosus, rheumatoid arthritis, celiac disease, Crohn's disease, ulcerative colitis, multiple sclerosis, Sjogren's syndrome, autoimmune sclerosis, antiphospholipid syndrome, myasthenia gravis, and type 1 diabetes. Mechanistic investigations of probiotics have linked them to at least four metabolic pathways, suggesting that the interaction between microbes and the immune system is related to signaling for the production of signaling for the metabolism of tryptophan, signaling for the activation of aryl hydrocarbon receptors in the intestine, signaling for the release of histamine in the intestine, and so on. Several randomized controlled trials have demonstrated that the administration of probiotics resulted in the improvement of symptoms related to gastrointestinal and multi-organ inflammation in patients diagnosed with rheumatoid arthritis (RA), ulcerative colitis, and multiple sclerosis [54]. Probiotics have the potential to confer health benefits through the modulation of immunological function [55].

8.6.3 Curative Role of Probiotics Against *Helicobacter pylori*

Helicobacter pylori is a Gram-negative bacterium that lives on the luminal side of the stomach epithelium. It moves about using flagella. Since substantial side effects and disturbances in the natural microbiota have been identified with the use of standard triple therapy (STT), it has been found to be less effective in treating *Helicobacter pylori* infection [56]. Since probiotics enhance immune-regulation capabilities, they work as a preventative measure against *Helicobacter pylori* infection. Supplemental usage of probiotics has been shown to improve the eradication rates of *Helicobacter pylori*, lessen the severity of treatment-related side effects, and lessen the inflammation of the gastrointestinal tract caused by the bacteria. Although probiotics have shown promise in treating *Helicobacter pylori*, conflicting test results have made it hard to draw firm conclusions. It is possible that *Lactobacillus* is the only probiotic capable of completely eliminating *Helicobacter pylori*. Several variables contribute to probiotics' success in treating *Helicobacter pylori*. Examples of probiotic bacteria include those that have both an immunological and a non-immunological effect on *Helicobacter pylori*. Proposed benefits for these substances include direct bacteriostatic or bactericidal activities, reduced cytokine generation, enhanced stomach defense, engagement with the host's innate immune response, and others [56]. A vaccination against *Helicobacter pylori* was created by engineering *L. lactis* to synthesize the lipoprotein Lpp20 of *Helicobacter pylori*. When tested *in vivo*, this tailored probiotic triggered an immunological response in the animals. Another study used genetic engineering to make *Bacillus subtilis* spores that surface-expressed the *Helicobacter pylori* urease B protein [9].

8.6.4 Role of Probiotics Against Inflammatory Bowel Disease

Inflammatory bowel diseases (IBDs) like Crohn's disease and ulcerative colitis are characterized by gastrointestinal mucosal irritation, intestinal permeability, and dysbiotic microbiota [9]. Millions of people suffer from the debilitating effects of irritable bowel syndrome (IBS), putting a load on healthcare systems throughout. A higher permeability defect in the stomach was seen in IBD patients compared to controls. The consumption of lymphocyte-specific elimination agents has been shown to inhibit lymphocyte growth and the generation of inflammatory cytokines [57]. Recent research has posited conflicting theories on which microorganisms play a role in the onset of IBD. The immune system's reaction to the good bacteria normally found in the gut is the first problem. Modifications to the microbiota of the gut or the mucosal protective barrier, both of which activate detrimental immune responses towards the mucosa, are the second probable cause in the development of inflammatory bowel disease (IBD). The approach may include prescription medications, dietary adjustments, and complementary therapies like probiotics [58]. *Lactobacillus paracasei* L74 CBA-containing fermented milk reduces the production of pro-inflammatory cytokines. *Streptococcus salivarius* has anti-inflammatory characteristics due to its ability to inhibit NF-Kb activation. According to Fernandez *et al.*, the protective effects of *Lactobacillus salvarius*, which were previously connected to local synthesis of IL-10, which has anti-inflammatory characteristics, require Nucleotide Binding Oligomerization (NOD2) receptors. When tested in a mouse model of colitis, the probiotic strain *Lactobacillus plantarum* Lp91 was found to increase IL-10 expression and decrease levels of tumor necrosis factor (TNF) and cyclooxygenase-2 (COX-2). In the context of managing inflammatory bowel disease, it is worth noting that probiotic lactic acid bacteria, which are commonly employed as probiotics, have demonstrated potential benefits [59]. Specifically, these microorganisms have the ability to generate substantial quantities of antioxidant enzymes, thereby potentially aiding in the treatment of this condition. Studies have demonstrated that augmenting the population of advantageous probiotic microorganisms within the gastrointestinal tract and colon can lead to enhanced gastrointestinal well-being among individuals afflicted with a condition known as inflammatory bowel disease [60].

Diarrhea is widely recognized as the prevailing adverse response to antimicrobial therapy. Antibiotics with broad-spectrum activity, such as amino-penicillin, cephalosporin, or clindamycin, have an increased propensity to induce antibiotic-associated diarrhea. The administration of

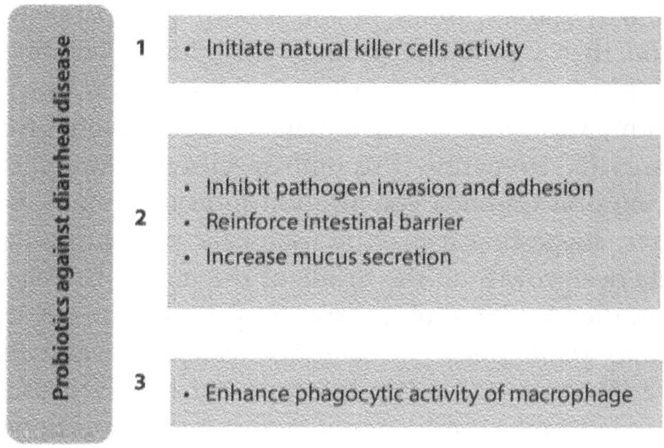

Figure 8.5 The potential mode of action of the probiotic for the treatment of diarrhea.

antibiotics has the potential to disrupt the equilibrium of the indigenous enteric microbiota, resulting in a reduction in the population of endogenous microorganisms residing in the gastrointestinal tract. There are many ways in which the host suffers when the population of these symbiotic bacteria is reduced [61]. Probiotics have been demonstrated to be useful for a variety of illnesses, including the treatment of antibiotic-associated diarrhea, the aversion of traveler's diarrhea, and the treatment of acute infectious diarrhea. Probiotics therapy over a longer length of time was more effective at repairing the gut microbiota, restoring tissue architecture, and decreasing systemic inflammation than probiotics therapy over a shorter period of time. Long-term probiotic treatment normalized levels of IL-10, as well as all of the major fecal bacteria. Figure 8.5 displays the nature of the probiotics' effectiveness in treating antibiotic-associated diarrhea [62].

8.6.5 Role of Probiotics in the Treatment of Oral Candidiasis Colonization (OCC) in Denture Wearers

Denture wearers are more likely to contract angular cheilitis and Down syndrome (DS), two conditions linked to the fungus *Candida* sp. Candida infections are typically associated with systemic illnesses and immune inadequacy, in addition to local causes such as inappropriate occlusal vertical dimension (OVD), trauma induced by an ill-fitting denture, and poor denture washing, among other variables. Research has demonstrated that certain strains of the probiotic bacterium Lactobacillus, including *Lacticaseibacillus rhamnosus* and *Limosilactobacillus reuteri*, can decrease the amount of Candida found in the mouth [63]. Ishikawa and colleagues found that a probiotic product including a combination of *B. longum, L. bulgaris*, and *S. thermophilus* in addition to oral local antifungal drugs (nystatin) was more effective than the conventional treatment for Candida-associated stomatitis. This multispecies probiotic showed promise in reducing oral Candida colonization in asymptomatic seniors who wore dentures, suggesting it may be useful for preventing oral candidiasis. Candida overgrowth on the inside of the mouth has been linked to an increase in the likelihood of invasive fungal infections. A combination of *Lactobacillus rhamnosus*, *Lactobacillus acidophilus*, and *Bifidum* spp. was effective in treating *Candida* infections in elderly denture wearers. Probiotics can lessen the prevalence of opportunistic infections because of their ability to strengthen the immune system [64].

8.6.6 Role of Probiotics in the Treatment of Migraines

Studies have demonstrated that avoiding certain meals can reduce the occurrence and severity of migraines. Increased level of immunoglobulin G (IgG) specific to food may be an indicator of food hypersensitivity. Consequently, avoiding migraines by consuming IgG-free meals may be helpful. There is an increasing body of research suggesting that inflammation has an effect in the onset of migraines. Improved intestinal barrier function via the gut–brain axis may be one mechanism by which probiotics alleviate migraine symptoms. Multiple studies point to serotonin as a crucial component of the brain–gut axis. However, the central nervous system only contains 3% of an individual's total serotonin. The rest stays in the digestive tract. It has been found that bacteria in the digestive tract play a part in serotonin production. Therefore, using the correct probiotics to alter gastrointestinal function may be an approach to aiding migraine sufferers [65].

8.6.7 Role of Probiotics for the Treatment Diabetes

Diabetes is a set of metabolic diseases characterized by sustained elevation of blood sugar. Rabiee and Babajafari (2018) state that probiotics are becoming increasingly well-liked as supplements for health among both the general public and the medical establishment [66]. This is perhaps because of the potential importance of these microbes in enhancing general health, especially in the control of diabetes. Since probiotics can alter gut flora, they can play a role in warding off metabolic disorders like diabetes and obesity in children [67].

In people with type 1 diabetes, the immune system mistakenly attacks and kills off their own pancreatic beta (β)-cells, making the disease an autoimmune disorder. The onset of type 1 diabetes is due to both genetic predisposition and environmental factors. Beta-cell mortality and decreased insulin production in type 1 diabetes are preceded by the development of autoantibodies against beta-cell antigens and insulitis (pancreatic inflammation) [66, 68]. Probiotic strains, when consumed, were found to decrease levels of IL-6, IL-1, and TNF-α and increase levels of TGF-β and IL-10, all of which are anti-inflammatory cytokines. This suggests that probiotics may be useful in preventing the onset of type 1 diabetes [68]. Oral probiotic supplementation was studied for its potential to protect non-obese diabetic mice from developing spontaneous autoimmune diabetes. Finally, they found that an oral probiotic combination including Bifidobacteria (*B. longum, B. infantis,* and *B. breve*) and Lactobacilli (*L. acidophilus, L. casei, L. delbrueckii* subsp. *L. bulgaricus,* and *L. plantarum*) displayed antidiabetic properties. There is evidence that exposing healthy patients to *L. plantarum* or another member of the *L.* genus can delay or prevent the onset of autoimmune diabetes [66].

Disabling consequences associated with Type 2 Diabetes (T2DM) pose a significant threat to public health and the economy. Many factors, including the gut microbiota's influence on energy expenditure, inflammation, appetite regulation, fullness, and lipid and glucose metabolism, have been linked to obesity. Alterations in the intestinal flora have been linked to type 2 diabetes. Several chemicals produced by the gut microbiota have been linked to the regulation of glucose levels in the body. Many of these chemicals, such as secondary bile acids, amino acid derivatives, and short-chain fatty acids, have been linked to effects on metabolic processes and immunologic processes [69, 70]. As a whole, probiotics may boost intestinal barrier function, reduce systemic LPS levels, raise Incretins, lessen ER stress, and enhance insulin sensitivity in the periphery. Improved glucose tolerance, lipid metabolism, antioxidant status, gut flora, and SCFA composition are all potential mechanisms by which probiotics combat diabetes. In addition, probiotics reduce the body's reaction to oxidative stress, autoimmune attacks, and inflammation [71]. Both *L. plantarum* and *B. lactis* are examples of probiotic bacteria that have been shown to mitigate the side effects of high-fat diets and even alter the immunological responses that can lead to inflammatory diseases. Probiotics have been shown to be effective in treating diabetes [72], and a combination of *L. rhamnosus, L. gasseri,* and *B. lactis* has been shown to reduce weight gain in people, particularly in the form of fat tissue mass adiposity [7].

8.6.8 Role of Probiotic in the Treatment of Autism Spectrum Condition

Recent research suggests that an imbalance in the gut's flora may have a role in autism spectrum disorder (ASD). Psychobiotics, a novel classification of probiotics, refer to living organisms that confer advantageous impacts on mental well-being. A number of mechanisms are used by probiotics to alter the make-up and/or function of the gut microbiota, including bacteriocins and metabolites including lactic, propionic, and acetic acids [73]. Probiotic treatment has been shown to alter the microbiota–gut–brain axis in people with ASD. Probiotics may aid ASD rehabilitation by preserving or enhancing gut barrier integrity, which is achieved through the promotion of "leaky gut" healing [74]. By increasing BDNF and stimulating the formation of regulatory T cells while decreasing LPS and giving tolerogenic signals, probiotics have been shown to have a suppressive effect on inflammatory pathways in the gut and central nervous system (CNS) [75]. Probiotic immunomodulation may lead to an increase in the production of anti-inflammatory mediators like interleukin-10 (IL-10) and transforming growth factor beta (β-TFGF), and a decrease in the production of pro-inflammatory cytokines like interleukin-12 (IL-12), tumor necrosis factor alpha (TNF-α), and interferon gamma (IFN- γ). Because it decreases inflammation and dysregulated immune responses in the gut, probiotic medication may be effective for resolving both gut microbial and behavioral issues in ASD [75, 76]. Finally, it has been suggested that the vagus nerve-mediated gut–brain axis (GBA) may be involved in the effects of various probiotic strains on neurotransmission and emotional states [75]. These strains include *Bifidobacterium* sp. (*B. longum, B. breve, B. infantis,* and *B. bifidum*) and *Lactobacillus* sp.

8.7 Recent Advancement in Probiotics

Probiotics have been shown in recent research to alter the make-up of the gut microbiota, which may have important health implications. It has yet to be fully clarified, however, what mechanism is responsible for this modulation. Oral administration of probiotics reduced the severity of the condition by boosting the proportion of the intestinal microbiota that generate SCFA or therapeutic metabolites. These microbes include Bifidobacterium, Faecalibacterium, and Akkermansia, among others [77]. The importance of novel probiotics and prebiotics has increased as the pace of human microbiome scientific research has increased. Depending on where you reside, there may be different regulations in place for the introduction of a brand-new probiotic or prebiotic. In today's world, modern probiotics are frequently used to facilitate digestion. Recent developments in the field of cultivomics have made it possible to create new host-specific probiotic treatments by utilizing enhanced techniques and gnotobiotic animal models. These developments were made possible by recent progress in cultivomics. Despite the bright outlook for the future of modern probiotics, more regulations are needed to create authentic probiotic products and define novel probiotics through state-of-the-art research and technology [78].

8.7.1 Recent Progress in Treatment of Acute Pancreatitis

Acute pancreatitis is rapidly becoming the most common reason for gastrointestinal hospitalization worldwide. The etiology of acute pancreatitis (AP) is complicated, involving inflammation of acinar cells, immune system involvement, and systemic clinical repercussions. The use of probiotics in postoperative patients has shown encouraging results [79]. In patients who have been diagnosed with severe acute pancreatitis, the use of probiotics is now discouraged according to current recommendations. Patients with acute pancreatitis may benefit from the administration of probiotics because of their potential to increase upper gastrointestinal motility, decrease intestinal permeability by strengthening the gastrointestinal barrier, and trigger a response that is anti-inflammatory by activating anti-inflammatory cytokines. For one thing, probiotics can strengthen the digestive tract's protective lining, which, in turn, decreases intestinal permeability "(just like interleukin-10)". Streptococcus, Lactobacillus, and Bifidobacterium were administered to a mouse model of acute pancreatitis to reduce bacterial translocation and enhance clinical outcome [80, 81].

8.7.2 Autophagy: A Novel Mechanism for the Anti-Inflammatory Effects of Probiotics

Autophagy is a cellular metabolic process that is essential to a wide range of physiologic functions. Autophagy serves a crucial role in preserving cellular homeostasis during periods of stress, and any aberrations or disturbances in its regulation may potentially contribute to the development of human diseases such as cancer [82]. The autophagosome is generated through the engulfment of cargo by the phagophore, a compartment consisting of two membranes. Subsequently, the phagophore undergoes closure, leading to the fusion of lysosomes and subsequent degradation of the cargo by lysosomal hydrolases [83]. According to genome-wide association studies (GWAS), around two hundred genetic risk loci for inflammatory bowel disease, also known as IBD, have been discovered. Many of these genes have been linked to innate immunity, mucosal barrier function, and bacterial identification [84].

Intestinal diseases can develop as a result of dysfunctional epithelium cells. Toll-like receptors, C-type lectin receptors, retinoic acid-inducible gene-I like receptors, and nucleotide binding oligomerization domain 2-like receptor gene defects all cause mucosal dysfunction and induce intestinal inflammation [85]. It has been well established that the nucleotide binding oligomerization domain-2 receptor gene is the primary susceptibility gene for chronic illness. This receptor recognizes muramyl dipeptides, a conserved pattern present in the peptidoglycan of both Gram-negative and Gram-positive bacteria. While *Bifidobacterium bifidum* PI22 showed the greatest protection in an acute colitis model, it showed only modest protection in a chronic colitis model, and *Bifidobacterium lactis* LA804, which showed the least protection in the acute model, showed the greatest protection in the chronic type. Although *Lactobacillus helveticus* PI5 was not soothing *in vitro*, it was the most effective in establishing the epithelial barrier and reducing the degree of inflammation of acute colitis in mice. Interestingly, mice were significantly protected from both types of colitis by *Lactobacillus salivarius* LA307 [85, 86].

8.7.3 Synthetic Engineering of Probiotics for Therapeutic and Preventative Purposes

The development of innovative therapeutic modalities for disease prevention and therapy is being aided by breakthroughs in gut bacteria engineering [87]. These techniques are designed to improve outcomes by rewiring host microbiome relationships like the host's metabolism and immune system. According to research published in Pharmaceutical Strategies for Targeting the Colon, a robust population of beneficial bacteria known as probiotics is crucial to the long-term wellness and efficient functioning of the digestive system. You can get your hands on probiotics as a nutraceutical product in both the standard pharmaceutical dose form and in typical non-standard food products thanks to inclusion complexes with cyclodextrins that allow for efficient, sustained release tablets through solid dispersion technology [88].

There are a number of benefits to using engineered probiotics, including increased specificity, controlled release of antimicrobial drugs, and reduced risk of antibiotic resistance (Figure 8.6).

a) Antimicrobials and other therapeutic substances are secreted by probiotics in response to pathogen-released molecules, which are then used to treat ailments.
b) Asthmatics can benefit from the use of modified probiotics that deliver antigens for the purpose of treating and preventing allergies.
c) Upon recognizing quorum sensing signals, engineered probiotics linked to tumor tissue can secrete anti-tumor substances.
d) With the help of rational design, probiotics can be outfitted with complicated circuits that allow them to carry out their duties more precisely. Only by detecting two compounds at a time are engineered probiotics able to produce the necessary medicines.

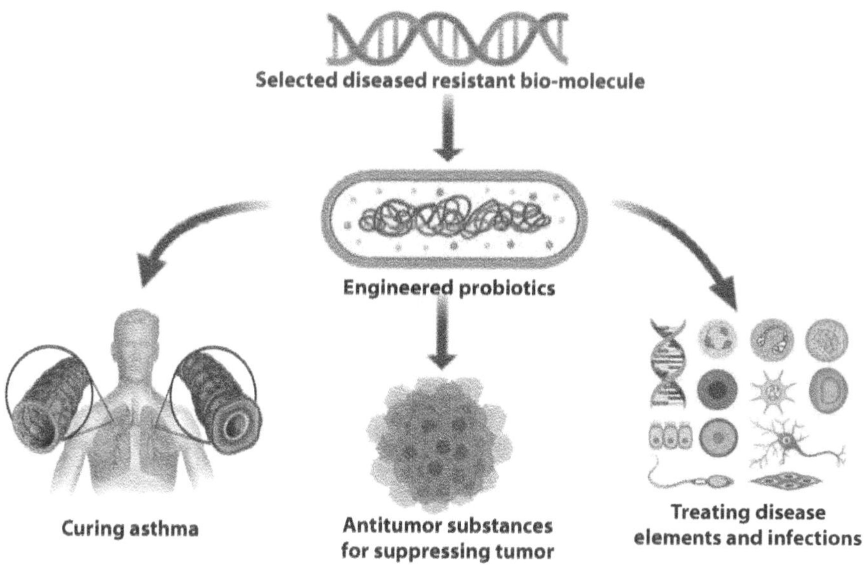

Figure 8.6 A schematic depicting the therapeutic potential of genetically engineered probiotics for treating disease.

L. lactis has been modified genetically to combat *Enterococcus* spp. that has developed resistance to multiple antibiotics. The *E. faecalis* pheromone was the impetus for creating the modified *L. lactis*. *C. difficile*, the most frequent pathogen in rich nations and a leading cause of antibiotic-associated diarrhea, has also been targeted by customized probiotics. *C. difficile* infection is best treated by fecal microbiota transplantation (FMT). Probiotics expressing specific antigens have been developed to combat pathogens such *Vibrio cholerae, Shigella dysenteriae,* and *Pseudomonas aeruginosa* [89, 90].

8.7.4 Probiotics in Anti-Oxidation Process

The beneficial effects of probiotics are thought to result from their Nobel Prize-winning methods of action in autoxidation, which are depicted in Figure 8.7. Multiple pathways have been identified by which probiotic bacteria exhibit potent antioxidant capabilities *in vivo* and *in vitro*: (1) Probiotics can remove metal ions through chelation. (2) Probiotics generate antioxidant enzymes. (3). Probiotics generate antioxidant metabolites,

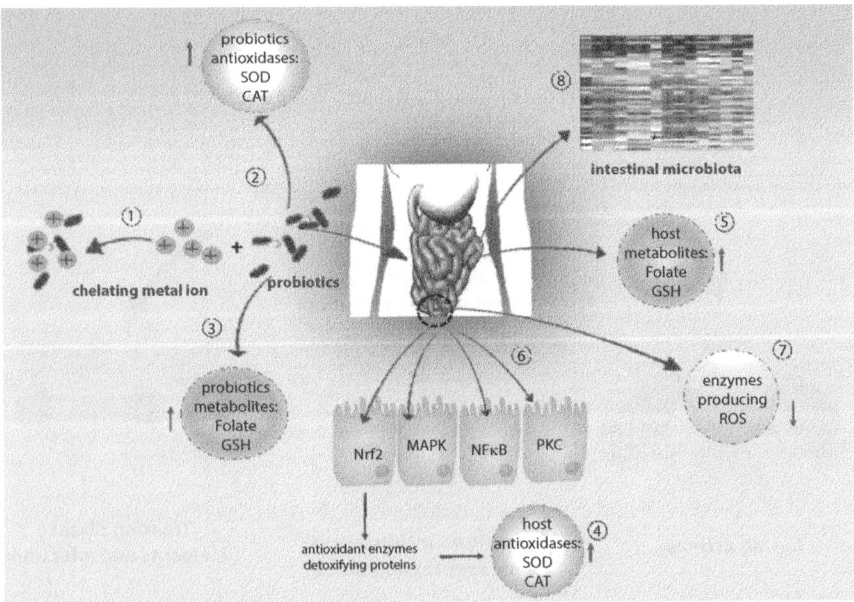

Figure 8.7 Probiotics' Nobel mechanisms of action in antioxidation Figure adopted from Wang *et al.* [91] and it is open article (copyright © 2017 by authors) distributed under the terms and condition of the Creative Commons Attribution (CC BY) License.

which is a huge plus. (4) Probiotics boost the antioxidant capacity of the host. (5) Antioxidant metabolite levels in the host are increased by probiotics. (6) Probiotics can control signaling pathways. (7) Probiotics reduce the production of reactive oxygen species (ROS) by blocking the action of enzymes. (8) Intestinal microorganisms are affected by probiotics. When human subjects were subjected to intense physical stress, *Lacticaseibacillus rhamnosus* showed significant antioxidant activity. Athletes experiencing oxidative stress may benefit from taking *L. rhamnosus* due to its capacity to increase antioxidant levels and reduce the harmful impact of reactive oxygen species. Studies conducted over the past few decades have uncovered numerous mechanisms through which various strains of probiotic bacteria can exert antioxidant activity. However, there has been a dearth of reviews examining the evidence for probiotic antioxidant pathways [91].

8.7.5 Preventing Protozoal Infections and Harmful Element Levels in Hair

Use of probiotics may prevent or delay parasite invasion. Recent advances in the field of fecal microbiota transplantation (FMT) have made it a promising option for treating recurrent *Clostridium difficile* infection. It is possible that fecal microbiota transplantation will one day be used to treat neuropsychiatric diseases in addition to inflammatory bowel disorder, type 2 diabetes, and metabolic disease. Donor excrement or feces is combined with a saline or other solution, strained, and infused into a recipient during FMT [92].

Toxins are typically removed from the body with the help of chelating agents like ethylene diamine tetra acetic acid (EDTA), meso-2,3-dimercaptosuccinic acid (DMSA), and 2,3-dimercapto-1-propanesulfonic acid (DMPS). However, it may cause significant side effects such as dehydration, failure of the liver and kidney, hypoglycemia, and coagulation insufficiency, and may lower levels of other essential minerals (such as Ca, Fe, and Zn). Evidence suggests that *Limosilactobacillus reuteri* Cadmium (Cd)70-13 and Led (Pb)71-1 may adsorb Cd (25% removal) and Pb (59% removal) from Rogosa and Sharpe (MRS) culture mediums, respectively. This finding extends previous *in vitro* findings that show that *Lactobacillus gasseri* and *Limosilactobacillus reuteri* can bind dangerous ions like Cd. *Lacticaseibacillus rhamnosus* may be able to remove Al3 and Cd2 from water by using its exopolysaccharide (EPS). Copper and lead are also removed from water by *Bifidobacterium logum* (*B. longum*). To sum up the function of probiotics in lowering hair toxicity, fewer patients had abnormal concentrations of Hg and Be after taking probiotics. Taking probiotics for a minimum of six months has been shown to dramatically reduce Hg and Be levels [93].

8.8 Conclusion and Recommendation

Due to the spread of multidrug-resistant infections, the prevalence of serious adverse effects, the development of resistance, and the depletion of antibiotics in the environment, it is vital that new methods be developed to battle deadly germs. In addition, noncommunicable diseases and those with limited therapeutic choices require beneficial or antagonistic microorganisms worldwide, especially in wealthier countries. In order to combat this issue, probiotics have been identified as one of the most promising biotherapies due to their ability to increase specificity, control the release of antimicrobial agents, and reduce the likelihood of antibiotic resistance. They have been shown to have host immune-modulatory effects, which makes them a potentially useful therapeutic and preventative measure against a wide range of disorders, including inflammatory diseases. Probiotics are strains of bacteria that have been carefully chosen for their ability to (1) benefit the host, (2) survive the journey through the intestines, (3) bind to the intestinal epithelial cell membrane, (4) produce antibiotic compounds to fight infections, and (5) stabilize the intestinal microbiota. Because probiotic effects vary with dose, setting, and strain, knowing the species and genus of a probiotic is crucial for eliciting the desired effects in the host. Although probiotics have been proven to have favorable effects on health, unique microbes created from our microbial symbionts need further investigation into the systemic pathways and components underlying the beneficial functions. Researchers from the fields of microbiology and pharmacology are encouraged to create modified probiotics in the near future to treat diseases that cannot be managed with existing conventional drugs.

Acknowledgments

Dr. Yugal Kishore Mohanta and Mr. Jibanjyoti Panda are highly indebted and give their sincere thanks to SERB-DST, Government of India for financial support for the research facilities (Nano-biotechnology and Translational Knowledge Laboratory) through research grant no. SRG/2022/000641.

References

1. Pancu, D.F., Scurtu, A., Macasoi, I.G., Marti, D., Mioc, M., Soica, C., Coricovac, D., Horhat, D., Poenaru, M., Dehelean, C., Antibiotics: Conventional Therapy and Natural Compounds with Antibacterial Activity-A Pharmaco-Toxicological Screening. *Antibiotics (Basel, Switzerland)*, 10, 4, 401, 2021.
2. Patangia, D.V., Anthony Ryan, C., Dempsey, E., Paul Ross, R., Stanton, C., Impact of antibiotics on the human microbiome and consequences for host health. *Microbiologyopen*, 11, 1, e1260, 2022.
3. Noce, A., Marrone, G., Di Daniele, F., Ottaviani, E., Jones, G.W., Bernini, R., Romani, A., Rovella, V., Impact of gut microbiota composition on onset and progression of chronic non-communicable diseases. *Nutrients*, 11, 5, 1073, 2019.
4. Sarkar, A., Yoo, J.Y., Dutra, S.V.O., Morgan, K.H., Groer, M., The association between early-life gut microbiota and long-term health and diseases. *J. Clin. Med.*, 10, 3, 1–24, 2021.
5. Banik, S., Nath, P.C., Roy, R., Microbiome and gut-brain axis affecting stress behavior. *Am. J. Appl. Bio-Technol. Res.*, 3, 4, 17–34, 2023.
6. Yang, J., Pu, J., Lu, S., Bai, X., Wu, Y., Jin, D., Cheng, Y., Zhang, G., Zhu, W., Luo, X., Rosselló-Móra, R., Xu, J., Species-Level Analysis of Human Gut Microbiota With Metataxonomics. *Front. Microbiol.*, 11, 3389, 2020.
7. Mazloom, K., Siddiqi, I., Covasa, M., Probiotics: How effective are they in the fight against obesity? *Nutrients*, 11, 2, 258, 2019.
8. Ferchichi, M., Sebei, K., Boukerb, A.M., Karray-Bouraoui, N., Chevalier, S., Feuilloley, M.G.J., Connil, N., Zommiti, M., Enterococcus spp.: Is it a bad choice for a good use—a conundrum to solve? *Microorganisms*, 9, 11, 2222, 2021.
9. Zheng, D., Liwinski, T., Elinav, E., Interaction between microbiota and immunity in health and disease. *Cell Res.*, 30, 6, 492–506, 2020.
10. Teneva, D. and Denev, P., Biologically Active Compounds from Probiotic Microorganisms and Plant Extracts Used as Biopreservatives. *Microorganisms*, 11, 8, 1896, 2023.
11. Borse, S., Singh, D., Upadhyay, D., Sharma, V., Nivsarkar, M., Probiotic use in the management of hypertension: A new era of therapeutic management. *Indian J. Health Sci. Biomed. Res.*, 11, 3, 207, 2018.
12. Herath, M., Hosie, S., Bornstein, J.C., Franks, A.E., Hill-Yardin, E.L., The Role of the Gastrointestinal Mucus System in Intestinal Homeostasis: Implications for Neurological Disorders. *Front. Cell. Infect. Microbiol.*, 10, 00248, 2020.
13. Tegegne, B.A. and Kebede, B., Probiotics, their prophylactic and therapeutic applications in human health development: A review of the literature. *Heliyon*, 8, 6, e09725, 2022.
14. Binda, S., Hill, C., Johansen, E., Obis, D., Pot, B., Sanders, M.E., Tremblay, A., Ouwehand, A.C., Criteria to Qualify Microorganisms as "Probiotic" in Foods and Dietary Supplements. *Front. Microbiol.*, 11, 1662, 2020.

15. Markowiak-Kopeć, P. and Śliżewska, K., The Effect of Probiotics on the Production of Short-Chain Fatty Acids by Human Intestinal Microbiome. *Nutrients*, 12, 4, 1107, 2020.
16. Zheng, J., Wittouck, S., Salvetti, E., Franz, C.M.A.P., Harris, H.M.B., Mattarelli, P., O'toole, P.W., Pot, B., Vandamme, P., Walter, J., Watanabe, K., Wuyts, S., Felis, G.E., Gänzle, M.G., Lebeer, S., A taxonomic note on the genus Lactobacillus: Description of 23 novel genera, emended description of the genus Lactobacillus beijerinck 1901, and union of Lactobacillaceae and Leuconostocaceae. *Int. J. Syst. Evol. Microbiol.*, 70, 4, 2782–2858, 2020.
17. Plaza-Diaz, J., Ruiz-Ojeda, F.J., Gil-Campos, M., Gil, A., Mechanisms of Action of Probiotics. *Adv. Nutr.*, 10, suppl_1, S49–S66, 2019.
18. Milner, E., Stevens, B., An, M., Lam, V., Ainsworth, M., Dihle, P., Stearns, J., Dombrowski, A., Rego, D., Segars, K., Utilizing Probiotics for the Prevention and Treatment of Gastrointestinal Diseases. *Front. Microbiol.*, 12, 689958, 2021.
19. Latif, A., Shehzad, A., Niazi, S., Zahid, A., Ashraf, W., Iqbal, M.W., Rehman, A., Riaz, T., Aadil, R.M., Khan, I.M., Özogul, F., Rocha, J.M., Esatbeyoglu, T., Korma, S.A., Probiotics: mechanism of action, health benefits and their application in food industries. *Front. Microbiol.*, 14, 1216674, 2023.
20. Tsaousis, G.N., Papadopoulou, E., Apessos, A., Agiannitopoulos, K., Pepe, G., Kampouri, S., Diamantopoulos, N., Floros, T., Iosifidou, R., Katopodi, O., Koumarianou, A., Markopoulos, C., Papazisis, K., Venizelos, V., Xanthakis, I., Xepapadakis, G., Banu, E., Eniu, D.T., Negru, S., Stanculeanu, D.L., Ungureanu, A., Ozmen, V., Tansan, S., Tekinel, M., Yalcin, S., Nasioulas, G., Analysis of hereditary cancer syndromes by using a panel of genes: novel and multiple pathogenic mutations. *BMC Cancer*, 19, 1, 535, 2019.
21. Śliżewska, K., Markowiak-Kopeć, P., Śliżewska, W., The Role of Probiotics in Cancer Prevention. *Cancers (Basel)*, 13, 1, 20, 2020.
22. Nazir, Y., Hussain, S.A., Abdul Hamid, A., Song, Y., Probiotics and Their Potential Preventive and Therapeutic Role for Cancer, High Serum Cholesterol, and Allergic and HIV Diseases. *Biomed. Res. Int.*, 2018, 3428437, 2018.
23. Bourrie, B.C.T., Willing, B.P., Cotter, P.D., The Microbiota and Health Promoting Characteristics of the Fermented Beverage Kefir. *Front. Microbiol.*, 7, 00647, 2016.
24. Nami, Y., Haghshenas, B., Haghshenas, M., Abdullah, N., Khosroushahi, A.Y., The Prophylactic effect of probiotic Enterococcus lactis IW5 against different human cancer cells. *Front. Microbiol.*, 6, 1317, 2015.
25. Liu, Z.H., Huang, M.J., Zhang, X.W., Wang, L., Huang, N.Q., Peng, H., Lan, P., Peng, J.S., Yang, Z., Xia, Y., Liu, W.J., Yang, J., Qin, H.L., Wang, J.P., The effects of perioperative probiotic treatment on serum zonulin concentration and subsequent postoperative infectious complications after colorectal cancer surgery: A double-center and double-blind randomized clinical trial. *Am. J. Clin. Nutr.*, 97, 1, 117–126, 2013.

26. El-Nezami, H.S., Polychronaki, N.N., Ma, J., Zhu, H., Ling, W., Salminen, E.K., Juvonen, R.O., Salminen, S.J., Poussa, T., Mykkänen, H.M., Probiotic supplementation reduces a biomarker for increased risk of liver cancer in young men from Southern China. *Am. J. Clin. Nutr.*, 83, 5, 1199–1203, 2006.
27. Toi, M., Hirota, S., Tomotaki, A., Sato, N., Hozumi, Y., Anan, K., Nagashima, T., Tokuda, Y., Masuda, N., Ohsumi, S., Ohno, S., Takahashi, M., Hayashi, H., Yamamoto, S., Ohashi, Y., Probiotic Beverage with Soy Isoflavone Consumption for Breast Cancer Prevention: A Case-control Study. *Curr. Nutr. Food Sci.*, 9, 3, 194–200, 2013.
28. Ohashi, Y., Nakai, S., Tsukamoto, T., Masumori, N., Akaza, H., Miyanaga, N., Kitamura, T., Kawabe, K., Kotake, T., Kuroda, M., Naito, S., Koga, H., Saito, Y., Nomata, K., Kitagawa, M., Aso, Y., Habitual intake of lactic acid bacteria and risk reduction of bladder cancer. *Urol. Int.*, 68, 4, 273–280, 2002.
29. Ma, E.L., Choi, Y.J., Choi, J., Pothoulakis, C., Rhee, S.H., Im, E., The anti-cancer effect of probiotic Bacillus polyfermenticus on human colon cancer cells is mediated through ErbB2 and ErbB3 inhibition. *Int. J. Cancer*, 127, 4, 780–790, 2010.
30. Rebersek, M., Gut microbiome and its role in colorectal cancer. *BMC Cancer*, 21, 1, 1325, 2021.
31. Górska, A., Przystupski, D., Niemczura, M.J., Kulbacka, J., Probiotic Bacteria: A Promising Tool in Cancer Prevention and Therapy. *Curr. Microbiol.*, 76, 8, 939–949, 2019.
32. Banerjee, D., Jain, T., Bose, S., Bhosale, V., Importance of probiotics in human health. *Funct. Food Hum. Health*, 539–554, 2018.
33. Sharma, A., Importance of probiotics in cancer prevention and treatment, in: *Recent Developments in Applied Microbiology and Biochemistry*, pp. 33–45, Elsevier, Academic Press, Netherlands, 2018.
34. Eslami, M., Abdolshahi, A., Emadi, A., Yousefi, B., Importance of Probiotics in the Prevention and Treatment of Covid-19. *J. Microbiol. Biotechnol. Food Sci.*, 12, 3, 17127–17143, 2022.
35. Libertucci, J. and Young, V.B., The role of the microbiota in infectious diseases. *Nat. Microbiol.*, 4, 1, 35–45, 2019.
36. Ranjbar, S., Seyednejad, S.A., Azimi, H., Rezaeizadeh, H., Rahimi, R., Emerging Roles of Probiotics in Prevention and Treatment of Breast Cancer: A Comprehensive Review of Their Therapeutic Potential. *Nutr. Cancer*, 71, 1, 1–12, 2019.
37. Molska, M. and Reguła, J., Potential mechanisms of probiotics action in the prevention and treatment of colorectal cancer. *Nutrients*, 11, 10, 2453, 2019.
38. Dasari, S., Kathera, C., Janardhan, A., Praveen Kumar, A., Viswanath, B., Surfacing role of probiotics in cancer prophylaxis and therapy: A systematic review. *Clin. Nutr.*, 36, 6, 1465–1472, 2017.
39. Sharaf, L.K., Sharma, M., Chandel, D., Shukla, G., Prophylactic intervention of probiotics (L.acidophilus, L.rhamnosus GG) and celecoxib modulate

Bax-mediated apoptosis in 1,2-dimethylhydrazine-induced experimental colon carcinogenesis. *BMC Cancer*, 18, 1, 1111, 2018.

40. Qorri, B., Harless, W., Szewczuk, M.R., Novel molecular mechanism of aspirin and celecoxib targeting Mammalian neuraminidase-1 impedes epidermal growth factor receptor signaling axis and induces apoptosis in pancreatic cancer cells. *Drug Des. Dev. Ther.*, 14, 4149–4167, 2020.
41. Van Baal, M.C., Van Rens, M.J., Geven, C.B., Van De Pol, F.M., Van Den Brink, I.W., Hannink, G., Nagtegaal, I.D., Peters, W.H., Rijkers, G.T., Gooszen, H.G., Association between probiotics and enteral nutrition in an experimental acute pancreatitis model in rats. *Pancreatology*, 14, 6, 470–477, 2014.
42. Tewari, V.V., Dubey, S.K., Gupta, G., Bacillus clausii for Prevention of Late-onset Sepsis in Preterm Infants: A Randomized Controlled Trial. *J. Trop. Pediatr.*, 61, 5, 377–384, 2015.
43. Rodriguez, H. and Miller, J.E., Do prophylactic probiotics prevent the incidence of Clostridium difficile colitis infection in hospitalized patients? *J. Okla. State Med. Assoc.*, 112, 1, 18–19, 2019.
44. Franko, J., Raman, S., Krishnan, N., Frankova, D., Tee, M.C., Brahmbhatt, R., Goldman, C.D., Weigel, R.J., Randomized Trial of Perioperative Probiotics Among Patients Undergoing Major Abdominal Operation. *J. Am. Coll. Surg.*, 229, 6, 533–540.e1, 2019.
45. DeVeaux, A., Ryou, J., Dantas, G., Warner, B.B., Tarr, P.I., Microbiome-targeting therapies in the neonatal intensive care unit: safety and efficacy. *Gut Microbes*, 15, 1, 2221758, 2023.
46. Dong, Y., Glaser, K., Speer, C.P., Late-onset sepsis caused by Gram-negative bacteria in very low birth weight infants: a systematic review. *Expert Rev. Anti. Infect. Ther.*, 17, 3, 177–188, 2019.
47. Underwood, M.A., Umberger, E., Patel, R.M., Safety and efficacy of probiotic administration to preterm infants: ten common questions. *Pediatr. Res.*, 88, Suppl 1, 48–55, 2020.
48. Zbinden, A., Zbinden, R., Berger, C., Arlettaz, R., Case series of Bifidobacterium longum bacteremia in three preterm infants on probiotic therapy. *Neonatology*, 107, 1, 56–59, 2015.
49. Beghetti, I., Panizza, D., Lenzi, J., Gori, D., Martini, S., Corvaglia, L., Aceti, A., Probiotics for Preventing Necrotizing Enterocolitis in Preterm Infants: A Network Meta-Analysis. *Nutrients*, 13, 1, 192, 2021.
50. Iqbal, Z., Ahmed, S., Tabassum, N., Bhattacharya, R., Bose, D., Role of probiotics in prevention and treatment of enteric infections: a comprehensive review. *3 Biotech.*, 11, 5, 242, 2021.
51. Sun, Y.-C., Wang, C.-Y., Wang, H.-L., Yuan, Y., Lu, J.-H., Zhong, L., Probiotic in the prevention of ventilator-associated pneumonia in critically ill patients: evidence from meta-analysis and trial sequential analysis of randomized clinical trials. *BMC Pulm. Med.*, 22, 1, 168, 2022.

52. Han, S., Lu, Y., Xie, J., Fei, Y., Zheng, G., Wang, Z., Liu, J., Lv, L., Ling, Z., Berglund, B., Yao, M., Li, L., Probiotic Gastrointestinal Transit and Colonization After Oral Administration: A Long Journey. *Front. Cell. Infect. Microbiol.*, 11, 609722, 2021.
53. Yang, T., Santisteban, M.M., Rodriguez, V., Li, E., Ahmari, N., Carvajal, J.M., Zadeh, M., Gong, M., Qi, Y., Zubcevic, J., Sahay, B., Pepine, C.J., Raizada, M.K., Mohamadzadeh, M., Gut dysbiosis is linked to hypertension. *Hypertension (Dallas, Tex. 1979)*, 65, 6, 1331–1340, 2015.
54. Song, H.Y., Zhou, L., Liu, D.Y., Yao, X.J., Li, Y., What roles do probiotics play in the eradication of helicobacter pylori? current knowledge and ongoing research. *Gastroenterol. Res. Pract.*, 2018, 9379480, 2018.
55. Zhao, Y., Dong, B.R., Hao, Q., Probiotics for preventing acute upper respiratory tract infections. *Cochrane Database Syst. Rev.*, 2022, 8, CD006895, 2022.
56. Roberts, L.T., Issa, P.P., Sinnathamby, E.S., Granier, M., Mayeux, H., Eubanks, T.N., Malone, K., Ahmadzadeh, S., Cornett, E.M., Shekoohi, S., Kaye, A.D., Helicobacter Pylori: A Review of Current Treatment Options in Clinical Practice. *Life (Basel, Switzerland)*, 12, 12, 2038, 2022.
57. Chong, P.P., Chin, V.K., Looi, C.Y., Wong, W.F., Madhavan, P., Yong, V.C., The Microbiome and Irritable Bowel Syndrome - A Review on the Pathophysiology, Current Research and Future Therapy. *Front. Microbiol.*, 10, 1136, 2019.
58. Qiu, P., Ishimoto, T., Fu, L., Zhang, J., Zhang, Z., Liu, Y., The Gut Microbiota in Inflammatory Bowel Disease. *Front. Cell. Infect. Microbiol.*, 12, 733992, 2022.
59. Banoth, D., Wali, M.H., Bekova, K., Abdulla, N., Gurugubelli, S., Lin, Y.M., Khan, S., The Role of Oral Probiotics in Alleviating Inflammation, Symptom Relief, and Postoperative Recurrence and Their Side Effects in Adults With Crohn's Disease: A Systematic Review. *Cureus*, 15, 12, e50901, 2023.
60. Dempsey, E. and Corr, S.C., Lactobacillus spp. for Gastrointestinal Health: Current and Future Perspectives. *Front. Immunol.*, 13, 840245, 2022.
61. Kesavelu, D. and Jog, P., Current understanding of antibiotic-associated dysbiosis and approaches for its management. *Ther. Adv. Infect. Dis.*, 10, 20499361231154444, 2023.
62. Kopacz, K. and Phadtare, S., Probiotics for the Prevention of Antibiotic-Associated Diarrhea. *Healthcare (Basel, Switzerland)*, 10, 8, 1450, 2022.
63. Hu, L., Zhou, M., Young, A., Zhao, W., Yan, Z., *In vivo* effectiveness and safety of probiotics on prophylaxis and treatment of oral candidiasis: A systematic review and meta-analysis. *BMC Oral. Health*, 19, 1, 140, 2019.
64. Mundula, T., Ricci, F., Barbetta, B., Baccini, M., Amedei, A., Effect of Probiotics on Oral Candidiasis: A Systematic Review and Meta-Analysis. *Nutrients*, 11, 10, 2449, 2019.
65. Xie, Y., Zhou, G., Xu, Y., He, B., Wang, Y., Ma, R., Chang, Y., He, D., Xu, C., Xiao, Z., Effects of Diet Based on IgG Elimination Combined with Probiotics

on Migraine Plus Irritable Bowel Syndrome. *Pain Res. Manage.*, 2019, 7890461, 2019.

66. Rabiee, M.R. and Babajafari, S., Probiotics and diabetes: A Review. *Int. J. Nutr. Sci.*, 3, 2, 73–81, 2018.
67. Vyas, N., Nair, S., Rao, M., Miraj, S.S., Childhood Obesity and Diabetes: Role of Probiotics and Prebiotics. *Glob. Perspect. Child. Obes.*, 363–376, 2019.
68. Mishra, S., Wang, S., Nagpal, R., Miller, B., Singh, R., Taraphder, S., Yadav, H., Probiotics and prebiotics for the amelioration of type 1 diabetes: Present and future perspectives. *Microorganisms*, 7, 3, 67, 2019.
69. Caesar, R., Pharmacologic and Nonpharmacologic Therapies for the Gut Microbiota in Type 2 Diabetes. *Can. J. Diabetes*, 43, 3, 224–231, 2019.
70. Okesene-Gafa, K.A.M., Brown, J., Mccowan, L., Crowther, C.A., Probiotics for treating women with gestational diabetes for improving maternal and fetal health and well-being. *Cochrane Database Syst. Rev.*, 2018, 2, CD012970, 2018.
71. Sun, Z., Sun, X., Li, J., Li, Z., Hu, Q., Li, L., Hao, X., Song, M., Li, C., Using probiotics for type 2 diabetes mellitus intervention: Advances, questions, and potential. *Crit. Rev. Food Sci. Nutr.*, 60, 4, 670–683, 2020.
72. Ayesha, I.E., Monson, N.R., Klair, N., Patel, U., Saxena, A., Patel, D., Venugopal, S., Probiotics and Their Role in the Management of Type 2 Diabetes Mellitus (Short-Term Versus Long-Term Effect): A Systematic Review and Meta-Analysis. *Cureus*, 15, 10, e46741, 2023.
73. Del Toro-Barbosa, M., Hurtado-Romero, A., Garcia-Amezquita, L.E., García-Cayuela, T., Psychobiotics: Mechanisms of Action, Evaluation Methods and Effectiveness in Applications with Food Products. *Nutrients*, 12, 12, 3896, 2020.
74. Cristofori, F., Dargenio, V.N., Dargenio, C., Miniello, V.L., Barone, M., Francavilla, R., Anti-Inflammatory and Immunomodulatory Effects of Probiotics in Gut Inflammation: A Door to the Body. *Front. Immunol.*, 12, 578386, 2021.
75. El Khatib, R., Karam-Sarkis, D., Waligora-Dupriet, A.-J., Butel, M.-J., Could Gut Modulation through Probiotic Supplementation Be Beneficial in Autism Spectrum Disorder?, in: *Prebiotics and Probiotics-Potential Benefits in Human Nutrition and Health*, 2020.
76. Morshedi, M., Hashemi, R., Moazzen, S., Sahebkar, A., Hosseinifard, E.S., Immunomodulatory and anti-inflammatory effects of probiotics in multiple sclerosis: a systematic review. *J. Neuroinflammation*, 16, 1, 231, 2019.
77. Choi, H.-J., Recent advances on next generation probiotics linked to the gut microbiome. *Food Sci. Ind.*, 52, 3, 261–271, 2019.
78. Kumar, R., Sood, U., Gupta, V., Singh, M., Scaria, J., Lal, R., Recent Advancements in the Development of Modern Probiotics for Restoring Human Gut Microbiome Dysbiosis. *Indian J. Microbiol.*, 60, 1, 12–25, 2020.

79. Chen, C., Wen, T., Zhao, Q., Probiotics Used for Postoperative Infections in Patients Undergoing Colorectal Cancer Surgery. *BioMed. Res. Int.*, 2020, 5734718, 2020.
80. Pan, L.L., Li, J., Shamoon, M., Bhatia, M., Sun, J., Corrigendum: Recent advances on nutrition in treatment of Acute Pancreatitis. *Front. Immunol.*, 9, 849, 2018, DOI: 10.3389/fimmu.2017.00762.
81. Van Den Nieuwboer, M. and Claassen, E., Dealing with the remaining controversies of probiotic safety. *Benef. Microbes*, 10, 6, 605–616, 2019.
82. Kim, S., Eun, H.S., Jo, E.K., Roles of autophagy-related genes in the pathogenesis of inflammatory bowel disease. *Cells*, 8, 1, 77, 2019.
83. Zaylaa, M., Al Kassaa, J.A.I., Boutillier, V.P.D., Desramaut, J., Rosenstiel, P., Nguyen, H.T.T., Dabboussi, F., Pot, B., Grangette, C., Autophagy: A novel mechanism involved in the anti-inflammatory abilities of probiotics. *Cell Physiol. Biochem.*, 53, 5, 774–793, 2019.
84. El Hadad, J., Schreiner, P., Vavricka, S.R., Greuter, T., The Genetics of Inflammatory Bowel Disease. *Mol. Diagn. Ther.*, 28, 1, 27–35, 2024.
85. Fujimoto, K. and Uematsu, S., *Innate immunity at mucosal surfaces*, Elsevier, Academic Press, Netherlands, 2019.
86. Alard, J., Peucelle, V., Boutillier, D., Breton, J., Kuylle, S., Pot, B., Holowacz, S., Grangette, C., New probiotic strains for inflammatory bowel disease management identified by combining *in vitro* and *in vivo* approaches. *Benef. Microbes*, 9, 2, 317–331, 2018.
87. Hwang, I.Y. and Chang, M.W., Engineering commensal bacteria to rewire host–microbiome interactions. *Curr. Opin. Biotechnol.*, 62, 116–122, 2020.
88. Baral, K.C., Bajracharya, R., Lee, S.H., Han, H.-K., Advancements in the Pharmaceutical Applications of Probiotics: Dosage Forms and Formulation Technology. *Int. J. Nanomed.*, 16, 7535–7556, 2021.
89. Chua, K.J., Kwok, W.C., Aggarwal, N., Sun, T., Chang, M.W., Designer probiotics for the prevention and treatment of human diseases. *Curr. Opin. Chem. Biol.*, 40, 8–16, 2017.
90. Sanap, D., Garje, M., Godge, G., Probiotics, their Health Benefits and Applications for Development of Human Health: A Review Introduction. *J. Drug Deliv. Ther.*, 9, 4-s, 631–640, 2019.
91. Wang, Y., Wu, Y., Wang, Y., Xu, H., Mei, X., Yu, D., Wang, Y., Li, W., Antioxidant properties of probiotic bacteria. *Nutrients*, 9, 5, 521, 2017.
92. Stensvold, C.R. and van der Giezen, M., Associations between Gut Microbiota and Common Luminal Intestinal Parasites. *Trends Parasitol.*, 34, 5, 369–377, 2018.
93. Fang, S.T., Cheng, D.E., Huang, Y.T., Hsu, T.Y., Horng Shing Lu, H., A Pilot Study of the Influence of Probiotics on Hair Toxic Element Levels After Long-Term Supplement with Different Lactic Acid Bacteria Strains. *J. Probiotics Health*, 06, 01, 1000203, 2018.

9

Probiotics in the Prevention and Treatment of Psoriasis

Prativa Biswasroy[1], Deepak Pradhan[1], Dilip Kumar Pradhan[2], Goutam Rath[1*] and Goutam Ghosh[1†]

[1]*Department of Herbal Nanotechnology, School of Pharmaceutical Sciences, Siksha O Anusandhan (Deemed to be University), Bhubaneswar, Odisha, India*
[2]*Department of Medicine, Pandit Raghunath Murmu Medical College and Hospital, Baripada, Odisha, India*

Abstract

Psoriasis is a T-cell-mediated chronic inflammatory skin disease with pathogenic autoimmune features associated with hyperproliferation (uncontrolled keratinocyte growth) and poor epidermal keratinocyte differentiation. The underlying issue of psoriasis is clinically unidentified, affecting an estimated 2%–3% of the global population. Besides the triggers that influence an autoimmune response, recent evidence suggests that skin and gut flora compositions are significantly associated with inflammatory response and cause clinical severity in diseases like psoriasis. Despite its overall therapeutic strategy, clinical limitations confine its utility. Currently, there seems to be a demand for natural products as alternatives to synthetic drugs that enhance the existence of a gut–skin axis, which could reflect a novel treatment method for psoriatic patients. In this regard, oral supplements like probiotics might be simple, reliable, and inexpensive for preventing and treating skin inflammation. Very few studies were conducted to investigate how probiotic supplements such as *Lactobacillus pentosus* GMNL-77, *Lactobacillus sakei* proBio-65, *Lactobacillus sporogenes*, and others affect gut flora and regulate the immune response. Several types of research indicate that the *in vivo* effect of a potentially probiotic strain in imiquimod-induced psoriasis-like skin inflammation in BALB/c mice significantly reduced proinflammatory cytokines such as tumor necrosis factor-alpha, interleukin (IL)-6, and IL23/IL-17A axis-associated

**Corresponding author*: goutamrath@soa.ac.in
†*Corresponding author*: goutamghosh@soa.ac.in

Hrudayanath Thatoi, Sonali Mohapatra, Swagat Kumar Das and Sukanta Kumar Pradhan (eds.) Applied Biotechnology and Bioinformatics: Agriculture, Pharmaceutical Research and Environment, (217–234)

cytokines. Furthermore, it reduced the amount of IL-17- and IL-22-producing CD4 T cells in the splenocytes of the imiquimod-treated group. Moreover, evidence from preclinical and clinical studies reveals that probiotics reduce the psoriasis area and severity index (PASI), inhibit the inflammatory reaction, regulate the immune system, and control microbial diversity. This chapter outlines the scientific evidence regarding the microbiome of psoriatic patients, as well as deep insights into microbiome-targeted potential treatments for psoriasis.

***Keywords*:** Psoriasis, microbiome, gut–skin axis, probiotics, inflammation

9.1 Introduction

Psoriasis is a potentially life-threatening immune-mediated inflammatory skin condition marked by scaly-red patches, hyperactivation of T cells, and hyperproliferation of keratinocytes, with a global incidence of 2–3% [1]. As per clinicians, the underlying cause of psoriasis is unspecified, but it is figured to be a complicated autoimmune disease influenced mainly through internal and external signals (trauma, physical and mental stress, specific drugs, and infections), and has a significant impact on a patient's self-esteem [2]. In humans, abnormal dendritic cell (cDC) production in the epidermis causes complex relationships between the immune cells and keratinocytes, triggering a series of cascade inflammatory processes initiated by stimulation of T cell subsets (Th1and Th17), which release cytokines such as interleukin (IL)-1, 6, 17, 22, and 23, and tumor necrosis factor-alpha (TNF-α) [3]. Despite the current large inventory of therapeutic strategies, such as topical, systemic, and phototherapy, the drug hierarchy remains based on symptomatic treatments. The most clinical challenges associated with drug therapy are hepatotoxicity, nephrotoxicity, teratogenicity, and so on. Furthermore, a recent study revealed that dysbiosis of the gut microbiome, microbial production, intestinal immune cell migration, and systemic cytokine production disrupt skin immune homeostasis in individuals [4]. According to recent data, a lower number of specific bacteria, like *Faecalibacterium prausnitzii* and *Akkermansia muciniphila* (potentially manage short-chain fatty acid (SCFA), and Th-17 cellular function) cause dysbiosis of the microbial community that may aggravate psoriasis, as the interactions of various microbial population in the epidermis and intestinal microbiota (gut–skin axis) perform a significant role in the development of psoriasis [5]. To address these concerns, researchers investigated a possible connection between the bacterial communities of the intestines and the epidermis and its implications for psoriasis. For regulating microbial diversity and enhancing psoriasis manifestations,

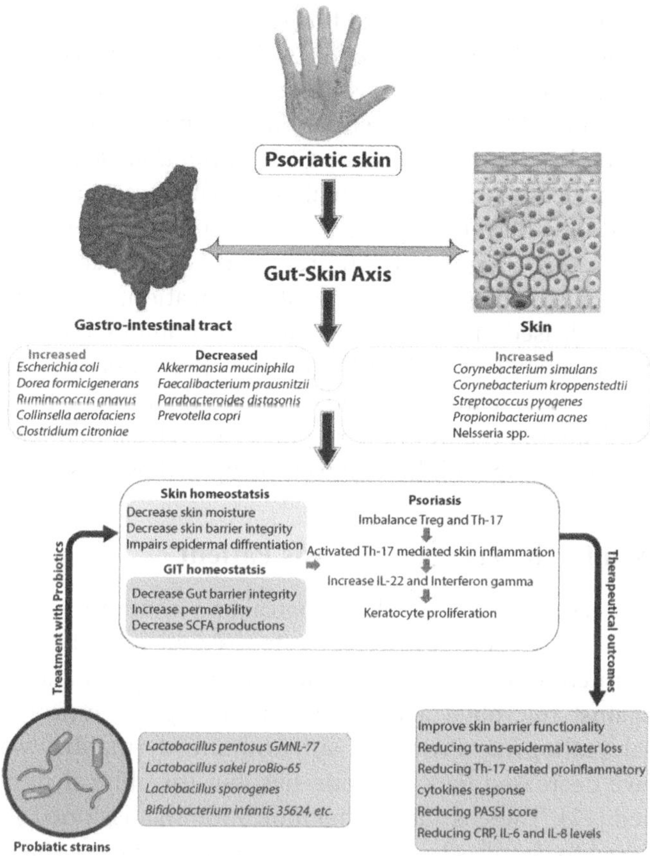

Figure 9.1 Gut–skin axis microbiota.

"probiotics" are an effective approach [6]. Evidence from preclinical and clinical trials illustrates that probiotics mitigate psoriasis area and severity index (PASI), suppress psoriasis inflammatory response, regulate immune cells, and control microbial community. The interaction of gut–skin axis microbiota is represented in Figure 9.1.

9.2 Interruption of the Microbiome: A Pathogenic Effect in Psoriasis

Psoriasis is thought to be caused primarily by abnormal activation of the microbial community blooms of pathogenic organisms that ultimately harm skin function and integrity observed in affected psoriasis patients with priority to the development of the gut–skin axis [7]. The by-products of the aromatic amino group, phenol, and p-cresol can reach the bloodstream and remain on the skin's surface, interrupting epidermal barrier integrity and epithelial cell differentiation [8]. These metabolites are generated by microbes (pathogenic) like *Clostridium difficile* and are thought to be a risk in the disturbing intestinal flora, which can have serious consequences [9]. Furthermore, elevated plasma p-cresol levels are linked to decreased moisture retention (skin area) and keratinization. As a result, intestinal dysbiosis increases epithelial permeation, causing cytotoxic T cells to be released and distorting their harmonious relationship with immunomodulatory T cells. The following is a discussion about current findings in the skin and gut microbiota, such as their interactions with cellular metabolism and the immune response, as well as their structure and functional relationship with human well-being and psoriasis.

9.2.1 Skin Dysbiosis in a Psoriasis Patient

Skin, which serves mostly as a physiological and immunological barrier to foreign threats, is closely restricted by its resident microbial community [10, 11]. Skin commensal bacteria shield the skin by enhancing immune systems against microbial agents and improving immune resilience to commensal pathogens. Psoriasis, autoimmune dermatitis, was discovered to have decreased barrier properties, but the clinical significance is unidentified [12]. There is substantial indication that lesioned skin in psoriasis is linked to changes in skin microflora, which play a vital role in the pathogenesis of this illness [13]. Repetitive skin illness due to microbiota dysbiosis is linked to the occurrence and prognosis of autoimmune diseases like psoriasis [14]. It was also identified that skin-resident innate b lymphocytes (ILCs) could control the microflora to maintain homeostasis in the sebaceous gland. Lack of ILCs causes sebaceous gland hyperplasia, which is followed by an increase in antibacterial lipids and a constraint of Gram-positive symbiotic organisms. Skin-resident microflora and epidermal inflammatory cytokines collaborate to maintain the epidermis in form and composition. Prolonged treatment with chemotherapeutic agents may result in the disruption of microflora diversity and triggering the systemic inflammatory response [15]. Furthermore, interruption of the structure and functions of the skin's resident microflora, like certain skin fungal and bacterial flora (*Candida albicans* and *Corynebacterium* species), upregulates Th17 cell in the skin that is statistically significant to psoriasis. Many research findings have demonstrated that the existence of Propionibacterium and Actinobacteria organisms in lower amounts in these patient populations causes a shift towards an immune disorder because the presence of these microbes generally fosters a Th2-mediated pathway, which further downregulates the Th1 response, which is closely linked to autoimmune actions [16]. Besides, other microorganism species like Acidobacteria, Streptococcaceae, Schlegelella, Proteobacteria, Rhodobacteraceae, Moraxellaceae, Campylobacteraceae, and Firmicutes are more abundant in psoriatic skin in comparison to normal skin. From the swab samples analysis report, it was observed that the total concentration of *Streptococcus* increases significantly in psoriasis patients as compared with healthy controls. *Streptococcus pneumoniae* super-antigen induces T lymphocytes to target skin-specific microorganisms via molecular receptors, leading to severe inflammatory diseases, according to research on the M protein of *Streptococcus* pyrogen, which shares homology with the 50-kDa type I epithelial tissue (keratin) [17]. As per the survey report, Chang *et al.* distinguish clinical manifestations (n = 26 healthy participants and 28 psoriatic lesioned participants) based on the amplification area of the 16S rRNA V1–V3 region. Remarkably, *Staphylococcus* was not significantly associated with any skin disease, whereas the quantity of *Staphylococcus* across all specimens is linked to a distinct disease condition. Highlighting the inter-microbe connection with several *Staphylococcus* that may contribute to the numerous microorganisms linked to healthy and psoriatic skin patients [18]. Assarsson *et al.* experimented on 30 plaque psoriasis patients treated with nbUVB therapies. The 16S rRNA gene V3–V4 was targeted for MiSeq sequencing in this study, and the analysis was performed using the CLC technique. Therapeutically, there was a rise in Gardnerella, Staphylococcus, Anaerococcus, Conchiformibius, Finegoldia, Prevotella, and Clostridium species whereas a decrease in Staphylococcus and Firmicutes species was observed in psoriasis [19]. Consequently, in another method developed by Assarsson *et al.* on the variable region (16S rRNA V3–V4), the amplification area indicates a larger number of bacterial organisms such as Firmicutes and Proteobacteria with a lower number of Fusobacteria and Cyanobacteria. The authors observed the variation of microbes in plaque psoriasis (n = 50) and healthy controls (n = 77) in these experimental studies and allowed to treat with UVB (narrowband ultraviolent B). Quan and his colleagues concluded that systemic inflammation was linked to a notable increase in Corynebacterium and a decrease in Cutibacterium by comparing 27 psoriasis patients to 19 healthy people using a variable region (16S rRNA V3–V4) [20]. Furthermore, Fyhrquist and fellow researchers identify the skin pathogen related to atopic dermatitis (n = 82), psoriasis (n = 119), and healthy individuals (n = 115), and revealed that a lower proportion of Corynebacterium members may perform a key role in the regulatory inflammatory response initiated in psoriasis [21]. Corynebacterium species have frequently been assigned as an opportunistic infection-causing microbial community, and *Corynebacterium kroppenstedtii* is more abundant among them and occurred in the skin flora of psoriasis patients compared with individuals, usually linked to human infectious agents, largely with granulomatous mastitis and breast inflammations [22].

After interpreting the majority of research on the relationship between skin microbial species and psoriasis, results in numerous kinds of literature indicate an increase in the abundance of Firmicutes, Actinobacteria, and Proteobacteria, triggering an inflammatory cascade. Additionally, there was not any evidence that the diversity of the bacterial communities on psoriatic lesioned skin was lower than on healthy skin.

9.2.2 Gut Dysbiosis in a Psoriasis Patient

A widespread perception governs that gut dysbiosis is linked to the development of psoriasis as well as other illnesses and is closely correlated with an impaired intestinal epithelial barrier [23]. According to current research, a decrease in intestinal microbiota resulting from various factors such as gut barrier integrity, stimulation of regulatory immune cells, hyperproduction of cytokine, and altered carbohydrate, cobalamin, and iron metabolism is linked to a higher Firmicutes-to-Bacteroidetes ratio (F/B ratio) and loss of certain bacteria species Ruminococcus and Megasphaera [24, 25]. According to Codoner and colleagues, there has been a significant rise in the population of Streptococcus, Bifidobacterium, Akkermansia, Akkermonsia spp., and Faecalibacterium and decreases in Bacteroides in the psoriatic patient. Experimentally, the author has taken 52 patients and compared them to healthy controls [26]. Scher *et al.* found that the psoriasis group's microbiome was less diverse than that of controls. Clinically, there was a difference in diversification between these 2 experiments, psoriasis ($n = 16$) and psoriatic arthritis ($n = 15$), which may be attributed to differences in specimen collection, DNA analysis, sequential library preparation, and/or data evaluation compared to 17 healthy controls [27]. Overall, there has been a decrease in Akkermansia, Ruminococcus, Akkermansia, and Coprococcus, with an increase in Bacteroidetes species. Chen *et al.* found a significantly high abundance of Ruminococcaceae and Lachnospiraceae with a lowered abundance of Bacteroidaceae and Prevotellaceae (family level). Further, the results show enriched Ruminococcus and Megasphaera (genus level), and reduced levels of Bacteroidetes (phylum level) in psoriasis patients. In clinical trials, 32 psoriasis and 64 non-psoriasis subjects were studied, and it was noticed that dysbiosis of intestinal flora suppressed cobalamin and iron transport [28]. Huang *et al.* studied 35 psoriasis patients and discovered that Bacteroidetes, Bacillus, Bacteroides, Sutterella, Lactococcus, and Parabacteroides are major contributors to microbiota dysregulation, whereas Firmicutes, Thermus, Streptococcus, Rothia, Granulicatella, and Gordonibacter are significant contributors to microbiota distribution in healthy subjects. The experimental findings also revealed that the Firmicutes/Bacteroidetes ratio was altered in psoriasis patients According to the results of Shapiro *et al.*'s experiment, psoriasis patients (n = 24) had significantly lower levels of Lachnospira, Collinsella, Sutterella, and Actinomyces species and significantly higher levels of microbial pathogens (Firmicutes, Blautia, Faecalibacterium, Ruminococcus, Prevotella, Coprococcus, and Christensen) than the control group (n = 22). The impairment of the intestinal barrier is linked to a steady increase in blood I-FABP (intestinal fatty acid receptor) and bacterial DNA polymerase scales [29].

However, survey results suggest that gut flora homeostasis is necessary to sustain skin barrier integrity by regulating the systemic immune response.

9.3 Therapeutic Effect of Probiotics for Psoriasis

Current studies have highlighted the impact of the gut–skin axis on the onset and progression of psoriasis, emphasizing the importance of using probiotics in disease management [30]. To enhance the clinical signs and pathogenesis of psoriasis, "probiotic" has a diverse range of impacts on homeostasis and immune regulation of both mucous membranes and systemic immunity by limiting microbial counts and antagonizing symbiotic relationships [31]. Despite the limited evidence on probiotic supplementation in the treatment of psoriasis, positive results have been reported. The majority of probiotic microorganisms are members of the lactate-synthesizing genera *Lactobacillus* and *Bifidobacterium*. Such genera, when consumed orally, regulate inflammatory mediators like IL-17 and -22 cells as well as anti-inflammatory Foxp3+ binding site T cells [32]. One of the evidence-based data is suggested by Chen and colleagues, in that orally ingesting *Lactobacillus pentosus* (*L. pentosus*) GMNL-77, a specific probiotic in BALB/c mice, produces interesting insights using the imiquimod (IMQ)-induced psoriasis approach. Experimental data show a significant result by alleviating erythematous scaling lesions, with epidermal hyperplasia, decreased proinflammatory associated cytokines (TNF-α, IL-6,23/17A, and IL-22), as well as diminished IL-17 and 22 generating CD4+ T cells in the spleen compared with the control group [33]. Irfan and coworkers show that an ethanolic extract (SEL001) prepared from the effective probiotic bacteria *L. sakei* proBio-65 has therapeutic potential in a mouse model of IMQ-treated psoriasis-like dermatitis. Six days of experimental work on ICR mice (n = 30) shows a reduction of IL-17A, IL-19, and IL-23 (psoriasis-related proinflammatory markers) and possesses a protective effect on erythema and scalability [34]. As per Lu and his team, seven different female BALB/c mouse groups (n = 60) were given a different probiotic strain (*L. reuteri* CCFM1132, *B. adolescentis* CCFM667, *L. paracasei* CCFM1074, and *B. breve* CCFM1078). Oral administration of strains increased short-chain fatty acid levels, which have been linked to cytokine production, as well as decreased histopathologic features by downregulating Th17 axis-related chemokines that cause inflammation. The variety of intestinal flora in a psoriasis-like murine model was observed to be reduced after 16S rRNA gene sequencing, but all effective strains altered the proportion of gut microbiome when particularly in comparison with ineffective strains [35]. The investigation by Ogawa and his research team reported that treating a *Leuconostoc mesenteroides* NTM048 for 21 days using an IMQ-induced mouse model reduced skin redness, scalability, and IL-17 levels, which may be related to plasma deoxycholic acid (DCA) levels obtained from the intestinal flora. A relatively high amount of Akkermansia and lowered Staphylococcus and Streptococcus indicate modifications in the intestinal flora [36]. Certain effective microbes, such as *Bifidobacteria infantis* 35624, have an effective immunoregulatory impact by modifying commensal–immune interconnection that has been demonstrated by Groeger *et al.* In this placebo-controlled study of psoriasis, the authors conclude that oral probiotic supplementation for 6–8 weeks shows a significant attenuation of TNF-α and IL-6 against healthy controls. Moreover, probiotic therapy was found to be effective in a patient with serious pustular psoriasis who had developed medication resistance like methotrexate (MTX), dapsone, and other medications [37]. Another clinical study conducted by Navarro-Lopez and his team of investigators claims to support the administration of a capsule of a probiotic strain containing a combination of three probiotic bacteria, *L. rhamnosus* CECT 8361, *B. lactis* CECT 8145, and *B. longum* CECT 7347 (ratio of 1:1:1) for 12 weeks to plaque psoriasis (18- to 70-year-old adults), which show a lower risk of relapse, i.e., PASI 75 in 66.7% in patients compared with that of the placebo group, 41.9% [38].

Moreover, a randomized controlled trial with 75 participants in Indonesia found that taking the probiotic strain *L. plantarum* IS-10506 orally at a dose of 2 ´ 10^{10} CFU for 12 weeks is an appropriate, reliable,

Table 9.1 Evidence of beneficial probiotic interventions in treating psoriasis.

Study design	Preclinical/clinical model	Probiotic	Dose/duration/ route	Outcome of interest	References
Preclinical model					
Mice treated with IMQ developed psoriasis lesioned skin	-BALB/c mice (20–25 g) -Age: 6–8 weeks	*L. pentosus GMNL-77*	Dose—(5×10^7 CFU/0.2 mL/day or 5×10^8 CFU/0.2 mL/day) Duration—for 6–7 days Route—Orally	-↓ Erythematous scaling - RT-PCR showed ↓ the level of mRNA including TNF-α, IL-6, 23, 17A/F, and 22 -↓ Th17/Th22 T cells and spleen weights	[33]
Protective effect of ethanol extract reduces mice treated with IMQ-developed psoriasis lesioned skin	-Thirty ICR mice -Age: 6 weeks	*L. sakei proBio-65*	Dose—(50 mg/cm^2 on back and 10 mg on the right ear). Duration—for 6–7 days Route—Topically	- ↓ Skin hardening, as well as improved erythema and scalability -↓ Inflammatory mediators IL-6, 23, 17A/F, and 22 and expression such as IL-19, 17A, and 23 mRNA - Improved clinical strategy	[34]

(Continued)

Table 9.1 Evidence of beneficial probiotic interventions in treating psoriasis. (*Continued*)

Study design	Preclinical/clinical model	Probiotic	Dose/duration/ route	Outcome of interest	References
Mice treated with IMQ developed psoriasis lesioned skin	-BALB/c mice -Age: 6–8 weeks	*E. coli Nissle* 1917	Dose—(-). Duration—for 24 days Route—Orally	-↑ lesioned skin and ↓skin surface area - ↑ clinical strategy -↓ serum IL-8, IL-23, IL-10, IL-17A, IL-17F, IL-23, and TNF-α with ↑ serum IL-10	[40]
Clinical model					
Purpose of taking 3 months completely random with double-blind to assess the effectiveness of a probiotic mixture in reducing the extent of psoriasis.	-Adults -Age: 18–70 years old	*B. longum* CECT 7347, *B. lactis* CECT 8145 and *L. rhamnosus* CECT 8361	Dose—(1×10^9 CFU/capsule formulated on maltodextrin) Duration—for 7 days Route—Orally	-3 months completely follow-up (probiotic group - 66.7% ↓ in PASI whereas, 41.9% is in the placebo group)	[41]

(*Continued*)

Table 9.1 Evidence of beneficial probiotic interventions in treating psoriasis. (*Continued*)

Study design	Preclinical/clinical model	Probiotic	Dose/duration/route	Outcome of interest	References
Female has pustular psoriasis everywhere around the body	-Female -Age: 47 years old	*L. sporogenes*	Dose—one sachet Duration—thrice daily with biotin 10 mg once daily Route—Orally	- Fever goes down with no lesion in fifteen days -↓ Skin inflammatory disease caused by T cells and improved detection of systemic inflammation	[42]
Random with double-blind	Psoriasis patients-26	*B. infantis* 35624	Dose—one sachet Duration—for 6–8 weeks Route—Orally	- Regulates host inflammatory reactions - Significantly↓ plasma CRP and IL-17A, IL-19, IL-23 in both gastro and non-gastrointestinal inflammatory disorders	[37]

Abbreviations: *L. pentosus—Lactobacillus, E. coli—Escherichia coli, B. longum—Bifidobacterium longum,* IL—interleukin, ↓ Decrease, ↑ Increase, PASI—Psoriasis Area and Severity Index.

Table 9.2 Clinical trial of probiotics for the treatment of psoriasis (https:// ClinicalTrials.gov/).

Probiotic	Subject/duration/route	Study title	Disease condition	Location	Study phase	Status	NCT no.
Dietary Supplement-Probiotic	Subject: 98 participants Duration: 12 weeks Route: orally (one capsule per day)	An intervention study to evaluate a probiotic in mild to moderate psoriasis patients	Psoriasis	Spain	No	Completed	NCT02576197
L. plantarum IS-10506	Subject: 75 participants Duration: 12 weeks Route: orally (dose of 2 ′ 10^{10} CFU)	Probiotic diet supplemented in Psoriasis Vulgaris: A Clinical Trial	Psoriasis vulgaris	Indonesia	Phase 1 Phase 2	Recruiting	NCT05254249
Dietary Supplement: Omnibiotic Stress Repair (OBSR)	Subject: 74 participants Duration: 3 months Route: orally (once daily, 3 g)	The MEDIPSA clinical trial	Psoriatic arthritis	Austria	No	Recruiting	NCT04588623
Dietary Supplement: Sublimated mare mil	Subject: 40 participants Duration: 3 months Route: orally	Sublimated mare milk consumption affects the intestinal flora in patients with psoriasis.	Psoriasis	Kazakhstan	No	Completed	NCT03594877

and cost-effective therapeutic option for psoriasis management, especially *Psoriasis vulgaris*. The administration of probiotics is controlled by altering the gut flora profile, interleukin mediators, and disease harshness in patients with psoriasis vulgaris [39]. Different varieties of preclinical and clinical studies of probiotics for the management of psoriasis are cited in Tables 9.1 and 9.2. Another case report in Kazakhstan explored 40 participants and allowed sublimated mare milk supplementation for 3 months for psoriasis patients. After three months of monitoring, the findings reveal a positive clinical relevance on the gut immunity and intestinal microbial proportions in patients with the psoriatic condition and healthy individuals.

Based on the survey, probiotics given by mouth or directly to the skin may exert a fruitful effect on the progression of dermatitis. Despite probiotics' beneficial effects, more clinical research is required to evaluate the success rate and long-term clinical applications of probiotics in the diagnosis of autoimmune diseases such as psoriasis.

9.4 Conclusion

The microbiome of skin and gut share many functional similarities; however, each area has its web of commensal bacteria counteracting the epidermal inflammatory processes. Proinflammatory markers like IL-23/IL-17 play an important role in both gut and skin inflammation. The gut flora can influence distant organ systems, along with skin via multifaceted inflammatory cascades. The skin and gut microbes of psoriasis patients differ markedly from those found in healthy volunteers. The meta-analysis approach signifies the alteration in the microbiome colonization of *C. albicans* and *Staphylococcus aureus* contributes a pathogenic role in the development of psoriasis. However, modulating the microflora population with probiotics may be an effective strategy for the therapeutic interventions of skin inflammatory conditions. Therefore, the researcher aims to demonstrate the multifaceted phenomenon involved in the gut–skin axis, with a specific emphasis on the pharmacological activity of long-term intestinal flora modulation via probiotic administration. The vast majority of probiotic strains contain lactic acid-yielding microbes that are neither pathogenic nor toxic. More than 70 scientific trials on foods containing microorganisms have been conducted to assess the potential risks of probiotic strains, with satisfactory results. Moreover, numerous flaws in the relevant work, as well as a lack of a summary of the aforementioned scientific studies were noticed. Meanwhile, a few clinical reports in different sampling, size estimation, blind technique, and research methodologies are insufficient to provide a concise database. Therefore, a potential comprehensive analysis of *in vivo* experimental (preclinical and clinical) evaluations of probiotic microorganisms in psoriasis therapy should be investigated. In this era of antibiotic resistance and side effects associated with chemotherapy, probiotics may offer a safe and well-tolerated therapeutic modality as an alternative approach.

References

1. Rendon, A. and Schäkel, K., Psoriasis Pathogenesis and Treatment. *Int. J. Mol. Sci.*, 20, 6, 1475, 2019.
2. Wintermann, G.-B., Bierling, A.L., Peters, E.M.J., Abraham, S., Beissert, S., Weidner, K., Childhood Trauma and Psychosocial Stress Affect Treatment Outcome in Patients With Psoriasis Starting a New Treatment Episode. *Front. Psychiatry*, 13, 848708, 2022.
3. Mohd Noor, A.A., Azlan, M., Mohd Redzwan, N., Orchestrated Cytokines Mediated by Biologics in Psoriasis and Its Mechanisms of Action. *Biomedicines*, 10, 2, 498, 2022.
4. Zheng, D., Liwinski, T., Elinav, E., Interaction between microbiota and immunity in health and disease. *Cell Res.*, 30, 6, 492–506, 2020.
5. Thye, A.Y.-K., Bah, Y.-R., Law, J.W.-F., Tan, L.T.-H., He, Y.-W., Wong, S.-H., Thurairajasingam, S., Chan, K.-G., Lee, L.-H., Letchumanan, V., Gut–Skin Axis: Unravelling the Connection between the Gut Microbiome and Psoriasis. *Biomedicines*, 10, 5, 1037, 2022.
6. Wang, X., Zhang, P., Zhang, X., Probiotics Regulate Gut Microbiota: An Effective Method to Improve Immunity. *Molecules*, 26, 19, 6076, 2021.
7. Camilleri, M., Leaky gut: mechanisms, measurement and clinical implications in humans. *Gut*, 68, 8, 1516–1526, 2019.
8. Mahmud, M.R., Akter, S., Tamanna, S.K., Mazumder, L., Esti, I.Z., Banerjee, S., Akter, S., Hasan, M.R., Acharjee, M., Hossain, M.S., Pirttilä, A.M., Impact of gut microbiome on skin health: gut-skin axis observed through the lenses of therapeutics and skin diseases. *Gut Microbes*, 14, 1, 2096995, 2022.
9. Bien, J., Palagani, V., Bozko, P., The intestinal microbiota dysbiosis and *Clostridium difficile* infection: is there a relationship with inflammatory bowel disease? *Ther. Adv. Gastroenterol.*, 6, 1, 53–68, 2013.
10. Stehlikova, Z., Kostovcik, M., Kostovcikova, K., Kverka, M., Juzlova, K., Rob, F., Hercogova, J., Bohac, P., Pinto, Y., Uzan, A., Koren, O., Tlaskalova-Hogenova, H., Jiraskova Zakostelska, Z., Dysbiosis of Skin Microbiota in Psoriatic Patients: Co-occurrence of Fungal and Bacterial Communities. *Front. Microbiol.*, 10, 438, 2019.

11. Olejniczak-Staruch, I., Ciążyńska, M., Sobolewska-Sztychny, D., Narbutt, J., Skibińska, M., Lesiak, A., Alterations of the Skin and Gut Microbiome in Psoriasis and Psoriatic Arthritis. *Int. J. Mol. Sci.*, 22, 8, 3998, 2021.
12. Griffiths, C.E.M., Armstrong, A.W., Gudjonsson, J.E., Barker, J.N.W.N., Psoriasis. *Lancet*, 397, 10281, 1301–1315, 2021.
13. Benhadou, F., Mintoff, D., delMarmol, V., Psoriasis: Keratinocytes or Immune Cells – Which Is the Trigger? *Dermatology*, 235, 2, 91–100, 2019.
14. Hurabielle, C., Link, V.M., Bouladoux, N., Han, S.-J., Merrill, E.D., Lightfoot, Y.L., Seto, N., Bleck, C.K.E., Smelkinson, M., Harrison, O.J., Linehan, J.L., Tamoutounour, S., Lionakis, M.S., Kaplan, M.J., Nakajima, S., Belkaid, Y., Immunity to commensal skin fungi promotes psoriasiform skin inflammation. *Proc. Natl. Acad. Sci. U.S.A.*, 117, 28, 16465–16474, 2020.
15. Kobayashi, T., Ricardo-Gonzalez, R.R., Moro, K., Skin-Resident Innate Lymphoid Cells – Cutaneous Innate Guardians and Regulators. *Trends Immunol.*, 41, 2, 100–112, 2020.
16. Catinean, A., Neag, M.A., Mitre, A.O., Bocsan, C.I., Buzoianu, A.D., Microbiota and Immune-Mediated Skin Diseases—An Overview. *Microorganisms*, 7, 9, 279, 2019.
17. Krakauer, T., Pradhan, K., Stiles, B.G., Staphylococcal Superantigens Spark Host-Mediated Danger Signals. *Front. Immunol.*, 7, 1–14, 2016.
18. Chen, L., Li, J., Zhu, W., Kuang, Y., Liu, T., Zhang, W., Chen, X., Peng, C., Skin and Gut Microbiome in Psoriasis: Gaining Insight Into the Pathophysiology of It and Finding Novel Therapeutic Strategies. *Front. Microbiol.*, 11, 589726, 2020.
19. Assarsson, M., Söderman, J., Duvetorp, A., Mrowietz, U., Skarstedt, M., Seifert, O., Narrowband UVB treatment induces expression of WNT7B, WNT10B and TCF7L2 in psoriasis skin. *Arch. Dermatol. Res.*, 311, 7, 535-544, 2019.
20. Quan, C., Chen, X.-Y., Li, X., Xue, F., Chen, L.-H., Liu, N., Wang, B., Wang, L.-Q., Wang, X.-P., Yang, H., Zheng, J., Psoriatic lesions are characterized by higher bacterial load and imbalance between Cutibacterium and Corynebacterium. *J. Am. Acad. Dermatol.*, 82, 4, 955–961, 2020.
21. Fyhrquist, N., Muirhead, G., Prast-Nielsen, S., Jeanmougin, M., Olah, P., Skoog, T., Jules-Clement, G., Feld, M., Barrientos-Somarribas, M., Sinkko, H., van den Bogaard, E.H., Zeeuwen, P.L.J.M., Rikken, G., Schalkwijk, J., Niehues, H., Däubener, W., Eller, S.K., Alexander, H., Pennino, D., Suomela, S., Tessas, I., Lybeck, E., Baran, A.M., Darban, H., Gangwar, R.S., Gerstel, U., Jahn, K., Karisola, P., Yan, L., Hansmann, B., Katayama, S., Meller, S., Bylesjö, M., Hupé, P., Levi-Schaffer, F., Greco, D., Ranki, A., Schröder, J.M., Barker, J., Kere, J., Tsoka, S., Lauerma, A., Soumelis, V., Nestle, F.O., Homey, B., Andersson, B., Alenius, H., Microbe-host interplay in atopic dermatitis and psoriasis. *Nat. Commun.*, 10, 1, 4703, 2019.

22. Tauch, A., Fernández-Natal, I., Soriano, F., A microbiological and clinical review on Corynebacterium kroppenstedtii. *Int. J. Infect. Dis.*, 48, 33–39, 2016.
23. De Pessemier, B., Grine, L., Debaere, M., Maes, A., Paetzold, B., Callewaert, C., Gut–Skin Axis: Current Knowledge of the Interrelationship between Microbial Dysbiosis and Skin Conditions. *Microorganisms*, 9, 2, 353, 2021.
24. Magne, F., Gotteland, M., Gauthier, L., Zazueta, A., Pesoa, S., Navarrete, P., Balamurugan, R., The Firmicutes/Bacteroidetes Ratio: A Relevant Marker of Gut Dysbiosis in Obese Patients? *Nutrients*, 12, 5, 1474, 2020.
25. Wang, J., Zhong, Y., Zhu, H., Mahgoub, O.K., Jian, Z., Gu, L., Xiong, X., Different gender-derived gut microbiota influence stroke outcomes by mitigating inflammation. *J. Neuroinflammation*, 19, 1, 245, 2022.
26. Codoñer, F.M., Ramírez-Bosca, A., Climent, E., Carrión-Gutierrez, M., Guerrero, M., Pérez-Orquín, J.M., Horga de la Parte, J., Genovés, S., Ramón, D., Navarro-López, V., Chenoll, E., Gut microbial composition in patients with psoriasis. *Sci. Rep.*, 8, 1, 3812, 2018.
27. Scher, J.U., Ubeda, C., Artacho, A., Attur, M., Isaac, S., Reddy, S.M., Marmon, S., Neimann, A., Brusca, S., Patel, T., Manasson, J., Pamer, E.G., Littman, D.R., Abramson, S.B., Decreased Bacterial Diversity Characterizes the Altered Gut Microbiota in Patients With Psoriatic Arthritis, Resembling Dysbiosis in Inflammatory Bowel Disease: Gut Microbiota in PsA. *Arthritis Rheumatol.*, 67, 1, 128–139, 2015.
28. Chen, Y.-T., Yang, N.-S., Lin, Y.-C., Ho, S.-T., Li, K.-Y., Lin, J.-S., Liu, J.-R., Chen, M.-J., A combination of Lactobacillus mali APS1 and dieting improved the efficacy of obesity treatment via manipulating gut microbiome in mice. *Sci. Rep.*, 8, 1, 6153, 2018.
29. Shapiro, J., Cohen, N.A., Shalev, V., Uzan, A., Koren, O., Maharshak, N., Psoriatic patients have a distinct structural and functional fecal microbiota compared with controls. *J. Dermatol.*, 46, 7, 595–603, 2019.
30. Ellis, S.R., Nguyen, M., Vaughn, A.R., Notay, M., Burney, W.A., Sandhu, S., Sivamani, R.K., The Skin and Gut Microbiome and Its Role in Common Dermatologic Conditions. *Microorganisms*, 7, 11, 550, 2019.
31. Wieërs, G., Verbelen, V., Van Den Driessche, M., Melnik, E., Vanheule, G., Marot, J.-C., Cani, P.D., Do Probiotics During In-Hospital Antibiotic Treatment Prevent Colonization of Gut Microbiota With Multi-Drug-Resistant Bacteria? A Randomized Placebo-Controlled Trial Comparing Saccharomyces to a Mixture of Lactobacillus, Bifidobacterium, and Saccharomyces. *Front. Public Health*, 8, 578089, 2021.
32. Pandiyan, P. and McCormick, T.S., Regulation of IL-17A–Producing Cells in Skin Inflammatory Disorders. *J. Invest. Dermatol.*, 142, 3, 867–875, 2022.
33. Chen, Y.-H., Wu, C.-S., Chao, Y.-H., Lin, C.-C., Tsai, H.-Y., Li, Y.-R., Chen, Y.-Z., Tsai, W.-H., Chen, Y.-K., Lactobacillus pentosus GMNL-77 inhibits skin lesions in imiquimod-induced psoriasis-like mice. *J. Food Drug Anal.*, 25, 3, 559–566, 2017.

34. Rather, I.A., Bajpai, V.K., Han, J., Nam, G.-J., Imiquimod-induced psoriasis-like skin inflammation in mouse model. *Bangladesh J. Pharmacol.*, 11, 4, 849, 2016.
35. Lu, W., Deng, Y., Fang, Z., Zhai, Q., Cui, S., Zhao, J., Chen, W., Zhang, H., Potential Role of Probiotics in Ameliorating Psoriasis by Modulating Gut Microbiota in Imiquimod-Induced Psoriasis-Like Mice. *Nutrients*, 13, 6, 2010, 2021.
36. Ogawa, C., Inoue, R., Yonejima, Y., Hisa, K., Yamamoto, Y., Suzuki, T., Supplemental *Leuconostoc mesenteroides* strain NTM048 attenuates imiquimod-induced psoriasis in mice. *J. Appl. Microbiol.*, 131, 6, 3043–3055, 2021.
37. Groeger, D., O'Mahony, L., Murphy, E.F., Bourke, J.F., Dinan, T.G., Kiely, B., Shanahan, F., Quigley, E.M.M., Bifidobacterium infantis 35624 modulates host inflammatory processes beyond the gut. *Gut Microbes*, 4, 4, 325–339, 2013.
38. Navarro-López, V., Martínez-Andrés, A., Ramírez-Boscá, A., Ruzafa-Costas, B., Núñez-Delegido, E., Carrión-Gutiérrez, M., Prieto-Merino, D., Codoñer-Cortés, F., Ramón-Vidal, D., Genovés-Martínez, S., Chenoll-Cuadros, E., Pérez-Orquín, J., Picó-Monllor, J., Chumillas-Lidón, S., Efficacy and Safety of Oral Administration of a Mixture of Probiotic Strains in Patients with Psoriasis: a Randomized Clinical Trial. *Acta Derm. Venerol.*, 99, 12, 1078–1084, 2019.
39. Prakoeswa, C.R.S., Bonita, L., Karim, A., Herwanto, N., Umborowati, M.A., Setyaningrum, T., Hidayati, A.N., Surono, I.S., Beneficial effect of *Lactobacillus plantarum* IS-10506 supplementation in adults with atopic dermatitis: a randomized controlled trial. *J. Dermatol. Treat.*, 33, 3, 1491–1498, 2022.
40. Dusek, O., Fajstova, A., Klimova, A., Svozilkova, P., Hrncir, T., Kverka, M., Coufal, S., Slemin, J., Tlaskalova-Hogenova, H., Forrester, J.V., Heissigerova, J., Severity of Experimental Autoimmune Uveitis Is Reduced by Pretreatment with Live Probiotic Escherichia coli Nissle 1917. *Cells*, 10, 1, 23, 2020.
41. Polak, K., Jobbágy, A., Muszyński, T., Wojciechowska, K., Frątczak, A., Bánvölgyi, A., Bergler-Czop, B., Kiss, N., Microbiome Modulation as a Therapeutic Approach in Chronic Skin Diseases. *Biomedicines*, 9, 10, 1436, 2021.
42. Vijayashankar, M. and Raghunath, N., Pustular psoriasis responding to Probiotics – a new insight. *Our Dermatol. Online*, 3, 4, 326–329, 2012.

10

A Gateway to Multi-Omics-Based Clinical Research

Ashutosh Sahoo*, Deepanshu Verma and Prajnadipta Panda

School of Biosciences and Bioengineering, Indian Institute of Technology Mandi, Kamand, Himachal Pradesh, India

Abstract

Modern clinical research, driven by hereditary factors, environmental variables, infections, aging, and sedentary lifestyles, seeks creative solutions to the plethora of ailments afflicting humanity. This chapter investigates the use of multi-omics data to overcome ethical constraints and improve clinical research efficacy. Multi-omics research, which includes genomes, transcriptomics, proteomics, and metabolomics, provides a thorough understanding of biological systems. Its benefits include a holistic perspective, the identification of molecular networks, the finding of biomarkers, and personalized medicine applications. The chapter categorizes omics researches, highlighting their importance in illness knowledge and non-destructive treatment exploration. Advances in whole-genome sequencing and analytical platforms allow for the investigation of biomolecule interactions, resulting in a better knowledge of infectious and lifestyle disorders. In this chapter, we also attempt to integrate currently used methodologies and tools in different omics platforms for interpreting clinical data acquired from suffering patients and populations. This will provide the reader a quick overview of the clinical research and its future directions.

***Keywords*:** Multi-omics, genomics, proteomics, metabolomics, clinical samples analysis, clinical omics

**Corresponding author*: ashutoshsahoo16@gmail.com

Hrudayanath Thatoi, Sonali Mohapatra, Swagat Kumar Das and Sukanta Kumar Pradhan (eds.) Applied Biotechnology and Bioinformatics: Agriculture, Pharmaceutical Research and Environment, (235–256)

10.1 Introduction

In this modern era, we are surrounded by numerous diseases that are the root cause of mortalities in humans. Genetic factors, environmental variables, numerous pathogenic agents, advancing age, and sedentary lifestyle are the causes of these disorders. Researchers from all across the globe are putting enormous efforts to combat these diseases. Most of the research is focused on finding the cause and cure for the same. Clinical research is one such area where the researchers and clinicians join their hands in the treatment. They use a variety of tools that help them in finding the root cause of disease. They generate data from clinical samples that help in deciding the treatment process. There are limitations too, which mainly include the ethical perspective. The advancement in technology and healthcare is proving to be a boon for researchers negating the limitations [1, 2].

Clinical research studies involve human participants to simulate real-world conditions in order to better understand diseases, discover new treatments, and improve patient care. It entails the examination of health and disease-related issues in real-world situations, usually with the participation of human volunteers or patients. It can mainly be divided into 4 phases from 1 to 4. Clinical research incorporates a variety of study types, such as observational studies, in which participants are observed in their natural settings [3], and interventional studies, in which participants are given a specific therapy [4]. These studies can look into a variety of medical therapies, such as pharmaceutical drugs, medical equipment, surgical procedures, behavioral interventions, and preventive measures [5]. The ultimate objective of clinical research is to generate knowledge supported by evidence that can contribute to developments in medicine, improve the health of patients, and strengthen medical practice. Throughout the investigation process, ethical issues and patient safety are crucial, and studies must comply with strict scientific standards and regulatory norms. In many cases, these ethical considerations somehow limit the extent of study [6]. To overcome this limitation and to make clinical research easy, researchers have integrated the multi-omics data with the clinical research. As we have an enormous amount of omics data, implementing it in clinical research will be challenging.

Multi-omics research is an interdisciplinary strategy to acquire a full understanding of biological systems by combining different high-throughput omics technologies. Omics refers to the large-scale investigation of numerous biological components such as genomics (the study of genes), transcriptomics (studying gene expression), proteomics (studying proteins), metabolomics (studying small molecules/metabolites), and others [7]. Researchers may discover complex links and interactions within biological systems by combining and analyzing data from multiple omics levels. Out of the many advantages, some advantages of multi-omics studies are Comprehensive Perspective, Identification of Molecular Networks, Biomarker Discovery, Uncovering Disease Mechanisms, Data Integration and Systems Biology, and Personalized Medicine Application.

The omics study has been categorized into studies where we study using genome, transcriptome, proteome, and metabolome, which are known as genomics, transcriptomics, proteomics, and metabolomics, respectively. This book chapter provides the reader a glimpse about different omics, mainly genomics, transcriptomics, proteomics, and metabolomics. The related current clinical studies in different fields help us in not only understanding the disease in a better way but also finding a cure in a very non-destructive way.

10.2 Importance of Multi-Omics

Research in systems biology often focuses on specific directions, with a specific methodology targeted at understanding individual processes or groups of activities within complex systems. However, this technique frequently overlooks vital information about the activities taking place in the environment. As a result, we miss several crucial connections between various systems, as well as their connections with environmental elements. Therefore, a comprehensive approach that considers every detail connecting the systems is required. To do this, it is critical to conduct investigations into the genome, transcriptome, proteome, thermodynamics, enzymatic, and metabolic properties. The suffix "omics" is widely used to encompass many characteristics of a system. It integrates -omes such as genomes, proteomes, transcriptomes, and metabolomes. The suffix "omics" is widely used to include many characteristics of a system [8–10]. DNA, RNA, proteins, histone modifications in chromosomes, protein post-translational additions, lipids, metabolites, genes, and other factors are all included. Incorporating multi-omics information allows us to comprehend the abilities and biological processes of a system in a comprehensive manner, rather than viewing each aspect in isolation. Furthermore, it contributes to the development of biological models that can anticipate and mimic distinct situations when exposed to other systems.

Advances in whole-genome sequencing and analytical platforms allow for the discovery of previously undiscovered interactions between various

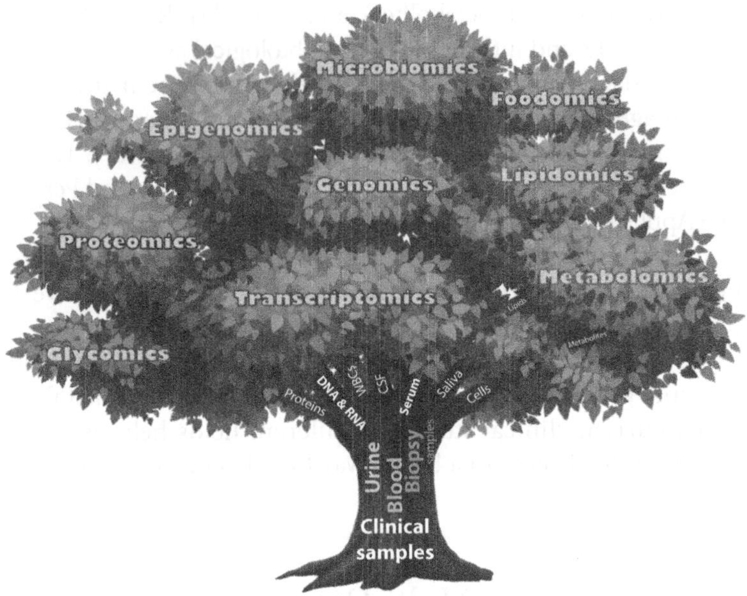

Figure 10.1 The relationship between numerous clinical samples and various domains of omics, as well as their key biomolecules obtained from corresponding clinical isolates.

biomolecules inside a system. We can derive significant insights into the systems and their functioning by combining powerful computing skills to process, filter, and analyze massive volumes of data provided by diverse biological samples. Researchers may now use this technique to analyze infectious and non-infectious illnesses that are quickly spreading throughout human populations by merging diverse omics investigations. Several bacteria, including *E. coli, S. aureus, K. pneumoniae, A. baumannii,* and *P. aeruginosa* (together known as the ESKAPEE group of pathogens), have evolved drug-resistant abilities, leading to an upsurge in infections, particularly in immunocompromised people. Understanding these infections' whole systems through a multi-omics approach might help forecast their spread [11, 12]. Aside from infectious illnesses, the human population has seen an increase in lifestyle disorders such as diabetes, hypertension, cardiovascular infections, various forms of malignancies, obesity, fatty liver diseases, and liver cirrhosis [13, 14]. It is challenging to identify the causal agents for these disorders because they affect a variety of organs, tissues, cell types, and dietary practices. In this case, omics approaches play a critical role by connecting the genome to the phenome, allowing researchers to examine gene expressions, metabolite use patterns, and latent talents that may not be visible in normal settings [15]. This aids in the prediction of genetic abnormalities and metabolic dysfunctions in order to avoid future difficulties [16]. Omics techniques also help to study pathogen and human evolution in response to changes in environmental circumstances, carbon sources, pH, temperature, atmosphere, humidity, and other physical characteristics that cause mutations. Omics approaches also aid in understanding the evolution of pathogens and humans in response to changes in environmental conditions, carbon sources, pH, temperature, atmosphere, humidity, and other physical parameters that induce mutations in organisms' basic coding sequences. As a result, large-scale genome sequencing of populations is now promoted to gain a holistic understanding of systems and design personalized medicines and population-based therapies worldwide [16]. Omics techniques can also help researchers understand how viruses and humans evolve in response to changes in environmental circumstances, carbon sources, pH, temperature, atmosphere, humidity, and other physical characteristics that cause mutations in organisms' basic coding sequences [17, 18]. As a result, large-scale genome sequencing of populations is now being advocated throughout the world in order to get a holistic understanding of systems and build personalized medications and population-based therapies [19, 20].

In addition to whole-genome sequencing, additional human body fluids such as urine and blood provide unique information about diseases and their causes. These fluids can be studied using high-throughput analytical technologies to find biomarkers for disorders like cancer, urinary tract infections, obesity, pregnancy, diabetes, renal disease, malaria, dengue, and chikungunya [21–23]. Population data guide us on a genome-to-metabolome trip of the systems, identifying reasons for process dysregulation and contributing to advances in medical research and tailored healthcare.

10.3 Genomics and Relevant Clinical Studies Along with Its Tools and Methods

Genomics is the study of an organism's genes and its chromosome. We know that every organism has a unique genetic material either in the form of DNA or RNA, which results in the phenotype of that organism. Understanding the genetic material of the organism and the information encoded in that can be helpful in understanding the system. After the completion of the Human genome project in the year 2003, 90% of the human genome was published. It revealed numerous facts such as the total number of genes encoded in a cell, proteins or enzymes encoded by them, and eventually related their function with the conclusion. It aided us in understanding the central dogma and the relationship between central genetic codes and the consequences in several cells, tissues, and organs of the human system. It offered the opportunity and means to comprehend other systems such as microbes, fungi, plants, and various other species living on Earth, all the way up to the complex mammalian system.

Genomics has been an important step to study human diseases. It has been divided into several branches, namely, structural, functional, comparative, and mutational.

10.3.1 Structural Genomics

It is the branch of genomics study where we determine and study the three-dimensional structures of proteins. Through this, we aim to determine the structure, function, and interactions of different genes in the genome. The three-dimensional structures of proteins are of utmost importance because they are directly related to their function. Understanding the structure can give insight into the interacting partners, mode of action, and the importance of the protein in different cellular processes. It also aids in drug discovery by helping in finding the new drug molecules. Structural genomics have become easier with the advancement in technologies like X-ray crystallography, NMR spectroscopy, and cryo-electron microscopy [24–26].

10.3.2 Functional Genomics

It is the branch of genomics in which we study about the function and interaction of genes and their product at the genomic level. It focuses on gene expression, their regulation, and their interaction within a biological system, which result in different cellular processes and phonotypes. It also leads to the discovery of novel genes, pathways, and regulatory mechanism. Understanding the function helps in understanding the diseases, development of drugs, and personalized medicine [27–30].

10.3.3 Comparative Genomics

It is a branch of genomics in which we study about the similarities and differences in the genomes of different species. It focuses on comparing the DNA sequences, gene structures, gene functions, and other genomic features of multiple organisms to gain insights into their evolutionary relationships, genetic diversity, and functional elements. It can help in constructing the phylogenetic trees, conserved genes, and regulatory elements among species. It also helps in understanding the genetic factors underlying the diseases by comparing the genome of healthy and diseased individuals. Knowing the genetic history of families can help in the early and easy diagnosis of diseases and ultimately cure of the same [31–34].

10.3.4 Mutational Genomics

It is a branch of genomics where we study genetic mutations and variations within the genome of the individual. We also investigate the identification, characterization, and analysis of different mutations that occur in DNA sequences. These can vary from point mutations to larger structural variations. Studying mutational genomics in clinical care can help us in diagnosing different traits, diseases, genetic history, and some other biological factors for treatment [35–37].

In clinical studies, genomics is a great tool to diagnose diseases that are left undiagnosed by conventional examination methods. It also plays an important role in improving the safety and efficacy of the drugs. It also helps in the better understanding of diseases, biomarker discoveries, personalized medicine, and targeted therapies [38–41].

There are many tools that are used to analyze and interpret genetic information in clinical research. The techniques include next-generation sequencing [42], microarrays [43], polymerase chain reaction [44], CRISPR-Cas9 [45], family health history, and SNP array genotyping [46]. In sequencing, whole genome sequencing, genome, and exome sequencing are widely used. As genomic research produces large amount of data, various online software are used to read that data. Also, cloud computing has been introduced for easy storage, analysis, and sharing of data. Clinicians are using electronic health records to consider genetic factors while diagnosing and treating patients. The availability of genomic data management system like the genomic analysis toolkit (GATK) and galaxy offer data management and analysis pipeline [47]. There are multiple tools that assess the significance of the genetic variants in clinical data, making the data actionable. Although there are advancements going on in the field of genomics in clinical care, as the amount of clinical data is increasing, there is a need for the development of the techniques that can analyze the data produced and give fruitful results.

10.4 Proteomics and Relevant Clinical Studies Along with Its Tools and Methods

Proteomics is a large-scale study of proteins within a biological system, encompassing their identification, quantification, and functional analysis. Proteomics plays a crucial role in understanding disease mechanisms, biomarker discovery, and personalized medicine by investigating the protein composition of cells, tissues, and body fluids. Based on the size of the protein examined by mass spectrometry (MS), most experimental strategies for MS-based proteomics can be split into two major categories: bottom-up and top-down [48]. Protein samples are proteolytically digested into peptides before being analyzed using a mass spectrometer in the more typical bottom-up approach [49]. In top-down proteomics, intact proteins are examined directly by MS [50, 51].

10.5 Sample Type and Acquisition

Clinical proteomics utilizes various sample types:

(a) **Tissues:** Biopsied tissues or surgically resected samples are commonly used to investigate proteomic changes associated with diseases, such as cancer.
(b) **Blood:** Serum and plasma samples provide valuable information regarding systemic protein changes and can be accessed through minimally invasive procedures.
(c) **Urine:** Urine proteomics has gained significance in non-invasive biomarker discovery and monitoring of certain diseases.
(d) **Cell Cultures:** Cellular proteomics is conducted using cell lines, primary cells, or patient-derived cells to understand cellular responses to stimuli or drug treatments.

10.6 Various Data Acquisition Methods for Proteomics Data Include the Following

(a) **Shotgun Proteomics:** Shotgun proteomics, also known as bottom-up proteomics, involves the enzymatic digestion of proteins into peptides, followed by their analysis using mass spectrometry (MS). This is a widely used approach for large-scale protein identification and quantification. Tandem MS (MS/MS) is often employed to sequence peptides and identify proteins with high sensitivity and accuracy [52].

(b) **Targeted Proteomics:** Targeted proteomics focuses on quantification of specific proteins or peptides of interest. Selected Reaction Monitoring (SRM) and Parallel Reaction Monitoring (PRM) are commonly used techniques for targeted protein quantification, offering higher sensitivity and reproducibility than shotgun proteomics [53].

(c) **Gel-Based Proteomics:** Gel-based methods, such as two-dimensional gel electrophoresis (2D-PAGE), separate proteins based on their isoelectric points and molecular weights. Spots corresponding to the individual proteins were excised from the gel and subjected to MS analysis. This approach is useful for analyzing changes in protein expression and post-translational modifications [54].

(d) **Label-Free Quantification:** Label-free quantification methods do not involve the chemical or isotopic labeling of peptides. Instead, they rely on MS signal intensities to measure the relative protein abundances between different samples. This approach is cost-effective and is suitable for comparative proteomic studies [55].

(e) **Isotope Labeling Techniques:** Stable isotope labeling methods, such as Stable Isotope Labeling with Amino Acids in Cell Culture (SILAC) and Isobaric Tags for Relative and Absolute Quantitation (iTRAQ), involve incorporating isotopically labeled amino acids or chemical tags into peptides. These labeled peptides were then quantified by MS, enabling the accurate relative quantification of protein expression changes between samples [56].

(f) **Data-Independent Acquisition (DIA):** DIA is an MS-based method that involves simultaneously acquiring the fragmentation spectra of all peptides in a sample, allowing comprehensive proteome coverage. This offers advantages in terms of reproducibility and data completeness for large-scale quantitative proteomic studies [57].

These data acquisition methods are integral to the success of proteomic studies and enable researchers to obtain comprehensive insights into the complex protein landscape of biological systems.

10.7 Techniques Used in Clinical Proteomics

(a) **SDS-PAGE (Sodium Dodecyl Sulfate Polyacrylamide Gel Electrophoresis):** SDS-PAGE separates proteins based on their molecular weights. It is widely employed in protein profiling and detection of major protein alterations in diseases [58].
(b) **2D-PAGE (Two-Dimensional Gel Electrophoresis):** 2D-PAGE combines isoelectric focusing and SDS-PAGE to separate proteins based on their isoelectric point and molecular weight. This facilitates the resolution of complex protein mixtures, aiding in the identification of differentially expressed proteins [59].
(c) **Liquid Chromatography-Mass Spectrometry/Mass Spectrometry (LC-MS/MS):** It is a powerful method for the identification and quantification of proteins. This involves liquid chromatography separation followed by tandem mass spectrometry analysis, enabling the identification of peptides and their fragmentation spectra [60].
(d) **Tandem Mass Spectroscopy: Tandem MS (MS/MS):** It involves the selection and fragmentation of specific precursor ions to obtain peptide sequence information. It aids peptide sequencing and protein identification with high sensitivity and accuracy [61].

10.8 Analysis Tools in Clinical Proteomics

(a) **Ion Fraction Analysis:** Software tools for ion fraction analysis assist in deconvoluting complex mass spectra and identifying individual peptide ions. These methods have improved the accuracy of peptide identification and quantification [62].
(b) **Protein Sequence Mutations:** Algorithms and databases for protein sequence mutation analysis can aid in identifying genetic variations associated with diseases. They contribute to the understanding of disease pathogenesis and the prediction of potential therapeutic targets [63].
(c) **Post-Translational Modifications (PTMs):** PTM analysis tools enable the identification and quantification of modified

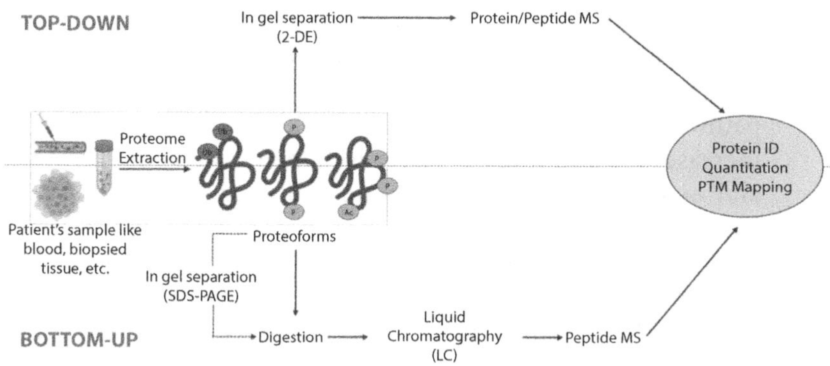

Figure 10.2 This figure depicts top-down and bottom-up approaches in proteomics (adapted from Carbonara *et al.*, 2021).

peptides, including phosphorylation, glycosylation, and acetylation. This information is crucial for the understanding of protein regulation and disease mechanisms [64].

10.9 Metabolomics and Relevant Clinical Studies Along with Its Tools and Methods

In a biological system, small molecules and metabolites play a key role in executing the phenotype of that system. They are controlled and influenced by the genetic, epigenetic, transcripts, proteomics, and environmental conditions. In case of single-cell organisms like bacteria, fungi, and algae, they have minimal regulations and effects on different regulatory steps. We can easily predict the fate of the metabolic phenotypes by analyzing their metabolic pathway networks. In case of higher organisms like mammals and plants, although the metabolic phenotype is largely dependent on the above-mentioned factors, a significant part of it is influenced by the nutrition availability, environment, and other symbiotic organisms [65, 66].

10.10 Different Types of Metabolomics

a) **Targeted metabolomics**—In this metabolomics study, metabolites and flux of a particular metabolic pathway is measured.
b) **Untargeted metabolomics**—Analyzing the metabolites of the entire system can reveal surprising results and novel

biomarkers. High-throughput techniques permit the extraction of the metabolome from clinical samples or organisms, as well as the detection and interpretation of the metabolic pathway activity of specific targets. This type of research increases the likelihood of discovering potential biomarkers and unique pathways for detection and drug development [67].

c) **Imaging metabolites**—Using imaging mass spectrometry techniques, such as matrix-assisted laser desorption ionization (MALDI), nanostructure-imaging mass spectrometry (NIMS), desorption electrospray ionization mass spectrometry (DESI), and secondary ion mass spectrometry (SIMS), among others, it is also possible to determine the localization of specific metabolites within a tissue sample. NIMS and DESI are particularly well-suited for analyzing minuscule molecules [68].

10.11 Techniques and Tools Used in Metabolomics

I. **Isotopic label feeding to models**—Now, the metabolomics study is expanded into a new direction to decode the environmental factors that influenced the metabolomics of an organism by tracer molecules leading towards mimicking the real environment unraveling true metabolic fluxes through the different nodes of the metabolic networks. In this method, we generally use the heavy isotopes of different molecules in the form of media components [69]. The metabolic enzymes do not discriminate between the common isotope of a molecule with the heavy isotope of that molecule and enable us to trace the flow of that molecule inside that system by using different analytical platforms like mass spectroscopy, NMR, and other highly sensitive techniques [70].

II. **Metabolites detection techniques**
NMR—Metabolomics uses NMR spectroscopy, a powerful analytical tool. It quantitatively analyzes complex biological material without any damage. The peculiar magnetic characteristics of atomic nuclei in metabolites allow NMR to identify and quantify several substances simultaneously. NMR is useful for researching metabolic pathways, biomarker

development, and biological systems' complex biochemical processes because of its structural insights and little sample preparation [71, 72].

HPLC, GC-MS, LC-MS—In metabolomics research, high-performance liquid chromatography (HPLC), gas chromatography-mass spectrometry (GC-MS), and liquid chromatography-mass spectrometry (LC-MS) are critical techniques. HPLC provides excellent precision and adaptability in the isolation and measurement of metabolites in complicated samples. Through the combination of gas chromatography and mass spectrometry, GC-MS excels in analyzing volatile and small-molecule metabolites, providing extensive structural information [72–74]. LC-MS, on the other hand, has a wide range of applications since it can analyze a wide variety of metabolites, including polar and non-polar molecules, allowing for complete profiling and identification. These strategies work together to analyze metabolic pathways, find biomarkers, and obtain a better understanding of the molecular composition of biological systems [75, 76].

Mass spectroscopy—Mass spectrometry (MS) is a fundamental method utilized in the field of metabolomics, providing exceptional sensitivity and specificity for the detection and measurement of metabolites in intricate biological samples. Mass spectrometry (MS) operates by inducing ionization of molecules, subsequently segregating them according to their mass-to-charge ratios, and ultimately discerning the resultant ions. GC-MS works well for volatile and small metabolites. Gas chromatography vaporizes and separates substances, then mass spectrometry ionizes and detects them. This approach is appropriate for primary metabolite analysis and substances with stable molecular structures [23, 67]. LC-MS can handle other metabolites, including polar and non-polar substances. Liquid chromatography separates sample chemicals before mass spectrometer ionization and detection. This method profiles complex metabolite mixtures to identify primary and secondary metabolites. Metabolomics using mass spectrometry reveals metabolic pathways, cellular responses to stimuli, disease biomarkers, and more [77–80]. By comparing mass spectra of known metabolites to sample data, researchers can

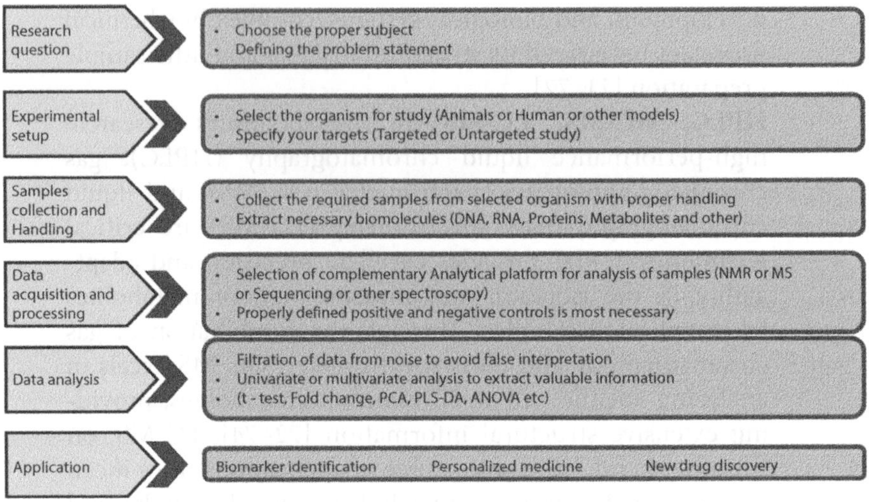

Figure 10.3 Describing about processes required to perform metabolic analysis on clinical samples obtained from patients, animals, or model organisms.

identify substances. Bioinformatics can also find patterns, correlations, and trends in complicated metabolomic datasets utilizing MS data.

10.12 Metabolite Databases

Metabolite databases contain detailed information on biological processes' chemical entities. These databases store metabolite chemical structures, characteristics, routes, and biological functions. Metabolomics research relies on them for metabolite identification, pathway analysis, and biomarker development [81]. Researchers can find chemical formulas, mass spectra, biological activities, and enzymes in metabolite databases including HMDB, KEGG, METLIN, and PubChem [82–85]. These websites allow scientists to compare experimental data to known metabolites, improving metabolite identification. Researchers use metabolite databases to find links between metabolites, pathways, and diseases to understand biochemical mechanisms. These datasets are expanding rapidly as metabolomics advances, offering a foundation for studying complex metabolic networks and their effects on health and illness [81, 86].

10.13 Data Analysis Tools and Software

R-based XCMS is commonly used for LC-MS data processing and peak detection. It aligns chromatographic peaks and does statistical analysis to find significant metabolite changes [87]. MZmine: Another open-source LC-MS data analysis software. It allows metabolite profiling through peak identification, alignment, deconvolution, and annotation. MetaboAnalyst: This web-based platform blends statistical and multivariate metabolomics data analysis capabilities [88]. Data preparation, normalization, statistical testing, pathway analysis, and visualization are supported. MetFrag: MetFrag identifies metabolites from MS/MS data. It suggests metabolite structures by comparing fragmentation patterns to databases. R package XCos does multivariate statistical analysis, data visualization, clustering, and classification to find patterns in complicated metabolomics datasets. MZmine GC-MS helps process and analyze gas chromatography-mass spectrometry data, like MZmine for LC-MS [89]. IDEOM: This user-friendly software automates NMR-based metabolomics data preparation, peak alignment, normalization, and statistical analysis. KNIME is a customizable data analytics platform for metabolomics workflows, incorporating preprocessing, analysis, and visualization tools and algorithms. MZCloud: This spectrum database compares experimental MS/MS spectra to reference spectra to detect unknown metabolites. Metabolomics research use Chenomx, TopSpin, and Bruker's NMR suite to analyse NMR spectra. These software tools help researchers identify biomarkers, analyze metabolic pathways, and understand biological systems from metabolomics data.

10.14 Application of Metabolomics in Clinical Studies

A clinical sample-based metabolomics study approach has been found to be useful in understanding nutrient utilization in a variety of scenarios, including cancer cells, drug-resistant pathogens, unculturable microbes, and pathogenic organisms. Furthermore, it aids researchers in understanding the dynamics of host–pathogen interactions as well as how organisms adapt to environmental challenges such as food shortages. Using 13C-labeled glutamic acid, for example, an important discovery in cancer research was the identification of glutamine as the preferred carbon source for cancer cells [21, 23]. Metabolomics has also been used to investigate real-time metabolic settings in infected patients through the analysis of bodily fluids such as serum, urine, and saliva. By comparing the levels of different metabolites in healthy and diseased circumstances, potential biomarkers for various metabolic syndromes that help in early illness diagnosis can be identified [20, 90, 91]. These biomarkers could be used to create simple bioassay kits for illness identification and monitoring. Biomarkers for diseases such as lung cancer have been identified using population-based clinical research.

Diagnostic facilities and clinical research organizations gather blood or urine samples directly from infected patients and extract metabolite information using LC-MS, GC-MS, and NMR. The collected data were then analyzed using high-throughput techniques, and their abundance fluctuations were tracked. Furthermore, healthy samples were collected from the same population to discover potential biomarkers, and changes were identified by processing the data in analytical tools such as MetaboAnalyst [92] and others. These metabolic data about the infected person can be used to create a supportive diet and medication to cure the medical condition. It also aided us in suggesting personalized treatment that is tailored to the patient's abilities.

10.15 Conclusion

Genomics has been an important tool used in the clinical research. Many of the genomics tools have facilitated in the improved care of patients. They have helped immensely in diagnosing the diseases and finding novel biomarkers for the disease and personalized therapies based on the patient genetic profile. With the advent of recent advancement in the sequencing technologies, they have provided the much-needed information in the case of diseases that are left undiagnosed by the conventional methods. One of the most difficult challenges is to develop precise analytic methodologies and standardize the deep mining processes, in order to meet clinical requirements of outcome quality. Proteomics has revolutionized clinical research by providing comprehensive insights into protein expression, post-translational modifications, and biomarker discovery in various diseases. The integration of multiple techniques, such as SDS-PAGE, 2D-PAGE, LC-MS/MS, and tandem mass spectroscopy, along with sophisticated analytical tools, has enhanced our understanding of disease mechanisms and facilitated the development of personalized medicine. As proteomics continues to evolve, it holds great promise for transforming patient care and advancing precision healthcare. Metabolomics of clinical patients also aids in understanding the metabolism of a certain medicine and its effectiveness within the body, which can be useful in changing the treatment in the event of resistance or negative effects. To summarize, metabolomics has become a vital tool for unravelling the complexities of biological systems, comprehending disease causes, and identifying valuable biomarkers for a variety of medical applications.

References

1. Augat, P., Simpson, H., Alt, V., Clinical research changing clinical practice. *Injury*, 54, 111023, 2023, doi: https://doi.org/10.1016/j.injury.2023.111023.
2. Wolinetz, C.D. and Tabak, L.A., Transforming Clinical Research to Meet Health Challenges. *JAMA*, 329, 20, 1740–1741, 2023.
3. Gilmartin-Thomas, J.F.M., Liew, D., Hopper, I., Observational studies and their utility for practice. *Aust. Prescr.*, 41, 3, 82, 2018.
4. Aggarwal, R. and Ranganathan, P., Study designs: Part 4–interventional studies. *Perspect. Clin. Res.*, 10, 3, 137, 2019.
5. Ligthelm, R.J., Borzì, V., Gumprecht, J., Kawamori, R., Wenying, Y., Valensi, P., Importance of observational studies in clinical practice. *Clin. Ther.*, 29, 6, 1284–1292, 2007.
6. Kandi, V. and Vadakedath, S., Ethical considerations in clinical research: a comprehensive review. *Am. J. Publ. Health Res.*, 10, 42–52, 2022.
7. Hasin, Y., Seldin, M., Lusis, A., Multi-omics approaches to disease. *Genome Biol.*, 18, 1, 1–15, 2017.
9. Quezada, H., Guzmán-Ortiz, A.L., Díaz-Sánchez, H., Valle-Rios, R., Aguirre-Hernández, J., Omics-based biomarkers: current status and potential use in the clinic. *Bol. Med. Hosp. Infant. Mex.*, 74, 3, 219–226, 2017, doi: https://doi.org/10.1016/j.bmhime.2017.11.030.
10. Hu, C. and Jia, W., Multi-omics profiling: the way toward precision medicine in metabolic diseases. *J. Mol. Cell Biol.*, 13, 8, 576–593, Aug. 2021, doi: 10.1093/jmcb/mjab051.
11. Priyamvada, P., Debroy, R., Anbarasu, A., Ramaiah, S., A comprehensive review on genomics, systems biology and structural biology approaches for combating antimicrobial resistance in ESKAPE pathogens: computational tools and recent advancements. *World J. Microbiol. Biotechnol.*, 38, 9, 153, 2022, doi: 10.1007/s11274-022-03343-z.
12. Ali, S., Alam, M., Hasan, G.M., Hassan, M., II, Potential therapeutic targets of Klebsiella pneumoniae: a multi-omics review perspective. *Brief Funct. Genomics*, 21, 2, 63–77, Mar. 2022, doi: 10.1093/bfgp/elab038.
13. Doran, S., Arif, M., Lam, S., Bayraktar, A., Turkez, H., Uhlen, M., Boren, J., & Mardinoglu, A. (2021). Multi-omics approaches for revealing the complexity of cardiovascular disease. *Brief. Bioinform.*, 22, 5, bbab061, 2021. https://doi.org/10.1093/bib/bbab061

14. van Karnebeek, C.D.M. *et al.*, The role of the clinician in the multi-omics era: are you ready? *J. Inherit Metab. Dis.*, 41, 3, 571–582, 2018, doi: 10.1007/s10545-017-0128-1.
15. Braggio, E. *et al.*, Genomic analysis of marginal zone and lymphoplasmacytic lymphomas identified common and disease-specific abnormalities. *Mod. Pathol.*, 25, 5, 651–660, 2012, doi: 10.1038/modpathol.2011.213.
16. Creasey, T. *et al.*, Genetic and genomic analysis of acute lymphoblastic leukemia in older adults reveals a distinct profile of abnormalities: analysis of 210 patients from the UKALL14 and UKALL60+ clinical trials. *Haematologica*, 107, 9, 2051–2063, Nov. 2021, doi: 10.3324/haematol.2021.279177.
17. Schiller, H.B., van Breugel, M., Nawijn, M.C., SARS-CoV-2-specific hotspots in virus–host interaction networks. *Nat. Immunol.*, 22, 7, 806–808, 2021, doi: 10.1038/s41590-021-00963-9.
18. Stare, T., Ramšak, Ž., Križnik, M., Gruden, K., Multiomics analysis of tolerant interaction of potato with potato virus Y. *Sci. Data*, 6, 1, 250, 2019, doi: 10.1038/s41597-019-0216-1.
19. Han, H.J., Ekweremadu, C., Patel, N., Advanced drug delivery system with nanomaterials for personalised medicine to treat breast cancer. *J. Drug Deliv. Sci. Technol.*, 52, 1051–1060, 2019, doi: https://doi.org/10.1016/j.jddst.2019.05.024.
20. Mur, L., Mironas, A., Paes, R., O'Shea, K., Lewis, K., Non-targeted integrative multi-omics approaches on sputum reveal potential diagnostic and monitoring biomarkers for respiratory diseases. *Eur. Respir. J.*, 50, 61, PA375, Sep. 2017, doi: 10.1183/1393003.congress-2017.PA375.
21. Ganti, S. and Weiss, R.H., Urine metabolomics for kidney cancer detection and biomarker discovery. *Urol. Oncol.*, 29, 5, 551–557, 2011, doi: https://doi.org/10.1016/j.urolonc.2011.05.013.
22. Yu, S. *et al.*, Integrative metabolomic characterization identifies plasma metabolomic signature in the diagnosis of papillary thyroid cancer. *Oncogene*, 41, 17, 2422–2430, 2022, doi: 10.1038/s41388-022-02254-5.
23. Wu, X., Ao, H., Gao, H., Zhu, Z., Metabolite biomarker discovery for human gastric cancer using dried blood spot mass spectrometry metabolomic approach. *Sci. Rep.*, 12, 1, 14632, 2022, doi: 10.1038/s41598-022-19061-3.
24. Shoemaker, S.C. and Ando, N., X-rays in the cryo-electron microscopy era: structural biology's dynamic future. *Biochemistry*, 57, 3, 277–285, 2018.
25. Chandonia, J.-M. and Brenner, S.E., The impact of structural genomics: expectations and outcomes. *Sci. (1979)*, 311, 5759, 347–351, 2006.
26. Brenner, S.E. and Levitt, M., Expectations from structural genomics. *Protein Sci.*, 9, 1, 197–200, 2000.
27. Hocquette, J.-F., Where are we in genomics? *J. Physiol. Pharmacol.*, 56, 37, 2005.
28. Kaur, S., Genomics, in: *Brenner's Encyclopedia of Genetics: Second Edition*, pp. 310–312, Jan. 2013, doi: 10.1016/B978-0-12-374984-0.00642-2.
29. Kramer, R. and Cohen, D., Functional genomics to new drug targets. *Nat. Rev. Drug Discov.*, 3, 11, 965–972, 2004.

30. Morozova, O. and Marra, M.A., Applications of next-generation sequencing technologies in functional genomics. *Genomics*, 92, 5, 255–264, 2008.
31. Lin, H., Yu, M., Wang, X., & Zhang, X.-H., Comparative genomic analysis reveals the evolution and environmental adaptation strategies of vibrios. *BMC Genomics*, 19, 1, 135, 2018. https://doi.org/10.1186/s12864-018-4531-2.
32. Miller, W., Makova, K.D., Nekrutenko, A., Hardison, R.C., Comparative genomics. *Annu. Rev. Genomics Hum. Genet.*, 5, 15–56, 2004.
33. Hardison, R.C., Comparative genomics. *PloS Biol.*, 1, 2, e58, 2003.
34. Sivashankari, S. and Shanmughavel, P., Comparative genomics-a perspective. *Bioinformation*, 1, 9, 376, 2007.
35. Alexandrov, L.B. and Stratton, M.R., Mutational signatures: the patterns of somatic mutations hidden in cancer genomes. *Curr. Opin. Genet. Dev.*, 24, 52–60, 2014.
36. Vijg, J., Somatic mutations, genome mosaicism, cancer and aging. *Curr. Opin. Genet. Dev.*, 26, 141–149, 2014.
37. Soskine, M. and Tawfik, D.S., Mutational effects and the evolution of new protein functions. *Nat. Rev. Genet.*, 11, 8, 572–582, 2010.
38. Cummings, B.B. *et al.*, Improving genetic diagnosis in Mendelian disease with transcriptome sequencing. *Sci. Transl. Med.*, 9, 386, eaal5209, 2017.
39. Kremer, L.S. *et al.*, Genetic diagnosis of Mendelian disorders via RNA sequencing. *Nat. Commun.*, 8, 1, 15824, 2017.
40. Jacob, H.J. *et al.*, Genomics in clinical practice: lessons from the front lines. *Sci. Transl. Med.*, 5, 194, 194cm5–194cm5, 2013.
41. Goetz, L. H., & Schork, N. J., Personalized medicine: motivation, challenges, and progress. *Fertil Steril.*, 109, 6, 952–963, 2018. https://doi.org/10.1016/j.fertnstert.2018.05.006.
42. Gargis, A.S. *et al.*, Assuring the quality of next-generation sequencing in clinical laboratory practice. *Nat. Biotechnol.*, 30, 11, 1033–1036, 2012.
43. Ramaswamy, S. and Golub, T.R., DNA microarrays in clinical oncology. *J. Clin. Oncol.*, 20, 7, 1932–1941, 2002.
44. Valones, M.A.A., Guimarães, R.L., Brandão, L.A.C., de Souza, P.R.E., Carvalho, A. de A. T., Crovela, S., Principles and applications of polymerase chain reaction in medical diagnostic fields: a review. *Braz. J. Microbiol.*, 40, 1–11, 2009.
45. Sharma, G., Sharma, A.R., Bhattacharya, M., Lee, S.-S., Chakraborty, C., CRISPR-Cas9: a preclinical and clinical perspective for the treatment of human diseases. *Mol. Ther.*, 29, 2, 571–586, 2021.
46. Carvalho, B., Bengtsson, H., Speed, T.P., Irizarry, R.A., Exploration, normalization, and genotype calls of high-density oligonucleotide SNP array data. *Biostatistics*, 8, 2, 485–499, 2007.
47. Van der Auwera, G.A. and O'Connor, B.D., *Genomics in the cloud: using Docker, GATK, and WDL in Terra*, O'Reilly Media, 2020.
48. Kar, U.K., Simonian, M., Whitelegge, J.P., Integral membrane proteins: bottom-up, top-down and structural proteomics. *Expert Rev. Proteomics*, 14, 8, 715–723, 2017.

49. Gillet, L.C., Leitner, A., Aebersold, R., Mass spectrometry applied to bottom-up proteomics: entering the high-throughput era for hypothesis testing. *Annu. Rev. Anal. Chem.*, 9, 449–472, 2016.
50. Donnelly, D.P. *et al.*, Best practices and benchmarks for intact protein analysis for top-down mass spectrometry. *Nat. Methods*, 16, 7, 587–594, 2019.
51. Toby, T.K., Fornelli, L., Kelleher, N.L., Progress in top-down proteomics and the analysis of proteoforms. *Annu. Rev. Anal. Chem.*, 9, 499–519, 2016.
52. Aebersold, R. and Mann, M., Mass spectrometry-based proteomics. *Nature*, 422, 6928, 198–207, 2003.
53. Lange, V., Picotti, P., Domon, B., Aebersold, R., Selected reaction monitoring for quantitative proteomics: a tutorial. *Mol. Syst. Biol.*, 4, 1, 222, 2008.
54. Görg, A., Drews, O., Lück, C., Weiland, F., Weiss, W., 2-DE with IPGs. *Electrophoresis*, 30, S1, S122–S132, 2009.
55. Zhang, B., VerBerkmoes, N.C., Langston, M.A., Uberbacher, E., Hettich, R.L., Samatova, N.F., Detecting differential and correlated protein expression in label-free shotgun proteomics. *J. Proteome Res.*, 5, 11, 2909–2918, 2006.
56. Ong, S.-E. and Mann, M., A practical recipe for stable isotope labeling by amino acids in cell culture (SILAC). *Nat. Protoc.*, 1, 6, 2650–2660, 2006.
57. Gillet, L. C., Navarro, P., Tate, S., Röst, H., Selevsek, N., Reiter, L., Bonner, R., & Aebersold, R. (2012). Targeted data extraction of the MS/MS spectra generated by data-independent acquisition: a new concept for consistent and accurate proteome analysis. *Mol. Cell. Proteomics: MCP*, 11, 6, O111.016717, 2012. https://doi.org/10.1074/mcp.O111.016717
58. Laemmli, U.K., Cleavage of structural proteins during the assembly of the head of bacteriophage T4. *Nature*, 227, 5259, 680–685, 1970.
59. O'Farrell, P.H., High resolution two-dimensional electrophoresis of proteins. *J. Biol. Chem.*, 250, 10, 4007–4021, 1975.
60. Yates, J.R., Eng, J.K., McCormack, A.L., Schieltz, D., Method to correlate tandem mass spectra of modified peptides to amino acid sequences in the protein database. *Anal. Chem.*, 67, 8, 1426–1436, 1995.
61. Clegg, G.A. and Dole, M., Molecular beams of macroions. III. Zein and polyvinylpyrrolidone. *Biopolymers*, 10, 5, 821–826, 1971.
62. Käll, L., Storey, J.D., MacCoss, M.J., Noble, W.S., Assigning significance to peptides identified by tandem mass spectrometry using decoy databases. *J. Proteome Res.*, 7, 01, 29–34, 2008.
63. Ng, P.C. and Henikoff, S., Predicting deleterious amino acid substitutions. *Genome Res.*, 11, 5, 863–874, 2001.
64. Olsen, J.V. and Mann, M., Status of large-scale analysis of post-translational modifications by mass spectrometry. *Mol. Cell. Proteomics*, 12, 12, 3444–3452, 2013.
65. Tan, S.Z., Begley, P., Mullard, G., Hollywood, K.A., Bishop, P.N., Introduction to metabolomics and its applications in ophthalmology. *Eye*, 30, 6, 773–783, 2016, doi: 10.1038/eye.2016.37.

66. Nielsen, J., & Oliver, S., The next wave in metabolome analysis. *Trends Biotechnol.*, 23, 11, 544–546, 2005. https://doi.org/10.1016/j.tibtech.2005.08.005.
67. Alonso, A., Marsal, S., & Julià, A., Analytical methods in untargeted metabolomics: state of the art in 2015. *Front. Bioeng. Biotechnol.*, 3, 23, 2015. https://doi.org/10.3389/fbioe.2015.00023.
68. Johnson, C.H., Ivanisevic, J., Siuzdak, G., Metabolomics: beyond biomarkers and towards mechanisms. *Nat. Rev. Mol. Cell Biol.*, 17, 7, 451–459, 2016, doi: 10.1038/nrm.2016.25.
69. Bueschl, C., Krska, R., Kluger, B., Schuhmacher, R., Isotopic labeling-assisted metabolomics using LC–MS. *Anal. Bioanal. Chem.*, 405, 1, 27–33, 2013, doi: 10.1007/s00216-012-6375-y.
70. Huang, X., Chen, Y.-J., Cho, K., Nikolskiy, I., Crawford, P.A., Patti, G.J., X13CMS: Global Tracking of Isotopic Labels in Untargeted Metabolomics. *Anal. Chem.*, 86, 3, 1632–1639, Feb. 2014, doi: 10.1021/ac403384n.
71. Wishart, D.S., NMR metabolomics: A look ahead. *J. Magn. Reson.*, 306, 155–161, 2019, doi: https://doi.org/10.1016/j.jmr.2019.07.013.
72. Hollywood, K., Brison, D.R., Goodacre, R., Metabolomics: Current technologies and future trends. *Proteomics*, 6, 17, 4716–4723, Sep. 2006, doi: https://doi.org/10.1002/pmic.200600106.
73. Jeong, J., Shi, X., Zhang, X., Kim, S., Shen, C., Model-based peak alignment of metabolomic profiling from comprehensive two-dimensional gas chromatography mass spectrometry. *BMC Bioinf.*, 13, 1, 27, 2012, doi: 10.1186/1471-2105-13-27.
74. Dunn, W.B. *et al.*, Procedures for large-scale metabolic profiling of serum and plasma using gas chromatography and liquid chromatography coupled to mass spectrometry. *Nat. Protoc.*, 6, 7, 1060–1083, 2011, doi: 10.1038/nprot.2011.335.
75. Yan, B. *et al.*, Serum metabolomic profiling in patients with systemic lupus erythematosus by GC/MS. *Mod. Rheumatol.*, 26, 6, 914–922, Nov. 2016, doi: 10.3109/14397595.2016.1158895.
76. Creek, D.J., Jankevics, A., Breitling, R., Watson, D.G., Barrett, M.P., Burgess, K.E.V., Toward Global Metabolomics Analysis with Hydrophilic Interaction Liquid Chromatography–Mass Spectrometry: Improved Metabolite Identification by Retention Time Prediction. *Anal. Chem.*, 83, 22, 8703–8710, Nov. 2011, doi: 10.1021/ac2021823.
77. Dettmer, K., Aronov, P.A., Hammock, B.D., Mass spectrometry-based metabolomics. *Mass Spectrom. Rev.*, 26, 1, 51–78, Jan. 2007, doi: https://doi.org/10.1002/mas.20108.
78. Brown, S.C., Kruppa, G., Dasseux, J.-L., Metabolomics applications of FT-ICR mass spectrometry. *Mass Spectrom. Rev.*, 24, 2, 223–231, Mar. 2005, doi: https://doi.org/10.1002/mas.20011.
79. Drolet, J., Tolstikov, V., Williams, B. A., Greenwood, B. P., Hill, C., Vishnudas, V. K., Sarangarajan, R., Narain, N. R., & Kiebish, M. A., Integrated

Metabolomics Assessment of Human Dried Blood Spots and Urine Strips. *Metabolites*, 7, 3, 35, 2017. https://doi.org/10.3390/metabo7030035.

80. Evans, E.D. *et al.*, Predicting human health from biofluid-based metabolomics using machine learning. *Sci. Rep.*, 10, 1, 1–13, 2020, doi: 10.1038/s41598-020-74823-1.
81. Moco, S., Vervoort, J., Moco, S., Bino, R.J., R.C.H., Bino, R., Metabolomics technologies and metabolite identification. *TrAC, Trends Anal. Chem.*, 26, 9, 855–866, De Vos2007, doi: https://doi.org/10.1016/j.trac.2007.08.003.
82. Wishart, D.S. *et al.*, HMDB: the Human Metabolome Database. *Nucleic Acids Res.*, 35, suppl_1, D521–D526, Jan. 2007, doi: 10.1093/nar/gkl923.
83. Wang, Y., Kora, G., Bowen, B.P., Pan, C., MIDAS: A Database-Searching Algorithm for Metabolite Identification in Metabolomics. *Anal. Chem.*, 86, 19, 9496–9503, Oct. 2014, doi: 10.1021/ac5014783.
84. Kanehisa, M., The KEGG Database, '*In Silico*' Simulation of Biological Processes, in: *Novartis Foundation Symposia*, pp. 91–103, 2002.
85. Smith, C. A., O'Maille, G., Want, E. J., Qin, C., Trauger, S. A., Brandon, T. R., Custodio, D. E., Abagyan, R., & Siuzdak, G., METLIN: a metabolite mass spectral database. *Ther. Drug Monit.*, 27, 6, 747–751, 2005. https://doi.org/10.1097/01.ftd.0000179845.53213.39.
86. Wishart, D.S., Advances in metabolite identification. *Bioanalysis*, 3, 15, 1769–1782, Aug. 2011, doi: 10.4155/bio.11.155.
87. Mahieu, N.G., Genenbacher, J.L., Patti, G.J., A roadmap for the XCMS family of software solutions in metabolomics. *Curr. Opin. Chem. Biol.*, 30, 87–93, 2016, doi: https://doi.org/10.1016/j.cbpa.2015.11.009.
88. Xia, J., Psychogios, N., Young, N., Wishart, D.S., MetaboAnalyst: a web server for metabolomic data analysis and interpretation. *Nucleic Acids Res.*, 37, suppl_2, W652–W660, Jul. 2009, doi: 10.1093/nar/gkp356.
89. Pluskal, T., Castillo, S., Villar-Briones, A., Orešič, M., MZmine 2: Modular framework for processing, visualizing, and analyzing mass spectrometry-based molecular profile data. *BMC Bioinf.*, 11, 1, 395, 2010, doi: 10.1186/1471-2105-11-395.
90. Dutta, N.K. *et al.*, Integration of metabolomics and transcriptomics reveals novel biomarkers in the blood for tuberculosis diagnosis in children. *Sci. Rep.*, 10, 1, 1–11, 2020, doi: 10.1038/s41598-020-75513-8.
91. Lewis, J.E. and Kemp, M.L., Integration of machine learning and genome-scale metabolic modeling identifies multi-omics biomarkers for radiation resistance. *Nat. Commun.*, 12, 1, 2700, 2021, doi: 10.1038/s41467-021-22989-1.
92. Pang, Z. *et al.*, MetaboAnalyst 5.0: narrowing the gap between raw spectra and functional insights. *Nucleic Acids Res.*, 49, W1, W388–W396, Jul. 2021, doi: 10.1093/nar/gkab382.
93. Carbonara K, Andonovski M, Coorssen JR. Proteomes Are of Proteoforms: Embracing the Complexity. *Proteomes.*, 9, 3, 38, 2021. https://doi.org/10.3390/proteomes9030038

11

Inherent Observation of Mucosal Non-Specific Immune Parameters in Indian Major Carps

Supriya Dash* **and Swagat Kumar Das**†

Department of Biotechnology, Odisha University of Technology and Research, Bhubaneswar, Odisha, India

Abstract

The study aims to explore the partial purification and evaluation of non-specific immune parameters like lysozyme, alkaline phosphatase, and protease enzyme activities from epidermal mucus of three Indian major carps (IMC) species, *Labeo rohita, Catla catla*, and *Cirrihinus mrigala*. The epidermal mucus extracts were subjected to ammonium sulfate salt precipitation followed by dialysis for partial purification of mucosal proteins. The dialyzed sample of *C. catla* exhibited the highest lysozyme and protease activities with a specific activity of 279.81 U/mg and 189.02 U/mg, respectively. However, the highest alkaline phosphatase activity was recorded in the dialyzed sample of *C. mrigala* with a specific activity of 3.02 U/mg. The SDS-PAGE analysis also revealed the presence of proteins of different molecular weights ranging between 31 and 97.4 kDa in the crude, ammonium sulfate precipitated and dialyzed samples of IMC. The present study could help us to decipher the magnitude and pattern of infection in three IMC species on the basis of their difference in innate immune parameters. However, further studies are required for the isolation of bioactive compounds in order to establish their innate immunological application.

Keywords: Mucus, lysozyme, alkaline phosphatase, protease, Indian major carps, immunity

**Corresponding author*: dashsupriya08@gmail.com

†*Corresponding author*: das.swagat@gmail.com

Hrudayanath Thatoi, Sonali Mohapatra, Swagat Kumar Das and Sukanta Kumar Pradhan (eds.) Applied Biotechnology and Bioinformatics: Agriculture, Pharmaceutical Research and Environment, (257–268) © 2025 Scrivener Publishing LLC

11.1 Introduction

Fishes living in aquatic environment are endowed with different pathogenic microbes. Despite that, they could maintain a healthy state because of a complex system of defense mechanisms [1]. The fish immune system consists of specific and non-specific immune components. The non-specific immune system includes physical barriers as well as cellular and humoral components among which gills, integument, mucus, complement, secretions of epidermal and epithelial cells, etc. are important [2]. The mucus being an innate immune system component is continuously produced and sloughed off, preventing adherence of pathogen [3]. Further, it contains a variety of bioactive substances that play a key role in proving innate immunity to the fish [4]. The mucus secreted by glands lining the nasal, esophageal, and other body cavities consists of mucins, electrolytes, water, epithelial cells, and leukocytes [5, 6]. The fish epidermal mucus produced by epidermal goblet or mucus cells [7] possesses several key components like lysozyme, proteolytic enzymes, alkaline phosphatase, immunoglobulins, complement proteins, lectins, C-reactive protein, etc. for innate immunity of fishes [8].

The present study is intended to investigate the activities of the innate immune specific enzymes such as lysozyme, alkaline phosphatase, and protease of different species of Indian major carps.

11.2 Materials and Methods

11.2.1 Collection and Preparation of Mucus Extracts

Labeo rohita, Catla catla, and *Cirrihinus mrigala* (IMC) fishes weighing approximately 200–500 g were collected from a local fish market in Khandagiri, Odisha, India. Mucus was carefully scraped from the dorsal body surface of fishes using a sterile spatula and were stored at -20°C till further use. Then, 1 ml of fish mucus was centrifuged at 10,000 rpm for 10 minutes at 4°C, and the collected supernatant was then stored in a refrigerator at 4°C.

11.2.2 Protein Precipitation

The protein from the crude mucus sample was precipitated by ammonium sulfate and stored at 4°C overnight [9].

11.2.3 Partial Protein Purification by Dialysis

The ammonium sulfate precipitated mucus sample was then dialyzed using phosphate buffered saline (PBS) for 48 hours at 4°C. The sample was collected and stored at 4°C till further use.

11.2.4 Protein Estimation

The total protein content of the mucus sample was determined by using the method of Lowry *et al.* (1951) [10].

11.2.5 Enzyme Activity

The crude, ammonium sulfate precipitated and dialyzed mucus extracts of *L. rohita, C. catla*, and *C. mrigala* fishes were estimated for their lysozyme [11], alkaline phosphatase [12], and protease activities [1].

11.2.6 Mucus Protein Profiling by SDS-PAGE

The fish mucus samples were analyzed by SDS-PAGE for the determination of molecular weight of proteins of mucus samples [13].

11.3 Results and Discussion

Crude proteins were first extracted from epidermal mucus of IMC and was purified using dialysis membrane. The protein content of the crude, ammonium sulfate precipitated and dialyzed mucus samples were recorded and tabulated in Table 11.1.

11.3.1 Protein Estimation

The epidermal fish mucus acts as a first line of defense against pathogens and offer a potential antimicrobial source [14]. The protein content in the crude mucus sample of *C. catla* was 1.65 mg/ml. It decreased to 0.35 mg/ml (21.21%) after precipitation with ammonium sulfate and then to 0.03 mg/ml (1.81%) after dialysis. For *L. rohita*, the protein content in the crude mucus sample was 1.48 mg/ml. It decreased to 0.45 mg/ml (30.40%) after precipitation with ammonium sulfate and then to 0.04 mg/ml (2.70%) after dialysis. Similarly, for *C. mrigala*, the protein content in the crude mucus sample was 1.49 mg/ml. It decreased to 0.21 mg/ml (14.09%) after

Table 11.1 Lysozyme activities of epidermal mucus extracts of *L. rohita, C. catla,* and *C. mrigala.*

Purification steps	Total volume (ml)	Protein concentration (mg/ml)	Activity (U/ml)	Specific activity (U/mg)	Yield (%)	Purification (fold)
L. rohita						
Crude	25	1.483	64.45	43.45	100	1
Ammonium sulfate precipitated	10	0.456	51.28	112.45	12.3	2.588
Dialyzed	5	0.04	7.39	150.7	0.66	3.468
C. catla						
Crude	25	1.65	94.59	56.79	100	1
Ammonium sulfate precipitated	10	0.348	30.53	87.72	8.4	1.54
Dialyzed	5	0.032	9.17	279.81	0.39	4.93
C. mrigala						
Crude	25	1.489	31.35	21.05	100	1
Ammonium sulfate precipitated	10	0.206	18.2	88.3	5.5	4.1
Dialyzed	5	0.072	8.46	116.8	0.97	5.43

precipitation with ammonium sulfate and then to 0.07 mg/ml (4.69%) after dialysis. The results revealed that crude mucus extracts contained high amount of proteins compared to other extracts [15]. It has been observed that the composition of mucus may differ on the basis of different ecological and physiological conditions like pH, stress, and stages of growth and maturity [16].

11.3.2 Enzyme Activities

The epidermal mucus of fish has numerous innate immune components that exert inhibitory or lytic action against different types of microbes [17].

11.3.2.1 Lysozyme Activity

Lysozyme, also known as N-acetyl muramide glycan hydrolase, exhibits antibacterial activity by hydrolyzing β-(1→4) linkage present in between N-acetyl muramic acid and N-acetyl-D-glucosamine residues of the peptidoglycan layer of the bacterial cell wall [18]. The lysozyme activity of the epidermal mucus extracts of *C. catla, L. rohita*, and *C. mrigala* was tabulated in Table 11.1. In all three fishes, the highest level of lysozyme activity was observed in the dialyzed sample followed by ammonium sulfate precipitated and crude mucus extracts. Among the three species, the highest specific lysozyme activity was observed in the dialyzed mucus sample of *C. catla* (279.81 U/mg) followed by *L. rohita* (150.7 U/mg) and *C. mrigala* (116.8 U/mg). The lysozyme activity of the crude mucus sample of *L. rohita* was 64.45 U/ml, which decreased to 51.28 U/ml and then to 7.39 U/ml after ammonium sulfate and dialysis, respectively. The specific activity of the crude mucus sample was 43.45 U/mg. It increased to 112.45 U/mg after ammonium sulfate precipitation and then to 150.7 U/mg after dialysis. The purification fold of the crude enzyme increased to 2.58 and 3.46 after fractionation with ammonium sulfate and dialysis, respectively. The recovery yield of the crude mucus sample (100%) was reduced to 12.3% and 0.66% after the ammonium sulfate and dialysis, respectively. The lysozyme activity of the crude mucus sample of *C. catla* was 94.59 U/ml, which decreased to 30.53 U/ml and then to 9.17 U/ml after ammonium sulfate and dialysis, respectively. The specific activity of the crude mucus sample was 56.79 U/mg. It increased to 87.72 U/mg after ammonium sulfate precipitation and then to 279.81 U/mg after dialysis. The purification fold of the crude enzyme increased to 1.54 and 4.93 after fractionation with ammonium sulfate and dialysis, respectively. The recovery yield of the crude mucus sample (100%) was reduced to 8.4% and 0.39% after ammonium sulfate and dialysis, respectively. The lysozyme activity of the crude mucus sample of *C. mrigala* was 31.35 U/ml, which decreased to 18.2 U/ml and then to 8.46 U/ml after ammonium sulfate and dialysis, respectively. The specific activity of the crude mucus sample was 21.05 U/mg. It increased to 88.3 U/mg after ammonium sulfate precipitation and then to 116.8 U/mg after dialysis. The purification fold of the crude enzyme increased to 4.1 and 5.43 after fractionation with ammonium sulfate and dialysis, respectively. The recovery yield of the crude mucus sample was reduced to 5.5% and 0.97% after ammonium sulfate and dialysis, respectively.

This is in contrast with the earlier study conducted by Nigam *et al.* (2012) [19] where the highest lysozyme activity was reported in the epidermal mucus extracts of *C. mrigala* followed by *C. catla* and *L. rohita.*

The variation in lysozyme activity can be attributed to factors like stress, maturity, diet, sex, species, genetic, habitat, spatial variations, etc. [20].

11.3.2.2 Alkaline Phosphatase Activity

Alkaline phosphatase (ALP), a hydrolase enzyme, can serve as a stress indicator and antibacterial agent [21]. It protects the fish during the initial stages of wound healing (Ingram, 1980).

The activity of ALP in the crude mucus sample of *L. rohita* was 1.003 U/ml, which decreased to 0.413 U/ml and then to 0.055 U/ml after ammonium sulfate precipitation and dialysis, respectively. The specific activity of the crude mucus sample was 0.676 U/mg, which further increased to 0.905 U/mg after ammonium sulfate precipitation and then to 1.122 U/mg after dialysis. The purification fold of the crude enzyme increased to 1.338 and 1.659 after fractionation with ammonium sulfate and dialysis, respectively. The recovery yield of the crude mucus sample was reduced to 16.47% and 0.6% after ammonium sulfate and dialysis, respectively. The ALP activity of the crude mucus sample of *C. catla* was 0.707 U/ml, which decreased to 0.613 U/ml and then to 0.063 U/ml after ammonium sulfate and dialysis, respectively. The specific activity of the crude mucus sample was 0.42 U/mg. It increased to 1.76 U/mg after ammonium sulfate precipitation and then to 1.92 U/mg after dialysis. The purification fold of the crude enzyme was increased to 4.19 and 4.57 after fractionation with ammonium sulfate and dialysis, respectively. The recovery yield of the crude mucus sample was reduced to 8.4% and 0.39% after ammonium sulfate and dialysis, respectively. The ALP activity of the crude mucus sample of *C. mrigala* was 1.324 U/ml, which decreased to 0.416 U/ml and then to 0.219 U/ml after ammonium sulfate and dialysis, respectively. The specific activity of the crude mucus sample was 0.88 U/mg. It increased to 1.99 U/mg after ammonium sulfate precipitation and then to 3.02 U/mg after dialysis. The purification fold of the crude enzyme increased to 2.26 and 3.43 after fractionation with ammonium sulfate and dialysis, respectively. The recovery yield of the crude mucus sample was reduced to 5.5% and 0.97% after ammonium sulfate and dialysis, respectively (Table 11.2).

11.3.2.3 Protease Activity

Proteases of different types like serine proteases, cysteine proteases, aspartic proteases, metallo-proteases, etc. have been reported in fish mucus, which may have direct involvement in the destruction of pathogen or the

Table 11.2 Alkaline phosphatase activities of epidermal mucus extracts of *L. rohita, C. catla*, and *C. mrigala*.

Purification steps	Total volume (ml)	Protein concentration (mg/ml)	Activity (U/ml)	Specific activity (U/mg)	Yield (%)	Purification (fold)
L. rohita						
Crude	25	1.483	1.003	0.676	100	1
Ammonium sulfate precipitated	10	0.456	0.413	0.905	16.47	1.338
Dialyzed	5	0.04	0.055	1.122	0.6	1.659
C. catla						
Crude	25	1.65	0.707	0.42	100	1
Ammonium sulfate precipitated	10	0.348	0.613	1.76	8.4	4.19
Dialyzed	5	0.032	0.063	1.92	0.39	4.57
C. mrigala						
Crude	25	1.489	1.324	0.88	100	1
Ammonium sulfate precipitated	10	0.206	0.416	1.99	5.5	2.26
Dialyzed	5	0.072	0.219	3.02	0.97	3.43

activation of other immunological components like antimicrobial peptides [22].

In the present study, the protease activity of the crude mucus sample of *L. rohita* was 111.72 U/ml, which decreased to 42.8 U/ml and then to 4.9 U/ml after ammonium sulfate precipitation and dialysis, respectively. The specific activity of the crude mucus sample was 75.31 U/mg, which further increased to 93.85 U/mg after ammonium sulfate precipitation and then to 100 U/mg after dialysis. The purification fold of the crude enzyme increased to 1.24 and 1.33 after fractionation with ammonium sulfate and dialysis, respectively. The recovery yield of the crude mucus sample was reduced to 12.3% and 0.66% after ammonium sulfate and dialysis, respectively. The protease activity of the crude mucus sample of *C. catla* was 125.9 U/ml, which decreased to 53.4 U/ml and then to 6.2 U/ml after ammonium sulfate and dialysis, respectively. The specific activity of the crude mucus sample was 75.63 U/mg. It increased to 153.44 U/mg after ammonium sulfate precipitation and then to 189.02 U/mg after dialysis. The purification fold of the crude enzyme was increased to 2.02 and 2.49 after fractionation with ammonium sulfate and dialysis, respectively. The recovery yield of the crude mucus sample was reduced to 8.4% and 0.39% after ammonium sulfate and dialysis, respectively. The lysozyme activity of the crude mucus sample of *C. mrigala* was 114.47 U/ml, which decreased to 18.5 U/ml and then to 7.24 U/ml after ammonium sulfate and dialysis, respectively. The specific activity of the crude mucus sample was 76.87 U/mg. It increased

Table 11.3 Protease activities of epidermal mucus extracts of *L. rohita, C. catla,* and *C. mrigala.*

Purification steps	**Total volume (ml)**	**Protein concentration (mg/ml)**	**Activity (U/ml)**	**Specific activity (U/mg)**	**Yield (%)**	**Purification (fold)**
L. rohita						
Crude	25	1.483	111.72	75.31	100	1
Ammonium sulfate precipitated	10	0.456	42.8	93.85	12.3	1.24
Dialyzed	5	0.04	4.9	100	0.66	1.33
C. catla						
Crude	25	1.65	125.29	75.63	100	1
Ammonium sulfate precipitated	10	0.348	53.4	153.44	8.4	2.02
Dialyzed	5	0.032	6.2	189.02	0.39	2.49
C. mrigala						
Crude	25	1.489	114.47	76.877	100	1
Ammonium sulfate precipitated	10	0.206	18.5	89.8	5.5	1.16
Dialyzed	5	0.072	7.24	100	0.97	1.3

to 88.8 U/mg after ammonium sulfate precipitation and then to 100 U/mg after dialysis. The purification fold of the crude enzyme was increased to 1.16 and 1.3 after fractionation with ammonium sulfate and dialysis, respectively. The recovery yield of the crude mucus sample was reduced to 5.5% and 0.97% after the ammonium sulfate and dialysis, respectively (Table 11.3).

11.3.3 SDS-PAGE

Electrophoretic comparison of the proteins of epidermal mucus extracts revealed the differences in the intensity of several protein bands in the three IMC fishes. The SDS-PAGE further demonstrated the variation in expression of similar proteins among *C. catla, L. rohita*, and *C. mrigala* fishes (Figure 11.1). The protein profile of mucus also varied from species to species, showing the protein bands that ranged between 31 and 97.4 kDa. SDS-PAGE revealed the presence of multiple high-intensity protein bands in the range of 31 to 97.4 kDa in the crude mucus extracts of the three fishes studied. The SDS-PAGE of ammonium salt precipitated epidermal mucus proteins demonstrated the presence of multiple bands in the range of 31 to 45 kDa and 66.2 to 97.4 kDa in all the three fishes. However, the protein profiling of dialyzed mucus samples from *C. catla, L. rohita*, and *C. mrigala* revealed the presence of a single prominent band of

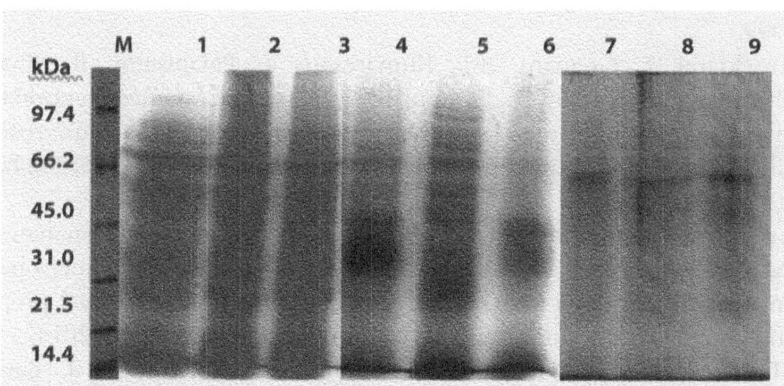

Figure 11.1 *SDS-PAGE analysis of mucus extracts of C. catla, L. rohita, and C. mrigala.* M: Molecular weight standards; 1: Crude sample of *C. catla*; 2: Crude sample of *L. rohita*; 3: Crude sample of *C. mrigala*; 4: *salt precipitated sample of C. catla*; 5: *Salt precipitated sample of L. rohita*; 6: *Salt precipitated sample of C. mrigala*; 7: Dialyzed sample of *C. catla*; 8: Dialyzed sample of *L. rohita*; 9: Dialyzed sample of *C. mrigala*.

66.2 kDa. Hence, there may be potent antimicrobial peptides falling in this range, which could be a novel source of antimicrobial peptides.

11.4 Conclusion

The innate immune factors are a prerequisite for the survival of fish in a harsh aquatic environment. Fish epidermal mucus containing different immune components showed a significant variation in the epidermal mucosal level of lysozymes, alkaline phosphatases, and proteases. In the present study, the immunologically important proteins from the epidermal mucus of IMC were isolated, partially purified, and characterized. However, further studies involving chromatographic separation and the identification of the purified protein is warranted to have a better understanding of the proteins involved in mucosal immune response in fishes. However, the variation in enzyme level among the different species of IMC could not be understood fully. It may be attributed to evolutionary, genetic, or environmental factors.

References

1. Dash, S., Samal, J., Thatoi, H., A comparative study on innate immunity parameters in the epidermal mucus of Indian major carps. *Aquac. Int.*, 22, 411–421, 2014.
2. Ellis, A.E., Immunity to bacteria in fish. *Fish Shellfish Immunol.*, 9, 291–308, 1999.
3. Arockiaraj, J., Gnanam, A.J., Kumaresan, V., Palanisamy, R., Bhatt, P., Thirumalai, M.K., Roy, A., Pasupuleti, M., Kasi, M., An unconventional antimicrobial protein histone from freshwater prawn Macrobrachium rosenbergii: Analysis of immune properties. *Fish Shellfish Immunol.*, 35, 1511–1522, 2013.
4. Arockiaraj, J., Kumaresan, V., Bhatt, P., Palanisamy, R., Gnanam, A.J., Pasupuleti, M., Kasi, M., Chaurasia, M.K., A novel single-domain peptide, anti-LPS factor from prawn: synthesis of peptide, antimicrobial properties and complete molecular characterization. *Peptides*, 53, 79–88, 2014.
5. Martínez-Antón, A., DeBolós, C., Garrido, M., Roca-Ferrer, J., Barranco, C., Alobid, I., Xaubet, A., Picado, C., Mullol, J., Mucin genes have different expression patterns in healthy and diseased upper airway mucosa. *Clin. Exp. Allergy*, 36, 448–457, 2006.

6. Knowles, M.R. and Boucher, R.C., Mucus clearance as a primary innate defense mechanism for mammalian airways. *J. Clin. Invest.*, 109, 571–577, 2002.
7. Ingram, G.A., Substances involved in the natural resistance of fish to infection–A review. *J. Fish Biol.*, 16, 23–60, 1980.
8. Dyson, H.J., Rance, M., Houghten, R.A., Wright, P.E., Lerner, R.A., Folding of immunogenic peptide fragments of proteins in water solution: II. The nascent helix. *J. Mol. Biol.*, 201, 201–217, 1988.
9. Loganathan, K., Arulprakash, A., Prakash, M., Senthilraja, P., Lysozyme, protease, alkaline phosphatase and esterase activity of epidermal skin mucus offreshwater snake head fish Channa striatus. *Int. J. Res. Pharm. Biosci.*, 3, 17–20, 2013.
10. Lowry, O., Rosebrough, N., Farr, A.L., Randall, R., Protein measurement with the folin phenol reagent. *J. Biol. Chem.*, 193, 265–275, 1951.
11. Sankaran, K. and Gurnani, S., On the variation in the catalytic activity of lysozyme in fishes. *Indian J. Biochem. Biophys.*, 9, 162–5, 1972.
12. Rosalki, S., Foo, A.Y., Burlina, A., Prellwitz, W., Stieber, P., Neumeier, D., Klein, G., Poppe, W., Bodenmueller, H., Multicenter evaluation of Iso-ALP test kit for measurement of bone alkaline phosphatase activity in serum and plasma. *Clin. Chem.*, 39, 648–652, 1993.
13. Laemmli, U.K., Cleavage of Structural Proteins during the Assembly of the Head of Bacteriophage T4. *Nature*, 227, 680–685, 1970.
14. Anbuchezhian, R., Gobinath, C., Samuthirapandian, R., Antimicrobial Peptide from the Epidermal Mucus of Some Estuarine Cat Fishes. *World Appl. Sci. J.*, 12, 256–260, 2011.
15. Rao, V., Marimuthu, K., Kupusamy, T., Rathinam, X., Valan Arasu, M., Al-Dhabi, N., Arockiaraj, J., Defense properties in the epidermal mucus of different freshwater fish species. *AACL Bioflux*, 8, 184–194, 2015.
16. Loganathan, K., Muniyan, M.K., Prakash, A.A., Raja, P.S., Prakash, M.S., Studies on the role of mucus from clarias batrachus (linn) agains selected microbes. *Int. J. Pharm. Appl.*, 2, 3, 202–206, 2011.
17. Shephard, K.L., Functions for fish mucus. *Rev. Fish. Biol. Fish*, 4, 401–429, 1994.
18. Ebran, N., Julien, S., Orange, N., Auperin, B., Molle, G., Isolation and characterization of novel glycoproteins from fish epidermal mucus: correlation between their pore-forming properties and their antibacterial activities. *Biochim. Biophys. Acta, Biomembr.*, 1467, 271–280, 2000.
19. Nigam, A.K., Kumari, U., Mittal, S., Mittal, A.K., Comparative analysis of innate immune parameters of the skin mucous secretions from certain freshwater teleosts, inhabiting different ecological niches. *Fish Physiol. Biochem.*, 38, 1245–1256, 2012.

20. Balfry, S.K. and Iwama, G.K., Observations on the inherent variability of measuring lysozyme activity in coho salmon (Oncorhynchus kisutch). *Comp. Biochem. Physiol. Part B Biochem. Mol. Biol.*, 138, 207–211, 2004.
21. Ross, N., Firth, K., Wang, A.P., Burka, F., Johnson, S., Changes in hydrolytic enzyme activities of naïve Atlantic salmon Salmo salar skin mucus due to infection with the salmon louse Lepeophtheirus salmonis and cortisol implantation. *Dis. Aquat. Organ.*, 41, 43–51, 2000.
22. Aranishi, F., Mano, N., Nakane, M., Hirose, H., Epidermal response of the Japanese eel to environmental stress. *Fish Physiol. Biochem.*, 19, 197–203, 1998.

Part III

ENVIRONMENT

12

Eco-Friendly Approaches for Converting Organic Waste to Bioenergy for Sustainable Development

Krishna Kumar Jaiswal[1*], Chandrama Roy Chowdhury[2], Deepti Yadav[3], Swapnamoy Dutta[4], Ishita Banerjee[5], Km Smriti Jaiswal[6], Arun Prasath Ramaswamy[6], Mrinal[1], Sangmesh B.[7], Amit K. Jaiswal[8], Vinod Kumar[9,10] and Krishnan Kanny[3]

[1]*Bioprocess Engineering Laboratory, Department of Green Energy Technology, Pondicherry University, Puducherry, India*
[2]*Department of Food Science, Czech University of Life Sciences, Prague, Czech Republic*
[3]*Composite Research Group, Durban University of Technology, Durban, South Africa*
[4]*Bredesen Center for Interdisciplinary Research and Graduate Education, University of Tennessee, Knoxville, TN, USA*
[5]*Department of Biochemistry & Cellular and Molecular Biology, The University of Tennessee, Knoxville, TN, USA*
[6]*Laboratory for Energy Materials and Sustainability, Department of Green Energy Technology, Pondicherry University, Puducherry, India*
[7]*School of Engineering, Jawaharlal Nehru University, New Delhi, India*
[8]*School of Food Science and Environmental Health, Faculty of Sciences and Health, Technological University Dublin – City Campus, Central Quad, Grangegorman, Dublin, Ireland*
[9]*Algal Research and Bioenergy Lab, Department of Food Science and Technology, Graphic Era (Deemed to Be University), Dehradun, Uttarakhand, India*
[10]*Peoples' Friendship University of Russia (RUDN University), Moscow, Russian Federation*

**Corresponding author*: kkjindia.get@pondiuni.ac.in

Hrudayanath Thatoi, Sonali Mohapatra, Swagat Kumar Das and Sukanta Kumar Pradhan (eds.) Applied Biotechnology and Bioinformatics: Agriculture, Pharmaceutical Research and Environment, (271–302) © 2025 Scrivener Publishing LLC

Abstract

The incessant generation of waste materials and their management affect the environment's safety and protection. Traditional waste management practices have shown a lack of proper measures. Thus, developing efficient technologies to convert waste into bioenergy through eco-friendly and sustainable approaches is vital. Due to climate change and carbon footprint, there is a great need to establish high-energy fuels such as biodiesel, bioethanol, and biogas from organic bio-wastes. This chapter demonstrates the potential for generating bioenergy/biofuels from various types of bio-waste through different thermochemical and biochemical routes. In addition, the chapter addresses the biochemical and thermochemical technologies to convert organic wastes into bioenergy for sustainable development. It has also conferred the origin of various organic wastes with organic waste-based biorefinery concepts concerning current and future uses. Thus, the intensification of strategies to convert organic waste into bioenergy production should be considered for economic, social, and environmental development.

***Keywords*:** Bio-waste, waste management, thermochemical conversion, biochemical processes, bioenergy and biofuels

12.1 Introduction

The increases in global population, suburbanization, and industrial development have resulted in an increase in energy demand and the search for alternative energy sources. Dependency on non-renewable fossil fuels such as coals, petroleum, and natural gases has increased manifold. This increased dependency resulted in the exhaustion of fossil fuels, a shortage of energy resources, and high fuel prices. It has called out for novel sustainable approaches for producing bioenergy resources [1, 2]. Biomass is a well-known renewable source to be utilized to produce solid, liquid, and gaseous states of renewable biofuels. Biomass has contributed to ~14% of the world's energy requirements. The concept of bioenergy, biorefinery, and zero waste discharge was introduced to minimize organic waste through the reuse and recycling of wastes for bioenergy production. Agricultural organic wastes can be converted into bioenergy forms through thermal or chemical conversion processes [3]. Organic wastes have been applied to thermal conversion processes, including conventional incineration and advanced thermal methods to produce bioenergy. Advanced thermal methods include pyrolysis, gasification, and plasma gasification.

Chemical processes in biorefinery include methanation. Other techniques involved in bioenergy production include fermentation, valorization, anaerobic digestion, combustion, and hydrothermal liquefaction [4].

Biofuels can be categorized as primary or secondary fuels. Primary fuels have been obtained from fuel wood and are primarily used for cooking and heating purposes. Biofuels are also categorized depending on different generations of sources of feedstock used to produce biofuel. First-generation bio-based fuels have been produced using agricultural crops like sugarcane, corn, rapeseed, etc. Second-generation bio-based fuels have been produced using non-food crops like lignocellulosic biomass. Third-generation bio-based fuels have been produced using algae biomass. Enormous amount of agricultural wastes are produced yearly, mostly removed by burning in an open field. It results in serious air pollution and environmental hazards. Nearly 5.5 billion liters of bioethanol are produced from agricultural residues [5].

Agricultural residues mainly used to produce biofuels include straw, sugar cane bagasse, rice straw, corn stover, sorghum, pistachios, almond and hazelnut shells, soybeans, cotton, tobacco, sunflower, peanut shells, hazelnuts, soybeans, rice, corn, and cattle manure. It also includes agricultural residues such as destroyed crops, fruits, and vegetables. Most fruits and vegetables are destroyed due to improper handling and packaging. These can also be used to produce biofuels. In some cases, apple and grape pomace can be used to obtain phenols as they are rich in phenols [6]. Similarly, orange waste is rich in oil content. Orange peels are rich in starch, soluble sugar, pectin, lignin, and cellulose, which favors a more outstanding bioethanol production. Additionally, different food wastes (such as cheese whey, canteen food waste, kitchen waste, wheat bran, bakery waste, cottonseed meal, corn cob waste, activated sludge waste, hydrolyzed food waste, molasses, pomegranate waste, coffee grounds, and bread residue) can also be used for biofuel production [7].

Moreover, sustainable bioenergy can also be produced by cultivating microalgae biomass. Microalgae are widely used in biorefinery due to their excellent carbon dioxide fixation capacity, simple extraction procedures, and high biomass productivity. Algal lipids have been extracted from microalgae such as *Botryococcus*, *Schizochytrium*, and *Nannochloropsis*, which can be converted into biodiesel. Since most organic waste is not reusable, recycling it through bioenergy production is essential. Bioenergy produced from organic waste can be used for fuel cells, electricity generation, micro-turbines, and cooking [8, 9]. Furthermore, the digestate obtained from biogas production can be used for sustainable farming and cultivation [10]. They act as excellent sources of fertilizers and contribute to greenhouse cultivation. They contribute to proper waste management by minimizing waste problems and reducing the volume of waste generated, the negative consequences of improper waste disposal, and bulk disposal in landfills. Specific challenges need to be addressed, including poor yield, substrate cost, the energy-intensive recovery process, and low productivity [11]. Bio-oils produced from organic waste are not well commercialized due to low fuel quality and high production cost. However, biomass yields and growth rate are quite high for microalgae biorefinery applications. Microalgae are grown and mixed with other flue gases that enhance the characteristics of various biofuels and bio-char production [12]. Other tetrameric challenges include technological, environmental, economic, and sustainability of biorefinery production techniques. For a better sustainable technological development, three elements must be addressed: economic growth, social development, and environmental development [13]. This chapter has conferred sustainable approaches to converting organic waste into bioenergy and biofuels. We have also discussed the origin of various organic wastes concerning current and future uses. Biochemical and thermochemical processes to produce biofuels have been explored in the organic waste-based biorefinery concept with success and challenges. Organic waste for bioenergy and biofuels displays promise for clean energy development and waste management.

12.2 Organic Waste in the Bioenergy Generation

Around the world, waste production is increasing severely, with the World Energy Council stating that >6 million tons of waste will be generated in 2025 [14]. When this huge volume of waste is discarded into the landfill, it produces various harmful greenhouse gases and ultimately affects global warming to climate change. Various biodegradable wastes have been conventionally discharged from multiple sources including households, food processing industries, and the hotel sector. Municipal solid waste can cause serious health complications and environmental problems [15]. Thus, there is an urgency to develop suitable waste management technologies to decrease the waste load. Along with that, to protect the environment from the effect of global warming, it is imperative to advance effective waste-to-energy techniques to generate an adequate level of renewable energy from waste material resources. Waste-to-energy technologies have the potential to reduce carbon footprint and address fuel protection and dependence on fossil fuel imports [16].

Researchers around the world are trying to develop various feasible and cost-effective methods to generate a large amount of renewable and

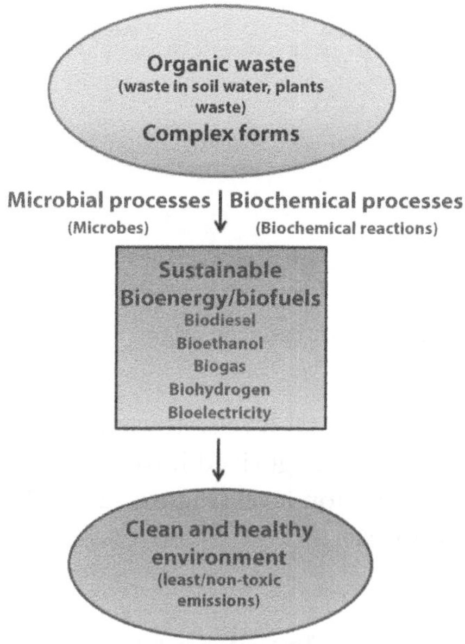

Figure 12.1 Organic waste in the production of different biofuels for a healthier environment.

sustainable energy and recover high-valued products from wastes. Many technical approaches and techniques exist to generate bioenergy from organic wastes (Figure 12.1). Some important technical routes to generate energy from waste are fermentation, transesterification, anaerobic digestion, hydrothermal liquefaction, direct combustion, pyrolysis, and thermal gasification [17]. Using the direct combustion technique, heat or electricity is generated from organic waste. In this route, the fuel achieved from the waste has been combusted in the presence of excessive levels of O_2 to generate electricity. In the gasification process, waste materials have been converted into syngas in the presence of excess O_2 and high temperatures. Pyrolysis of the bio-waste material is biochemical thermal disintegration in the lack of O_2. In the anaerobic digestion, organic waste material has been degraded using microbial metabolism in the lack of O_2, which commercially generates biogas and bio-fertilizer [18, 19].

12.3 Categories and Characteristics of Organic Waste

Organic waste originates from plant- or animal-based resources and is therefore biodegradable. These are generated at the domestic, commercial,

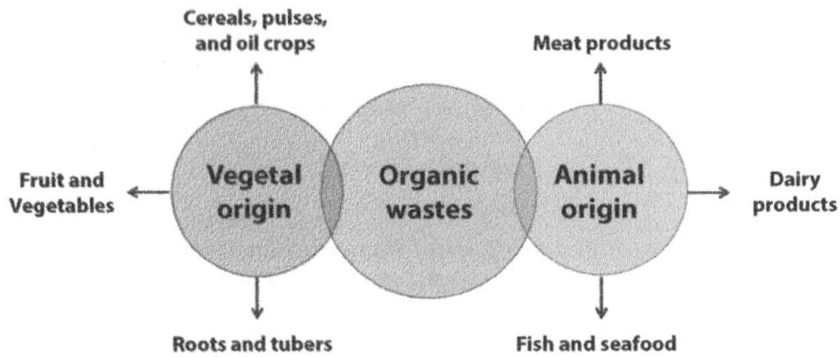

Figure 12.2 Sources of organic waste based on origin.

industrial, and even agricultural levels. These can be solid, liquid, or gaseous. Organic waste can be categorized into several groups depending on its origin, state, and production level (Figure 12.2). These can be described in general terms as follows:

12.4 Organic Waste Based on Origin

12.4.1 Plant-Based (Vegetal) Origin

Plant-based organic wastes belong to the plant or are generated from a part of a plant. These are also commonly known as green waste. Some common examples of plant waste are food waste, paper waste, bark, stems, leaves, gums, vegetable oils, and latex. Food waste typically includes vegetable/fruit peelings or food scraps. Vegetable oils such as sunflower, eucalyptus, orange, or jasmine can also be part of this waste. Also included in this category is debris generated from fields, such as grass clippings that are removed from the roots, bark, stems, or leaves. Products derived from paper, resin, wood, textiles (cotton), and medicinal herbs are also part of plant residues [20].

12.4.2 Animal-Based Origin

Organic wastes such as animal oil, lard, human or animal waste, skin, hair, fish waste, meat products, nails, fur, feathers, etc., belong to this category. They are usually generated at the household level, at small commercial levels (such as butcher shops, animal shelters, etc.), and in food industries that deal with animal products [21]. However, these may also include carcasses of animals intentionally euthanized or euthanized for disease control purposes. Others have shellfish shells, non-food products such as wool, leather, and cashmere, and leftovers from butcher shops [22, 23].

12.5 Organic Waste Based on the State of Matter

12.5.1 Solid Wastes

These include paper, wood, cardboard, food waste, vegetable/fruit peels, seeds, roots, stems, grass clippings, human/animal excrement, and other biological materials such as hair, skin, and yard waste. The spoiled or destroyed products derived from crop plants, such as stalks, straw, and cobs, are excellent examples of solid waste [24].

12.5.2 Liquid Wastes

These wastes include plant waste such as fruit juice, tree sap, vegetable oils, essential oils, tannins, resins, or other related organic acids. Animal wastes included in this category are urine and body fluids. Other than these, household organic waste includes sewage or water generated after cleaning and washing. These are also generated at the industrial level since most industries (such as dairy, lumber, pulp, textile, and sugar cane) use water for various purposes [25]. Water used for irrigation is also included in this category. One of the more specific liquid organic wastes is leachate. Leachate can be commonly defined as a collective source of contaminated fluid that has been generated due to the seepage of water through solid waste. It accumulates contaminants from all subsoil areas, resulting in a mixture of all the harmful elements in the garbage [26].

12.5.3 Gaseous Waste

Greenhouse gases are the most prominent form of organic waste. When organic matter rots, it emits methane and carbon dioxide. The gases produced in landfills include 98% and 90% carbon dioxide and methane. The rest comprises ammonia, oxygen, nitrogen, hydrogen, and sulfides. Ammonia and hydrogen sulfides are the main culprits of the unpleasant smell in landfills. Methane is naturally flammable, and its excessive production can also lead to landfill explosions. Factors such as oxygen content, type of waste, age of the landfill, temperature, and humidity play a massive role in landfill gas production. Gaseous household waste also contains trace amounts of chlorofluorocarbons, which are quickly emitted from air conditioners, microwave ovens, and refrigerators [27].

12.6 Organic Waste Based on the Level of Production

12.6.1 Agricultural Level

Common agricultural waste includes organic crop residues, manure, debris from other farms such as poultry and slaughterhouses, water runoff, soil, crop waste, silage waste, sawdust, and livestock waste.

12.6.2 Household Level

Most organic waste has been generated at home. Wastes regularly produced by everyday household chores include food waste, paper, cardboard, domestic sewage, sludge, and human waste. Food waste typically includes food scraps, fruit and vegetable scraps, fish or meat bones, food scraps, coffee grounds, tea bags, dairy products, or stale food.

12.6.3 Commercial Level

Some of the most common commercial waste is paper and cardboard from packaging, sewage, canals, remains of viscera from butcher shops, grass clippings, roots, leaves, and stems from plant-related outlets, food, and restaurant waste from food businesses.

12.6.4 Industrial Level

Industrial waste contains organic and inorganic residues. Some examples of industrial organic waste are wood and wood waste, gravel, soil, concrete and masonry, and plant and animal by-products.

12.7 Characteristics of Organic Waste

12.7.1 Biodegradability

Biodegradability is a notable characteristic for organic waste remediation. The term biodegradability of organic waste can be defined as the ability to biologically degrade or break down into basic substances (such as methane, carbon dioxide, water, etc.) through the action of live microorganisms (e.g., fungi and bacteria aerobically or anaerobically) while assimilating in nature. These products naturally break down into some usable forms and are returned to nature [28]. A biodegradability test can determine the biodegradability of a product. This test typically measures the amount of carbon dioxide released from the compost from a mixture of inoculum and organic waste over 45 days. It can also be tested through the amount of carbon dioxide generated over a set of time periods. Mathematically, biodegradability can be defined as the ratio of cumulative methane yield to theoretical methane yield as in Equation 12.1.

$$Biodegradability(\%) = \left(\frac{Cumulative\,methane\,yield}{Theoretical\,methane\,yield} \right) \times 100 \qquad (12.1)$$

According to Bayard (2015), biodegradability of solid waste materials is essential to control and optimize the high solid anaerobic digestion (HSAD) of bio-waste and other related municipal solid waste residues [29]. The batch mode of biochemical methane potential (BMP) is the usual mode to assess the anaerobic degradability and estimate methane generation rates/yields. However, the biggest drawback of this protocol is it being time-consuming (30–100 days). Some alternative approaches used are organic matter quantification and aerobic respirometry tests. Respirometry measures the respiration rate of live microorganisms (e.g., bacteria and fungi) by measuring their exchange rate of O_2 and CO_2. The quantification of organic matter quantifies the loss on ignition states in terms of volatile solids content (VS), chemical oxygen demand with chemical oxidation (COD_{tot}), and total organic carbon with combustion method (TOC_{solid}) [29].

12.8 Greenhouse Gases (GHGs)

Organic waste produces a large amount of GHGs that contribute to global warming. These GHGs include CO_2 (40–60%), CH_4 (45–60%), nitrogen, oxygen, sulfides, ammonia, carbon monoxide, hydrogen, and other related non-methane organic compounds (NMOCs). Gases such as benzene, propane, and butane are included in NMOCs. Odor is one of the typical characteristics of organic waste. The odor is usually a combination of ammonia and hydrogen sulfide. It has been found that there has been a close correlation between hydrogen sulfide and aromatics and the concentration of odorous gases. Other compounds like esters, aromatics, S-compounds, limonene, and alkanes are also responsible for the odor in organic waste. Scientists have discovered that these compounds are also responsible for the discomfort when one smells these odors. These can be harmful to human health. According to experiments performed by Di *et al.* (2013), it has been noticed that the gases generated from organic waste contain 49% aromatic compounds, 15% N compounds, 4% volatile fatty acids, 2% alkenes, 1% of each ester and S compounds, and 28% other compounds. In higher or lower concentrations, they found limonene, dimethylacetamide, naphthalene, 4-(para-tolyl)butyric acid, and docosane [30].

12.9 Benefits of Organic Waste

Organic wastes are harmless as they break down into essential elements available in nature in a certain period. These organic wastes can be useful for the environment as they can be used for different beneficial purposes for society. It has been proven that when adequately composted, organic waste can be used as fertilizer in fields and stabilizes organic matters [31]. They reduce moisture content and gradually increase plant nutrients' concentration while eradicating weed seeds and pathogens and reducing greenhouse gases. When disposed of in landfills, organic waste produces a large amount of methane. However, if used correctly through composting, organic waste can be converted into valuable compounds while reducing methane generation [32].

Moreover, a clear picture of compost maturity can be observed by mapping various physiological profiles of microbes during the composting. The temperature and C/N ratio of the composting mass signifies an expected reduction in microbial biomass. Then, during the cooling or maturation stage of waste composting, the microbial group is gradually reduced. In general terms, the development of an effective microbiota can be facilitated by certain process amendments that integrate fillers as supplements to the substrate (such as rice husk, sawdust, wood chips, etc.). This would maintain the quality of the compost and further optimize the C/N ratio. This provides a source of beneficial fertilizers while dropping the requirement for chemical fertilizers. Conversion might be the best way to reduce and use organic waste for the benefit of society.

12.10 Current and Prospective Use of Organic Waste

Inevitable waste accumulation prompts an inescapable problem affecting the soil, water, and air, or rather, every corner of the environment. In addition, the recent alarming concern about the scarcity of fossil fuel energy is why waste-to-energy techniques are prioritized. Here, we have mainly illustrated the potential utilization of organic waste, which includes industrial waste [banana peel, apple waste, newspaper, pine (softwood), potato peel waste, etc.], agricultural waste (barley straw, rapeseed straw, miscanthus, corn stover, sugarcane bagasse, etc.), and even urban waste (walnut shells, coastal bermudagrass, hotel waste, grasses, cattle manure, pig manure, etc.). Due to their useful composition, it is possible to use them in energy generation and the production of valuable products. Many traditional techniques have been used to convert this organic waste into bioproducts. The conversion of biomass to bioenergy is the highest priority at the moment. For example, biohydrogen, biodiesel, and bioethanol production have been considered an efficient option, since they can be easily obtained using various organic wastes as feedstocks [33]. It has the potential ability to replace the use of gasoline further. Currently, researchers have focused on developing efficient pretreatment and conversion techniques to obtain fuels from organic waste while maintaining their properties and appropriate performance.

As a result, biorefinery with advanced technologies is being built to meet the current demand. Countries such as Brazil and the United States have potentially improved their production rate in this regard. In addition to these fuels, recent studies have also addressed the development of microbial fuel cells using organic waste materials [34]. Waste materials such as chicken droppings, cow dung, food waste (rice and fruit residues), etc. have been employed in this technology [35]. Although this technology produces electricity, it can easily power devices such as phones, LED lamps, geo-environmental sensors, etc. However, a more compact technology is required to work on a large scale. In addition to the fuel perspective, various value-added products such as citric acid, acetone, cephalosporin c, glycerol carbonate, lactic acid, humic acid, succinic acid, 1,4-butanediol, and laccase are also produced using organic waste materials. For example, apple pomace residues have been used to produce citric acid through fermentation processes using enzymes from *Aspergillus niger* NRRL 567 and *Aspergillus niger* NRRL 2001 [36]. Using a similar waste feedstock, hydrolysis and fermentation techniques have also generated lactic acid [37].

Rice husk, rice straw, wheat straw, and wheat bran have also been used to produce laccase and lignin peroxidases, phytase, and xylanase-like products. In addition to feedstocks such as rice husk, corn husk/stalk, bamboo leaves, clover stalk, sewage sludge, and onion husks are used to develop biochar, activated carbon, and other carbon-based materials that have been established in applications for the removal of heavy metals from wastewater, formation of supercapacitors, and battery electrodes [38]. Thus, several applications have already been explored to utilize organic waste effectively. In the future, more can be examined with the help of advanced technologies considering its composition and economic importance.

12.11 Sustainable Bioenergy and Biofuels from Organic Waste

The global program is centered on the circular economy, focusing on the context of the utilization of organic waste, from which a clear path emerges for the use of viable bioenergy in the future. Organic waste has substantial untapped potential to provide clean energy around the world. Food waste occurs in large quantities throughout the global food supply chain. Conventional food waste disposal methods includes landfilling and incineration techniques that causes economic, social, and environmental problems. In terms of technological solutions, food waste can be converted into high-valued chemicals, which is a more viable and cost-effective management option [39, 40]. Waste from various food industry sectors can be categorized into two major groups and several subcategories: vegetable residues (including roots, tubers, cereals, fruits, vegetables, spices, oilseeds, cellars, and legumes) and residues from the animal feed industry (dairy, seafood, and meat) (Figure 12.2). These molecules obtained from this waste comprise a variety of functional groups, which is used to create a variety of valuable bio-based chemicals/materials (such as specialty, commodity, consumer, and niche) (Table 12.1). Sugars (e.g., glucose), organic acids (e.g., lactic acid), pectin, polysaccharides, polyphenols, and fatty acids have been shown to be synthesized from specific food wastes. The variety of constituents in food waste reveals the prospective influence on a wide range of chemical industries, which may partly substitute petroleum-based feedstocks with renewables. Increasing the use of food waste in chemical synthesis can alter the public perception of food waste and support to create renewable supply chains to achieve a closed-loop economy. A biorefinery

Table 12.1 Different sources of organic waste (food waste) and their potential applications.

Valuable components	Products	Sources	Applications	References
Commodity chemicals (biofuels)	Biodiesel	Waste cooking oil (consisting of triglycerides and free fatty acids), solid waste foods from restaurants, waste palm oil	Fuel in vehicles	[41]
	Ethanol, methane, volatile fatty acids, FAME, hydrogen, metabolites (acetate, butyrate, propionate, valerate, lactate, and ethanol)	Citrus waste, canteen-based food waste, mixed food waste, kitchen waste, cheese whey	Biogas	[42]
	Biochar	Vegetable waste, food waste, rice waste	Soil amendment to immobilize heavy metals, carbon-sequestering additive in the mortar	[43]

(*Continued*)

Table 12.1 Different sources of organic waste (food waste) and their potential applications. (*Continued*)

Valuable components	Products	Sources	Applications	References
Consumer chemicals	Volatile fatty acids	Canteen-based food waste	Dietary and medical applications	[44]
	Lactic acid	Waste-activated sludge, food waste, cottonseed meal, and corncob residues	Pharmaceutical and cosmetic applications	[45]
	Soluble sugars (glucose and fructose)	Kitchen waste	Platform chemicals	[46]
	Citric acid	Banana peels	Cosmetics, pharmaceuticals, dietary supplements, and foods	[47]
	Succinic acid	Bakery waste, including cakes and pastries	Polymers precursor, resins, solvents, food, and dietary supplement	[48]
	Ellagic acid	Fermented pomegranate wastes	Nutraceutical applications	[49]
Specialty chemicals	Invertase	Fruit peel waste	Catalytic enzyme for hydrolysis of sugar	[50]
	Cellulases	Banana waste	Lignolytic and cellulolytic enzymes to convert cellulose in soluble sugars	[51]

(*Continued*)

Table 12.1 Different sources of organic waste (food waste) and their potential applications. (*Continued*)

Valuable components	Products	Sources	Applications	References
	Essential oil (limonene)	Orange peel, citrus fruit	Essential oil with antimicrobial activity	[52]
	Ethanol	Citrus processing waste	Solvent	[53]
	Antioxidant molecules (carotenoids, chlorophyll, anthocyanins, betalains)	Grape pomace, pomegranate peel, tomato peel, orange peel, paprika leaves	Formulation of innovative packaging films and nutraceuticals	[54]
Niche chemicals	Chitosan	Shrimp waste	Antimicrobial packaging in the food and pharmaceutical industries	[55]
	Chitin	Shrimp shell	Bioactive material for antimicrobial activity	[56]
	Glucose, free amino nitrogen, and phosphate	Food waste containing rice, noodles, meat, and vegetables	Nutrient recovery or microalgae cultivation medium	[57]
	Carbohydrates and amino acids	Restaurant and bakery food waste	*Galdieria sulphuraria* 074G (Rhodophyta) growth for phycocyanin synthesis	[58]

based on food waste is predicted to play an essential role in the future, contributing to a greener and more sustainable society [39].

12.12 Conversion of Organic Waste into Bioenergy and High-Valued Products

Organic waste, such as agro-wastes, organic municipal solid waste, industrial waste, sewage sludge, and animal/livestock waste, is being investigated as a prospective source of bioenergy [3]. Tables 12.2 and 12.3 show the possible sources of organic waste for generating bioenergy and high-valued products.

12.13 Biofuels from Organic Waste: Biochemical and Thermochemical Processes

The use of organic waste in different domestic practices has been carried out since the dawn of civilization; however, there is a moment in which the growing research has managed to scientifically specify its terms and scope to make them more understandable. Various organic wastes, such as grasses, tree parts, agricultural crops, residues, and animal manure, are used practically instead of just being abandoned or landfilled. Starch, cellulose, sugars, triglycerides, alkaloids, resins, sterols, terpenoids, terpenes, and waxes are valuable and widely used in various fields [70]. This section emphasizes the biochemical and thermochemical pathways through which organic waste or specific biomass can be used favorably.

12.14 Fermentation

The biochemical conversion of biomass has been applied effectively in the generation of biofuels. Fermentation is one of the vital biochemical conversion techniques through which high fuel yield along with increased product selectivity can be achieved. In addition to process parameters such as pressure conditions, ambient temperature is also quite flexible in this biochemical technique. This technique mainly involves three successive steps such as pretreatment, hydrolysis (acid/enzymatic), and fermentation. Fermentation can be executed in both aerobic and anaerobic environments. In the first place, the pretreatment stage implies the biomass

Table 12.2 Organic waste as a potential source of bioenergy.

Type of organic waste	Components of organic waste	Bioenergy application	Applied methods	References
Palm oil mill effluent	Palm oil	Biogas	Pretreatment, mixing, anaerobic digestion	[59]
Animal waste	Blood, the rumen of pigs, goats, sheep, etc.	Biogas	Anaerobic digestion	[22]
Crop residues	Straws of wheat, barley, husk	Bioethanol	Pretreatment, hydrolysis, and fermentation	[60]
Fruit residues	Apple/grapes pomace, orange bagasse, peach seeds, pomegranate peel, date culls/pits	Bioethanol	Pretreatment, hydrolysis, and fermentation	[61]
Papermill effluent	Paper	Biomethane and biohydrogen	Two-stage sequencing system: pretreatment and dark fermentation	[62]
Food waste	Cottonseed meal, cheese whey, kitchen waste	Biohydrogen, biodiesel, biobutanol	Hydrolysis, fermentation, transesterification, enzymatic saccharification	[39]
Microbial biomass	Nannochloropsis	Biohydrogen	Pretreatment, fermentation, anaerobic digestion	[63]
Spent coffee grounds	Coffee beverage	Biogas, biodiesel, bioethanol	Anaerobic digestion, transesterification, enzymatic saccharification, fermentation	[64]
Textile waste	Wood, polyester fiber, cotton	Bio-oil	Two-stage pyrolysis	[65]

Table 12.3 Organic wastes as a potential source of other high-valued products.

Type of food wastes	Applied methods	High-valued products	Reference
Kitchen waste	Fermentation	Glucose and fructose	[46]
Curcuma longa waste	Saccharification and fermentation	L-lactic acid	[66]
Bakery waste, bread waste	Solid-state fermentation, bacterial fermentation	Succinic acid	[48]
Cottonseed meal and corncob residues	Enzymatic saccharification	D-lactic acid	[45]
Cheese whey and food waste	Acidogenic fermentation	Butyrate, valerate, ethanol, propionate, lactate, acetate, and hydrogen	[42]
Wheat bran	Enzymatic hydrolysis and fermentation	D-lactate	[67]
Food waste hydrolysate	Benchtop scale fermentation	Succinic acid	[68]
Canteen based food waste	Electro-fermentation	Volatile fatty acid, biohythane	[44]
Food waste/waste activated sludge	Cathodic electro-fermentation	Lactic acid	[69]
Fermented pomegranate waste	Yeast fermentation	Ellagic acid	[49]

processing by different methods (physical, chemical, physicochemical, biological, or a combination of them) or precisely the disruption of the cell wall, which affects the cellulose and hemicellulose upon subsequent hydrolysis. Subsequently, an acid/enzymatic hydrolysis is used to disrupt the pretreated biomass into monomers so that they can be favorable for fermentation. Finally, bacteria or yeasts have been used to convert the intermediate monomers to alcohols. Depending on the consolidation of the steps already mentioned, the ethanol production can be configured using the following routes: separate hydrolysis and fermentation, simultaneous saccharification and co-fermentation, simultaneous saccharification and fermentation, and consolidated bioprocessing [71]. Various fermentation approaches have also been explored in studies, such as the use of photo-fermentation involving various groups of organic substrates from photosynthetic bacteria that are used to generate biohydrogen in the incidence of light.

Similarly, dark fermentation can also be done following similar steps but without the presence of light. In recent studies, heterotrophic algae or yeasts are being used to convert sugars into lipids within their associated cells using appropriate solvents to disrupt the cells. Since then, the lipids produced have been purified and converted into liquid forms of transportation fuels by the hydrotreated vegetable oil diesel scenario, while genetically modified bacteria consume sugars and generate gaseous short-chain alkenes that can be recycled by oligomerization and hydrotreatment into jet/gasoline.

12.15 Anaerobic Digestion

Anaerobic digestion is an effective, less expensive, and environmentally sustainable biochemical approach to producing biofuels by using biomass as eligible feedstock. Generally, it requires an oxygen-free environment, in which, together with different anaerobic organisms, biodegradable organic substrates (e.g., chicken waste) are converted into renewable biogas [72]. This technique involves four steps: hydrolysis, acidogenesis, acetogenesis, and methanogenesis. The generated biogas typically contains methane (~50–75%), carbon dioxide (~30–50%), nitrogen (0–3%), H_2O (~6%), and O_2 (0–1%). Furthermore, terraces of trivial impurities such as hydrogen sulfide, ammonia, and water vapors have also originated [17]. In addition to producing biogas, the anaerobic digestion technique also contributes to greenhouse gas emission mitigation, nutrient recovery, dissolved oxygen depletion, etc. The factors affecting the anaerobic digestion process are temperature, pH, organic loading rate, alkalinity, hydraulic retention time, feedstock composition, volatile fatty acid concentration, etc. Various reactors have been used to process anaerobic digestion, such as fluidized beds, submerged packed beds, etc. These reactors are also suitable for wastewater's anaerobic digestion treatment process with high biochemical oxygen demand.

12.16 Combustion

Combustion is a centuries-old thermochemical conversion technique that can potentially be used to produce biofuels from biomass materials. Using this particular technique, electricity and heat are two primary forms of energy that can be produced by employing various forms of biomass. Generally, it explicitly involves an exothermic reaction between hydrocarbon and oxygen in the biomass, where they are oxidized into two main stable compounds: water and carbon dioxide. Combustion reactions generally consist of a few crucial stages: feedstock drying, pyrolysis, pre-combustion reaction, primary combustion, secondary combustion, and finally, effluent flue gas [73]. Recent studies have focused on the possible production of biofuels (both liquid and gaseous) from forest residues using combustion or co-combustion techniques [74]. It is important to note that density, ash percentage, the magnitude of biomass, specific gravity, elemental composition, and anatomical structure such as lignin, cellulose, hemicellulose, and moisture are essential factors of biomass that can have a potential impact on combustion efficiency.

12.17 Pyrolysis

Pyrolysis is an important direct thermochemical process that involves a medium-temperature range (500–1300°C) along with an oxygen-free reaction condition and is generally completed through three main conversion steps, which are pyrolysis, hydrothermal liquefaction, and solvolysis [75]. Not only liquid fuels but also this technique is used in the preparation of solid and gaseous fuels, bio-char, and value-added products from organic waste through intensive conversion pathways. Wood residues, agricultural residues, cotton gin residues, and peanut shells are waste materials containing cellulose and lignin frequently used for pyrolysis to obtain different fuels [73, 76]. Generally, this process takes place in the absence of air, and as a result, the waste biomass materials break down into a complex mixture of liquids, gases, and residual chars. The larger hydrocarbon molecules in the biomass have been broken down into smaller molecules through this process. In the case of liquid fuel processing, the fast pyrolysis technique has been used, while the slow pyrolysis technique has been used to produce gaseous fuels and solid carbon. This categorization of pyrolysis methods can be done based on their operating conditions. Mainly, parameters such as heating rate, process temperature, solids residence time, and biomass particle size are usually considered to divide the pyrolysis technique into several categories [77]. Slow pyrolysis has been carried out at low heating rates (0.1–1 K/s) and low temperatures (550–950°C). However, cracking of preliminary products occurs in this process due to increased residence time, which can negatively affect the quality and performance of the oil/fuel produced. Furthermore, the slow pyrolysis process in many cases requires additional energy input due to low heat transfer and longer residence time [78]. On the other hand, fast pyrolysis has been carried out with higher heating rates (10-200 K/s) and high-temperature ranges (850-1250°C). Fast pyrolysis is often used in studies as this technique is favorable for producing liquid fuels and various value-added chemicals. Advantages such as high energy efficiency and low investment costs have been observed using this process [79]. Apart from these two categories, flash pyrolysis is another technique considered suitable for converting biomass into liquid, solid, and gaseous fuels. This technique has resulted in a bio-oil conversion rate of almost 75% [80]. Under process temperatures between 450°C and 1000°C, high heating rate (>1000 K/s), and residence time (less than 0.5 s) together with an inert atmosphere, this process involves rapid devolatilization [81]. However, this technique has certain drawbacks, such as corrosiveness, solids in the oil, inadequate thermal stability, increase in viscosity over time due to carbon's catalytic action, and excessive water generation [82].

12.18 Gasification

The gasification technique can transform both fossil and non-fossil fuels (of the three types: liquid, solid, or gaseous) into valuable gases. A medium is required to carry out the gasification reactions, such as gas, steam, or subcritical or supercritical water. In general, a gaseous medium such as oxygen, air, or a mixture is frequently used. Using the gasification technique, biomass can be easily transformed into a multifunctional gaseous mixture, particularly called syngas or synthesis gas (mainly containing CO, H_2, CO_2, CH_4, and secondary components such as H_2O, H_2S, NH_3, tar) and the whole process is recommended to be done in the presence of a gasifying agent. This gas, as it is produced, is used to generate various forms of energy, such as the generation of heat and electricity. In addition, significant chemicals such as ammonia are also derived in many cases. In addition, after biomass conversion, a char-like solid residue has also been found [83, 84]. It is worth mentioning that the composition of syngas derived from biomass feedstock entirely depends on the type of feedstock, its characteristics, and parameters of the conversion process, such as the temperature and pressure inside the gasifier, the type of materials in the bed, the gasifying agent, and even the gasification technology (direct/indirect) also affect the final result [85]. In the direct process, direct exposure to hydrogen at high pressure has been applied, while in the indirect method, exposure to water vapor at higher pressure and temperature is carried out, where hydrogen acts as an intermediate product that is combined with the product. The indirect process also incorporates steam reforming. The high heating value, the increase of hydrogen to carbon, the removal of the produced fuel gas from sulfur, and the reduction of the oxygen content of the produced fuel are the few aspects that are controlled during this process. Thus, in conclusion, the production of H_2 and CO through biomass gasification offers a reasonable basis for generating liquid fuels for transportation, synthetic chemicals, methanol, and gasoline. Furthermore, methane has also been generated during gasification, which can be burned directly to produce energy [86].

12.19 Biorefinery Concept Based on Organic Waste for Clean Energy Management

Biorefinery can be defined as a series of processes and techniques used to convert biomass into usable materials, chemicals, and fuels. Organic waste biorefinery can be defined as techniques involved in recycling waste materials into dedicated biomass that can be used to produce bioenergy [87]. It also provides and introduces a management practice that improves the recovery of value-added products from organic waste. The economic and environmental sustainability offered by the biorefinery has increased its importance in various industries in recent years. The product range of a biorefinery receiving organic waste is restricted due to the variability of the waste stream. Organic waste can be a homogeneous waste, such as agro-industrial by-products or surplus materials that are refined as dedicated crops. A major aim of the waste biorefinery is to exploit the potential of organic waste from different resources to generate a range of sustainable bioenergy and biochemical biofuels. It is a platform for the integrated utilization of a wide range of organic waste resources to develop and implement environmental, economic, social, and political benefits [88]. A cascade principle is adopted for biorefinery, which is the flexible and sequential integration of different chemical, biological and thermal processes to develop a combination of biomolecules and biofuels to maximize production yields and revenue. Direct and reverse cascade approaches can be implemented based on bioenergy generation, downstream or upstream of biomaterial production. Economic sustainability, technical feasibility, environmental issues, and market conditions are vital in determining process integration for cascade approaches. There are some additional levels of complexity in waste biorefinery compared to conventional biorefinery due to variability, low purity of waste products, and heterogeneity. One of the significant challenges of the biorefinery is to mitigate the impacts of fluctuations in the characterization and composition of the waste since it determines the variety of techniques that would be required for the biorefinery [89]. For example, organic waste products consist of forest and agricultural waste, food waste, sludge, effluent from the food industry, and other organic household waste. These diversified organic wastes contain considerable amounts of sugars, proteins, lipids, vitamins, fibers, and bioactive compounds worth recovering. Therefore, specific combinations of techniques must be used to separate, treat, and purify any essential elements, such as pigments, organic acids, flavors, biopolymers, and other biofuels [87, 90]. Furthermore, unlike conventional wastes, food processing wastes are almost homogeneous but are affected by seasonal variability. Seasonal variability plays a crucial role in determining the biorefinery's strategies, timing, duration, logistics, and operational options. In such cases, seasonal flows can be buffered with the help of airtight storage tanks or containers and other conservation techniques, such as bio-drying or silage.

12.20 Success and Challenges of Organic Waste for Bioenergy

Bioenergy from organic waste becomes an enticing aspect of research to sustain the future energy system since natural reserves such as coal and oil have been depleted to a much lesser extent. Along with solar, wind, and geothermal energy as renewable sources, the conversion of organic waste into bioenergy has shown potential capacity. Advancing this aspect of energy generation focuses on a few particular elements, such as reducing waste, meeting energy demand, improving energy security, and reducing air pollution. In addition, parameters such as the handling, accessibility, and degradability of the generated fuels are also considered influential in their consumption and acceptability as a substitute fuel. Different biochemical and thermochemical methods have already been investigated, giving several valuable results, and some of them have already been implemented on a larger scale. However, some of those conversion techniques are centuries old, but the use of technology has improved bioenergy production systems [3, 91]. Nanomaterials play a vital role in catalyzing the procedures, such as pretreatment, hydrolysis, cellulose production, biofuel conversion efficiency, etc. However, with all the improvements, several drawbacks persist, such as a higher energy requirement, emission of dangerous gases, management and separation of mixed biomass for a specific purpose, fuel efficiency, presence of pollutants in the fuel produced, etc. For example, unwanted or polluting elements and other gases are often found in biogas, making it challenging to apply later. However, recent studies have reported a few additional pretreatment processes to remove hydrogen sulfide as unwanted gases and maximize fuel performance. Likewise, the generation of bio-oils and biodiesel from biomass must also be taken into account to optimize their characteristics and eliminate impurities as much as possible. Therefore, for bioenergy generation from different waste materials to be sustainable and economical, more advanced process design and optimization are needed to combine the benefit of a toxic-free environment and cleaner technology.

12.21 Conclusion and Recommendations

The increasing waste and its role in environmental pollution and increased energy requirements have become a global concern. To reduce organic waste, using waste materials to generate bioenergy/high-valued products is important for the modern age. The nexus between waste and energy has been based on the 5R principle, which forms the basis of sustainable development. The 5R stands for Reduce, Reuse, Recycle, Recover, and Restore. This helps to change the linear system of the make–use–dispose line to a realistic circular organization of the make–use–reuse–recycle line. The current state of biorefinery research revolves around real-time waste reduction and energy generation. Waste materials (e.g., agro-wastes, food processing waste, restaurant waste, organic household waste, municipal solid waste, forest waste, and other organic effluents) have been treated using various techniques. These techniques involve thermal methods such as pyrolysis, gasification, incineration, or biochemical processes such as aerobic composting, anaerobic digestion, fermentative biohydrogen production, bioethanol production, and other pretreatment techniques to offer the desired results. Future researchers are advised to affirm viable energy security in conjunction with an effective waste management system. Applying organic waste in bioenergy generation is one of the best ways to overcome this challenge. Bioenergy, such as bio-alcohol, biohydrogen, biodiesel, biogas, and bio-char, can play a vital role in reducing waste, improving the efficiency of waste management, and contributing to the fuel industry. Investigating the biohydrogen production process is necessary to increase yield and reduce the production rate. The circular economic concentration in the global plan for the use of organic waste should imply a sustainable future in the form of the bioenergy sector. Bioenergy sectors can improve fuel security, address air pollution, mitigate climate changes, and provide economic development in various countries. Measuring and recording carbon use and providing an effective framework policy to reduce carbon footprint are important to inhibit the expansion of fossil fuel production. Economic, technical, political, institutional, and social problems often create barriers to implementing bioenergy production. It is recommended to improve the cost of maintenance, investment costs, and operating costs required by the biorefinery. It is also recommended that the government provide the necessary support and policies to effectively implement bioenergy generation from organic waste.

References

1. Sahoo, M. and Sethi, N., Impact of industrialization, urbanization, and financial development on energy consumption: Empirical evidence from India. *J. Public Aff.*, 20, 3, e2089, 2020.
2. Correa, D.F., Beyer, H.L., Fargione, J.E., Hill, J.D., Possingham, H.P., Thomas-Hall, S.R., Schenk, P.M., Towards the implementation of sustainable biofuel production systems. *Renew. Sustain. Energy Rev.*, 107, 250–263, 2019.
3. Dhanya, B.S., Mishra, A., Chandel, A.K., Verma, M.L., Development of sustainable approaches for converting the organic waste to bioenergy. *Sci. Total Environ.*, 723, 138109, 2020.
4. Kamusoko, R., Jingura, R.M., Parawira, W., Chikwambi, Z., Strategies for valorization of crop residues into biofuels and other value-added products. *Biofuels, Bioprod. Biorefin.*, 15, 6, 1950–1964, 2021.
5. Nanda, S., Rana, R., Sarangi, P.K., Dalai, A.K., Kozinski, J.A., A broad introduction to first-, second-, and third-generation biofuels, in: *Recent advancements in biofuels and bioenergy utilization*, pp. 1–25, Springer, Singapore, 2018.
6. Ganesh, K.S., Sridhar, A., Vishali, S., Utilization of fruit and vegetable waste to produce value-added products: Conventional utilization and emerging opportunities-A review. *Chemosphere*, 287, 132221, 2022.
7. Chandra, M.S., Srinivasulu, M., Yadav, P.S., Ramesh, B., Narasimha, G., Chandrasekhar, T., Value Added Products from Agriculture, Paper and

Food Waste: A Source of Bioenergy Production, in: *Bioenergy Research: Commercial Opportunities & Challenges*, pp. 91–126, Springer, Singapore, 2021.

8. Jaiswal, K.K. and Prasath, A.R., Integrated growth potential of Chlorella pyrenoidosa using hostel mess wastewater and its biochemical analysis. *Int. J. Environ. Sci.*, 6, 5, 592–599, 2016.
9. Dahunsi, S.O., Electricity generation from food wastes and spent animal beddings with nutrients recirculation in catalytic fuel cell. *Sci. Rep.*, 10, 1, 1–14, 2020.
10. Kumar, V., Jaiswal, K.K., Tomar, M.S. *et al.* Production of high value-added biomolecules by microalgae cultivation in wastewater from anaerobic digestates of food waste: a review. *Biomass Conv. Bioref.* 13, 9625–9642, 2023. https://doi.org/10.1007/s13399-021-01906-y.
11. Ayilara, M.S., Olanrewaju, O.S., Babalola, O.O., Odeyemi, O., Waste management through composting: Challenges and potentials. *Sustainability*, 12, 4456, 2020.
12. Jaiswal, K.K., Kumar, V., Verma, R., Verma, M., Kumar, A., Vlaskin, M.S., Kim, H., Graphitic bio-char and bio-oil synthesis via hydrothermal carbonization-co-liquefaction of microalgae biomass (oiled/de-oiled) and multiple heavy metals remediations. *J. Hazard. Mater.*, 409, 124987, 2021.
13. Jaiswal, K.K., Dutta, S., Banerjee, I., Pohrmen, C.B., Kumar, V., Photosynthetic microalgae–based carbon sequestration and generation of biomass in biorefinery approach for renewable biofuels for a cleaner environment. *Biomass Convers. Biorefin.*, 1-19, 2021.
14. Liu, B., Zhang, N., Wang, L., Zhang, X., Electricity Generation Forecast of Shanghai Municipal Solid Waste Based on Bidirectional Long Short-Term Memory Model. *Int. J. Environ. Res. Public Health*, 19, 11, 6616, 2022.
15. Paravantis, J.A. and Kontoulis, N., Energy security and renewable energy: a geopolitical perspective, in: *Renewable Energy-Resources, Challenges and Applications*, IntechOpen, London, United Kingdom, 2020.
16. Li, Y., Liu, H., Yan, F., Su, D., Wang, Y., Zhou, H., High-calorific biogas production from anaerobic digestion of food waste using a two-phase pressurized biofilm (TPPB) system. *Bioresour. Technol.*, 224, 56–62, 2017.
17. Ma, Y. and Liu, Y., Turning food waste to energy and resources towards a great environmental and economic sustainability: An innovative integrated biological approach. *Biotechnol. Adv.*, 37, 7, 107414, 2019.
18. Patel, M., Zhang, X., Kumar, A., Techno-economic and life cycle assessment on lignocellulosic biomass thermochemical conversion technologies: A review. *Renew. Sustain. Energy Rev.*, 53, 1486–1499, 2016.
19. Jaiswal, K.K., Banerjee, I., Mayookha, V.P., Recent trends in the development and diversification of sericulture natural products for innovative and sustainable applications. *Bioresour. Technol. Rep.*, 13, 100614, 2021.

20. Kosseva, M.R., Sources, characteristics and treatment of plant-based food waste, in: *Food Industry Wastes*, pp. 37–66, Academic Press, London, United Kingdom, 2020.
21. Jaiswal, K.K., Jha, B., Prasath, R.A., Biodiesel production from discarded fish waste for sustainable clean energy development. *J. Chem. Pharm. Sci.*, 974, 2115, 2014.
22. Khalil, M., Berawi, M.A., Heryanto, R., Rizalie, A., Waste to energy technology: The potential of sustainable biogas production from animal waste in Indonesia. *Renew. Sustain. Energy Rev.*, 105, 323–331, 2019.
23. Jaiswal, K.K., Dutta, S., Banerjee, I., Mayookha, V.P., Bhushan, M., Lipid Extraction From Fish Processing Residues for Sustainable Biofuel Production, in: *Sustainable Fish Production and Processing*, pp. 293–319, Academic Press, London, United Kingdom, 2022.
24. Zhang, J., Kan, X., Shen, Y., Loh, K.C., Wang, C.H., Dai, Y., Tong, Y.W., A hybrid biological and thermal waste-to-energy system with heat energy recovery and utilization for solid organic waste treatment. *Energy*, 152, 214–222, 2018.
25. Kumar, V., Jaiswal, K. K., Vlaskin, M. S., Nanda, M., Tripathi, M. K., Gururani, P., Joshi, H. C., Hydrothermal liquefaction of municipal wastewater sludge and nutrient recovery from the aqueous phase. *Biofuels*, 13, 5, 657–662, 2021. https://doi.org/10.1080/17597269.2020.1863627.
26. Fazzino, F., Folino, A., Mauriello, F., Pedullà, A., Calabrò, P.S., Biofuel production from fruit and vegetable market waste and mature landfill leachate by an active filter-anaerobic digestion integrated system. *Energy Convers. Manage.: X*, 12, 100130, 2021.
27. Dhamodharan, K., Varma, V.S., Veluchamy, C., Pugazhendhi, A., Rajendran, K., Emission of volatile organic compounds from composting: A review on assessment, treatment and perspectives. *Sci. Total Environ.*, 695, 133725, 2019.
28. Paritosh, K., Yadav, M., Mathur, S., Balan, V., Liao, W., Pareek, N., Vivekanand, V., Organic fraction of municipal solid waste: overview of treatment methodologies to enhance anaerobic biodegradability. *Front. Energy Res.*, 6, 75, 2018.
29. Bayard, R., Gonzalez-Ramirez, L., Guendouz, J., Benbelkacem, H., Buffière, P., Gourdon, R., Statistical analysis to correlate bio-physical and chemical characteristics of organic wastes and digestates to their anaerobic biodegradability. *Waste Biomass Valorization*, 6, 5, 759–769, 2015.
30. Di, Y., Liu, J., Liu, J., Liu, S., Yan, L., Characteristic analysis for odor gas emitted from food waste anaerobic fermentation in the pretreatment workshop. *J. Air Waste Manage. Assoc.*, 63, 10, 1173–1181, 2013.
31. Kauser, H. and Khwairakpam, M., Organic waste management by two-stage composting process to decrease the time required for vermicomposting. *Environ. Technol. Innovation*, 25, 102193, 2022.

32. Usmani, Z., Sharma, M., Awasthi, A.K., Sharma, G.D., Cysneiros, D., Nayak, S.C., Gupta, V.K., Minimizing hazardous impact of food waste in a circular economy–Advances in resource recovery through green strategies. *J. Hazard. Mater.*, 416, 126154, 2021.
33. Callegari, A., Bolognesi, S., Cecconet, D., Capodaglio, A.G., Production technologies current role, and future prospects of biofuels feedstocks: a state-of-the-art review. *Crit. Rev. Environ. Sci. Technol.*, 50, 4, 384–436, 2020.
34. Santoro, C., Arbizzani, C., Erable, B., Ieropoulos, I., Microbial fuel cells: From fundamentals to applications - A review. *J. Power Sources*, 356, 225–244, 2017.
35. Awasthi, M.K., Sarsaiya, S., Patel, A., Juneja, A., Singh, R.P., Yan, B., Taherzadeh, M.J., Refining biomass residues for sustainable energy and bio-products: An assessment of technology, its importance, and strategic applications in circular bio-economy. *Renew. Sustain. Energy Rev.*, 127, 109876, 2020.
36. Sekoai, P.T., Ayeni, A.O., Daramola, M.O., Parametric optimization of citric acid production from apple pomace and corn steep liquor by a wild type strain of Aspergillus niger: A Response surface methodology approach. *Int. J. Eng. Res. Afr.*, 36, 98–113, Trans. Tech. Publications Ltd, 2018.
37. Kwan, T.H., Hu, Y., Lin, C.S.K., Valorisation of food waste via fungal hydrolysis and lactic acid fermentation with *Lactobacillus casei* Shirota. *Bioresour. Technol.*, 217, 129–136, 2016.
38. Yuan, X., Dissanayake, P.D., Gao, B., Liu, W.J., Lee, K.B., Ok, Y.S., Review on upgrading organic waste to value-added carbon materials for energy and environmental applications. *J. Environ. Manage.*, 296, 113128, 2021.
39. Xiong, X., Iris, K.M., Tsang, D.C., Bolan, N.S., Ok, Y.S., Igalavithana, A.D., Vikrant, K., Value-added chemicals from food supply chain wastes: State-of-the-art review and future prospects. *Chem. Eng. J.*, 375, 121983, 2019.
40. Jaiswal, K.K., Kumar, V., Vlaskin, M.S., Nanda, M., Verma, M., Ahmad, W., Kim, H., Hydropyrolysis of freshwater macroalgal bloom for bio-oil and bio-char production: Kinetics and isotherm for removal of multiple heavy metals. *Environ. Technol. Innovation*, 22, 101440, 2021.
41. Carmona-Cabello, M., Leiva-Candia, D., Castro-Cantarero, J.L., Pinzi, S., Dorado, M.P., Valorization of food waste from restaurants by transesterification of the lipid fraction. *Fuel*, 215, 492–498, 2018.
42. Girotto, F., Lavagnolo, M.C., Pivato, A., Cossu, R., Acidogenic fermentation of the organic fraction of municipal solid waste and cheese whey for bio-plastic precursors recovery–Effects of process conditions during batch tests. *Waste Manage.*, 70, 71–80, 2017.
43. Kumar, A., Ramamoorthy, D., Kumar, N. *et al.* Investigation on the potential of eco-friendly bio-char for amendment in serpentine soils and immobilization of heavy metals contaminants: a review. *Biomass Conv. Bioref.*, 13, 16585–16605, 2023. https://doi.org/10.1007/s13399-021-02257-4.

44. Sravan, J.S., Butti, S.K., Sarkar, O., Krishna, K.V., Mohan, S.V., Electrofermentation of food waste–regulating acidogenesis towards enhanced volatile fatty acids production. *Chem. Eng. J.*, 334, 1709–1718, 2018.
45. Bai, Z., Gao, Z., Sun, J., Wu, B., He, B., D-Lactic acid production by Sporolactobacillus inulinus YBS1-5 with simultaneous utilization of cottonseed meal and corncob residue. *Bioresour. Technol.*, 207, 346–352, 2016.
46. Vavouraki, A., II, Angelis, E.M., Kornaros, M., Optimization of thermo-chemical hydrolysis of kitchen wastes. *Waste Manage.*, 33, 3, 740–745, 2013.
47. Khairan, K., Makstum, A., Yulvizar, C., Utilization of banana peel waste for citric acid production by Aspergillus niger, in: *IOP Conference Series: Earth and Environmental Science*, vol. 364, IOP Publishing, p. 012005, 2019.
48. Zhang, A.Y.Z., Sun, Z., Leung, C.C.J., Han, W., Lau, K.Y., Li, M., Lin, C.S.K., Valorisation of bakery waste for succinic acid production. *Green Chem.*, 15, 3, 690–695, 2013.
49. Verotta, L., Panzella, L., Antenucci, S., Calvenzani, V., Tomay, F., Petroni, K., Napolitano, A., Fermented pomegranate wastes as sustainable source of ellagic acid: Antioxidant properties, anti-inflammatory action, and controlled release under simulated digestion conditions. *Food Chem.*, 246, 129–136, 2018.
50. Mehta, K. and Duhan, J.S., Production of invertase from Aspergillus niger using fruit peel waste as a substrate. *Intern. J. Pharm. Biol. Sci.*, 5, 2, B353–B360, 2014.
51. Dabhi, B.K., Vyas, R.V., Shelat, H.N., Use of banana waste for the production of cellulolytic enzymes under solid substrate fermentation using bacterial consortium. *Int. J. Curr. Microbiol. App. Sci.*, 3, 1, 337–346, 2014.
52. Espina, L., Somolinos, M., Lorán, S., Conchello, P., García, D., Pagán, R., Chemical composition of commercial citrus fruit essential oils and evaluation of their antimicrobial activity acting alone or in combined processes. *Food Control*, 22, 6, 896–902, 2011.
53. Lohrasbi, M., Pourbafrani, M., Niklasson, C., Taherzadeh, M.J., Process design and economic analysis of a citrus waste biorefinery with biofuels and limonene as products. *Bioresour. Technol.*, 101, 19, 7382–7388, 2010.
54. Benucci, I., Lombardelli, C., Mazzocchi, C., and Esti, M., Natural colorants from vegetable food waste: Recovery, regulatory aspects, and stability—A review. *Compr. Rev. Food Sci. Food Saf.*, 21, 3, 2715-2737, 2022.
55. Arancibia, M.Y., Alemán, A., López-Caballero, M.E., Gómez-Guillén, M.C., Montero, P., Development of active films of chitosan isolated by mild extraction with added protein concentrate from shrimp waste. *Food Hydrocoll.*, 43, 91–99, 2015.
56. Benhabiles, M.S., Salah, R., Lounici, H., Drouiche, N., Goosen, M.F.A., Mameri, N., Antibacterial activity of chitin, chitosan and its oligomers prepared from shrimp shell waste. *Food Hydrocoll.*, 29, 1, 48–56, 2012.

57. Pleissner, D., Lam, W.C., Sun, Z., Lin, C.S.K., Food waste as nutrient source in heterotrophic microalgae cultivation. *Bioresour. Technol.*, 137, 139–146, 2013.
58. Sloth, J.K., Jensen, H.C., Pleissner, D., Eriksen, N.T., Growth and phycocyanin synthesis in the heterotrophic microalga Galdieria sulphuraria on substrates made of food waste from restaurants and bakeries. *Bioresour. Technol.*, 238, 296–305, 2017.
59. Aziz, N.I.H.A. and Hanafiah, M.M., Life cycle analysis of biogas production from anaerobic digestion of palm oil mill effluent. *Renew. Energy*, 145, 847–857, 2020.
60. Panahi, H. K. S., Dehhaghi, M., Aghbashlo, M., Karimi, K., and Tabatabaei, M., Conversion of residues from agro-food industry into bioethanol in Iran: An under-valued biofuel additive to phase out MTBE in gasoline. *Renew. Energy*, 145, 699-710, 2020.
61. Rincon, L., Puri, M., Kojakovic, A., Maltsoglou, I., The contribution of sustainable bioenergy to renewable electricity generation in Turkey: Evidence based policy from an integrated energy and agriculture approach. *Energy Policy*, 130, 69–88, 2019.
62. Vaez, E. and Zilouei, H., Towards the development of biofuel production from paper mill effluent. *Renew. Energy*, 146, 1408–1415, 2020.
63. Choi, Y.Y., Patel, A.K., Hong, M.E., Chang, W.S., Sim, S.J., Microalgae Bioenergy with Carbon Capture and Storage (BECCS): An emerging sustainable bioprocess for reduced CO2 emission and biofuel production. *Bioresour. Technol. Rep.*, 7, 100270, 2019.
64. Atabani, A.E., Ala'a, H., Kumar, G., Saratale, G.D., Aslam, M., Khan, H.A., Mahmoud, E., Valorization of spent coffee grounds into biofuels and value-added products: Pathway towards integrated bio-refinery. *Fuel*, 254, 115640, 2019.
65. Yousef, S., Eimontas, J., Striūgas, N., Tatariants, M., Abdelnaby, M.A., Tuckute, S., Kliucininkas, L., A sustainable bioenergy conversion strategy for textile waste with self-catalysts using mini-pyrolysis plant. *Energy Convers. Manage.*, 196, 688–704, 2019.
66. Nguyen, C.M., Kim, J.S., Nguyen, T.N., Kim, S.K., Choi, G.J., Choi, Y.H., Kim, J.C., Production of L-and D-lactic acid from waste Curcuma longa biomass through simultaneous saccharification and cofermentation. *Bioresour. Technol.*, 146, 35–43, 2013.
67. Li, J., Sun, J., Wu, B., He, B., Combined utilization of nutrients and sugar derived from wheat bran for d-Lactate fermentation by *Sporolactobacillus inulinus* YBS1-5. *Bioresour. Technol.*, 229, 33–38, 2017.
68. Yang, X., Wang, H., Li, C., Lin, C.S.K., Restoring of glucose metabolism of engineered Yarrowia lipolytica for succinic acid production via a simple and efficient adaptive evolution strategy. *J. Agric. Food. Chem.*, 65, 20, 4133–4139, 2017.

69. Xue, G., Lai, S., Li, X., Zhang, W., You, J., Chen, H., Liu, Y., Efficient bioconversion of organic wastes to high optical activity of l-lactic acid stimulated by cathode in mixed microbial consortium. *Water Res.*, 131, 1–10, 2018.
70. Ruan, R., Zhang, Y., Chen, P., Liu, S., Fan, L., Zhou, N., Li, B., Biofuels: introduction, in: *Biofuels: Alternative feedstocks and conversion processes for the production of liquid and gaseous biofuels*, pp. 3–43, Academic Press, London, United Kingdom, 2019.
71. Cheng, F. and Brewer, C.E., Conversion of protein-rich lignocellulosic wastes to bio-energy: Review and recommendations for hydrolysis+ fermentation and anaerobic digestion. *Renew. Sustain. Energy Rev.*, 146, 111167, 2021.
72. Jha, B., Jaiswal, K.K., Ramaswamy, A.P., Impact of poly-aluminium chloride on foam suppression in a chicken waste-based biogas plant: A case study at KEPCO Kerala. *Int. J. Environ. Sci.*, 6, 6, 934–940, 2016.
73. Kamyab, H., Yuzir, A., Ashokkumar, V., Hosseini, S.E., Balasubramanian, B., Kirpichnikova, I., Review of the application of gasification and combustion technology and waste-to-energy technologies in sewage sludge treatment. *Fuel*, 316, 123199, 2022.
74. Nunes, L.J.R., Matias, J.C.O., Catalão, J.P.S., Mixed biomass pellets for thermal energy production: A review of combustion models. *Appl. Energy*, 127, 135–140, 2014.
75. Jaiswal, K. K., Kumar, V., Gururani, P., Vlaskin, M. S., Parveen, A., Nanda, M., and Grigorenko, A. V., Bio-flocculation of oleaginous microalgae integrated with municipal wastewater treatment and its hydrothermal liquefaction for biofuel production. *Environ. Technol. Innov.*, 26, 102340, 2022.
76. Sharma, N., Jaiswal, K.K., Kumar, V., Vlaskin, M.S., Nanda, M., Rautela, I., Ahmad, W., Effect of catalyst and temperature on the quality and productivity of HTL bio-oil from microalgae: A review. *Renew. Energy*, 174, 810–822, 2021.
77. Sharma, A., Pareek, V., Zhang, D., Biomass pyrolysis—A review of modelling, process parameters and catalytic studies. *Renew. Sustain. Energy Rev.*, 50, 1081–1096, 2015.
78. Demirbas, A.H., Yields and heating values of liquids and chars from spruce trunkbark pyrolysis. *Energy Sources*, 27, 14, 1367–1373, 2005.
79. Brammer, J.G., Lauer, M., Bridgwater, A.V., Opportunities for biomass-derived "bio-oil" in European heat and power markets. *Energy Policy*, 34, 17, 2871–2880, 2006.
80. Demirbas, A., Recent advances in biomass conversion technologies. *Energy Edu. Sci. Technol.*, 6, 19–41, 2000.
81. Aguado, R., Olazar, M., Gaisán, B., Prieto, R., Bilbao, J., Kinetic study of polyolefin pyrolysis in a conical spouted bed reactor. *Ind. Eng. Chem. Res.*, 41, 18, 4559–4566, 2002.
82. Cornelissen, T., Yperman, J., Reggers, G., Schreurs, S., Carleer, R., Flash co-pyrolysis of biomass with polylactic acid. Part 1: Influence on bio-oil yield and heating value. *Fuel*, 87, 7, 1031–1041, 2008.

83. Sikarwar, V.S., Zhao, M., Fennell, P.S., Shah, N., Anthony, E.J., Progress in biofuel production from gasification. *Prog. Energy Combust. Sci.*, 61, 189–248, 2017.
84. Soroush, S., Ronsse, F., Verberckmoes, A., Verpoort, F., Park, J., Wu, D., Heynderickx, P.M., Production of solid hydrochar from waste seaweed by hydrothermal carbonization: effect of process variables. *Biomass Convers. Biorefin.*, 14, 1, 183–197, 2024.
85. Molino, A., Larocca, V., Chianese, S., Musmarra, D., Biofuels production by biomass gasification: A review. *Energies*, 11, 4, 811, 2018.
86. Basu, P., *Biomass gasification, pyrolysis and torrefaction: practical design and theory*, Academic press, London, United Kingdom, 2018.
87. Kumar, V., Arora, N., Pandey, S. *et al.* Microwave-assisted pretreatment of harmful algal blooms for microbial oil-centered biorefinery approach. *Biomass Conv. Bioref.*, 12, 3097–3105, 2022. https://doi.org/10.1007/s13399-020-00941-5.
88. Tassoni, A., Tedeschi, T., Zurlini, C., Cigognini, I.M., Petrusan, J.I., Rodríguez, Ó., Corvini, P.F., State-of-the-art production chains for peas, beans and chickpeas—valorization of agro-industrial residues and applications of derived extracts. *Molecules*, 25, 6, 1383, 2020.
89. Awasthi, M.K., Ferreira, J.A., Sirohi, R., Sarsaiya, S., Khoshnevisan, B., Baladi, S., Taherzadeh, M.J., A critical review on the development stage of biorefinery systems towards the management of apple processing-derived waste. *Renew. Sustain. Energy Rev.*, 143, 110972, 2021.
90. Gururani, P., Bhatnagar, P., Bisht, B., Jaiswal, K.K., Kumar, V., Kumar, S., Rindin, K.G., Recent advances and viability in sustainable thermochemical conversion of sludge to bio-fuel production. *Fuel*, 316, 123351, 2022.
91. Arora, A., Nandal, P., Singh, J., Verma, M.L., Nanobiotechnological advancements in lignocellulosic biomass pretreatment. *Mater. Sci. Energy Technol.*, 3, 308–318, 2020.

13

Utilization of Food Waste for Bioenergy Production

Srutee Rout*, Rakesh Kumar Gupta, Sangeetha Karunanithi, Gnana Moorthy Eswaran U., Proshanta Guha and Prem Prakash Srivastav

Agricultural and Food Engineering Department, Indian Institute of Technology Kharagpur, West Bengal, India

Abstract

The importance of food waste has received little attention because of the lack of implementation of strategies and effective management policies. Enormous volumes of household waste have been generated at all points in the supply chain, including packaging, harvesting, consuming, and transporting. Due to its extensive production and inappropriate disposal practices, food waste (FW) has become a significant issue on a global scale. Recently, FW has typically ended up in landfills, which is the less desirable option. Organic material, which makes up the majority of FW and is easily converted into pollutants, has a negative impact on the ecosystem. Additionally, it has been established that FW landfill is a major contributor to the global warming trend. Methane and carbon dioxide are released in enormous quantities during the disposal of FW in landfills. Methane makes up between 40% and 70% of the biogas from landfills, and its warming impact of carbon dioxide is around 20–25 times higher. However, both the environment and human health are negatively impacted by these activities. Because FW contains a lot of water and other organic constituents, it can be recycled easily. This property of FW helps in the manufacture of bioenergy and other crucial chemicals used in industry, such as pigments, enzymes, organic acids, and essential oils. In addition to addressing FW disposal issues, the production of bioenergy from FW will help to reduce greenhouse gas emissions by displacing the use of coal, gasoline, and natural gas. Therefore, the generation of bioenergy, such as bioelectricity, biofuel, biodiesel, biobutanol, and biohythane, has been the main focus of this chapter.

**Corresponding author*: sruteerout1997@gmail.com

Hrudayanath Thatoi, Sonali Mohapatra, Swagat Kumar Das and Sukanta Kumar Pradhan (eds.) Applied Biotechnology and Bioinformatics: Agriculture, Pharmaceutical Research and Environment, (303–330) © 2025 Scrivener Publishing LLC

Keywords: Food waste, bioenergy, bioelectricity, biofuel, biohythane, biobutanol, biohydrogen

13.1 Introduction

One of the biggest challenges that developing countries are facing is the problem of waste management. Waste is often thought of as something that has lost its value. However, garbage's high organic content makes it suitable for energy recovery, helping to address issues with waste management, identify alternative energy sources, and lessen environmental contamination. There is a tremendous quantity of food waste produced globally, and effective solutions are required to turn this waste into resources. "The Food Waste Index Report 2021, from the United Nations Environment Programme (UNEP) said that around 931 million tonnes of food waste were generated in 2019, sixty-one per cent of which came from households, 26 per cent from food service and 13 per cent from retail" (United Nations, 2021). Various sources emit food waste into the environment, including home, organizations, and hospitals, and it might be precooked, expired, or leftover. The Food and Agriculture Organization (FAO) reveals that 1.5 billion tons of food are being lost from every trophic level of the food web, and this number has been rising annually [1]. Utilizing garbage as an input, the circular bioeconomy seeks to integrate green technology to produce goods that are commercially viable. Vegetables, dairy products, meat, fruits, and bakery products make up the majority of food loss sources. The organic proportion of municipal solid trash (OPMST), which consists primarily of food waste, is affected by water. This means that when food waste is dumped in a landfill, it decomposes and releases methane, while a greenhouse gas has a potential for global warming (GW) in a time period of 100 years. The appropriate treatment and disposal of biowaste is a global issue that is becoming more and more important, especially in developing nations, where the bulk of municipal solid waste (MSW) is made up of food waste.

Because both developed and developing nations are finding it difficult to handle the enormous volumes of food waste, different parts of the world have built frameworks for the proper management of this waste. According to studies, between 2005 and 2025, the amount of food waste could increase from 280 to 420 million tons [2]. The majority of carbon molecules in food waste are responsible for the climate change effect, and the release of hazardous gases into the atmosphere, namely, carbon dioxide, methane, and carbon monoxide, causes GW. The sequence of waste management includes prevention, conversion for human nutrition, use for livestock feed, use as a raw resource in manufacturing, biomass gasification or composting to create fertilizer, use as a sustainable source of energy, combustion, and disposal [3]. Municipal garbage was traditionally reduced through dumping or incineration, but these methods can affect human health and the environment [4]. Food waste needs to be handled to reduce ecological harm since it releases gas emissions that lead to climate change when it is dumped in landfills [5]. Dioxins are released during incineration, which harms the environment. They are part of the "dirty dozen," a collection of hazardous substances known as persistent organic pollutants (POPs). Dioxins have a high potential for being poisonous, which is a cause for concern. It has also been linked to an increase in cancer rates and problems in reproduction. It also prevents chemicals and nutrients from being extracted from the debris that is burnt [6]. To reduce the release of toxic gases into the environment, food waste is now pre-treated before being burned.

Food waste can be converted in a variety of ways to produce bioenergy, including thermochemical conversion, biochemical conversion, and transesterification [7, 8]. These different methods are explained below:

a) Thermochemical conversion: Food waste is preserved through dewatering in a process known as thermochemical degradation. The mass and volume of the original waste materials are significantly reduced as a result of this procedure, which may not produce a "new" food waste product but does lower hauling and application expenses. When organic material is treated to high temperatures in thermochemical processes, the bonds between the molecules are broken, and the result is oxygenated gas, bio-oil or charcoal, whereas in biochemical processes, microorganisms like bacteria or yeast transform the organic material into bioenergy.
b) Biochemical conversion: Biochemical pathways are achieved by utilizing specific microorganisms. By using different microorganisms, it is possible to produce several types of bioenergy, including biobutanol, biohydrogen, bioethanol, biofuel, etc. The two main processes of this conversion are anaerobic digestion and fermentation [9].
c) Transesterification: Triacylglycerides from a variety of feedstock, including non-edible oil seeds, animal fats, vegetable oils, used cooking oil, and microbial lipids are transformed

into fatty acid methyl esters (biodiesel) during transesterification (methanol or ethanol) [10].

The optimum technique for handling food waste has been demonstrated to be anaerobic digestion [11]. It is currently used for a number of purposes, such as recycling and the manufacture of biogas, biohythane, and biomethane. It recycles fertilizers again into the farmland by producing by-products like biofertilizers [12]. Anaerobic digestion and dark fermentation can both yield biohythane, a combination of methane and hydrogen. Fermentation can create biohydrogen and biomethane in a single stage or two stages. Because two-stage fermentation generates methane from anaerobic digestion from dark fermentation, it is preferred to use single-stage fermentation as it requires less pretreatment and shorter fermentation times [13]. Photo fermentation and biophotolysis depend on light whereas microbiological electrolysis cells and fermentation in the dark are both light-independent activities. There are numerous techniques to produce biodiesel, biobutanol, and bioelectricity, all of which support a circular economy. Fermentation of acetone butanol ethanol (ABE) by utilizing *Clostridium acetobutylicum* can produce biobutanol. By using base catalyzed transesterification, biodiesel can be produced. Alcohol and fat/oil (triglycerides) react, resulting in the production of esters and glycerol [14]. When opposed to the production of bioenergy using fossil fuels, bioenergy may be produced more cheaply and sustainably using renewable resources. In order to ensure the proper handling of food waste, it is crucial to use good food management procedures. As a result, harnessing renewable bioenergy has drawn greater attention recently, and new technologies are in the process for using bioenergy to synthesize numerous goods. Due to all the aforementioned environmental and financial advantages, this study discusses several methods, criteria, and potential outcomes for producing bioenergy from food waste using effective and affordable technologies.

13.2 Potential of Food Waste for Bioenergy Production

The literature claims that food waste generated as consumption grew, which significantly influenced the value of GDP (Gross Domestic Product) per capita. Canada produced maximum food waste at the rate of 0.8 kg/capita/day, followed by the US at 0.5 kg/capita/day. Germany, Japan, and England generate remarkable food waste around 12.258 MT, 23 MT, and 14.257 MT

Table 13.1 Source of food waste for generating energy.

Energy production	Process	Food waste	Yield	Reference
Syngas	Gasification	Sugarcane bagasse	13.3 mmol/g	[17]
		Aloe vera rind	11.5 mmol/g	
		Banana peel	15.0 mmol/g	
		Coconut shell	11.8 mmol/g	
		Lemon peel	12.8 mmol/g	
		Orange peel	12.3 mmol/g	
		Pineapple peel	14.8 mmol/g	
Bioethanol	Simultaneous saccharification and fermentation	Instant noodle waste	96.8%	[18]
Ethanol production	Acidic hydrolysis or enzymatic hydrolysis	Potato peel waste	7.58 g/L	[19]
Ethanol production	Enzymatic hydrolysis	Mixed food waste	65%	[20]
Ethanol production	Enzymatic saccharification	Household food waste	40.81%	[21]
Ethanol production	Hydrolysis	Mixed food waste	93.86 g/L	[22]
Biohydrogen	Anaerobic fermentation	Palm oil mill seed sludge	79 mmol H_2	[23]
Biochar	Hydrothermal carbonization	Raw potato, creamy yogurt, and raw onion	63%	[24]
Biodiesel	Fermentation	Waste cooking oil	64%	[25]
Biodiesel	Extraction and transesterification	Spent coffee grounds	10–15%	[26]
Biobutanol	Fermentation	Cane molasses	10 g/L	[27]

per year, respectively [15]. In the same way, the developing nations produced 56 kg/capita of food waste. The improper food management plan leads to generation of more food waste.

China and India are the world's top countries in food waste generation due to their vast populations, ineffective harvesting methods, improper handling, as well as poor marketing strategy. The food waste generation in underdeveloped nations is due to inadequate awareness about environmental issues and food management skills. Traditionally, the generated food waste has been mixed with MSW that is discarded in empty lands, which causes major environmental pollution and can be prevented by adopting novel technology used to convert those waste to produce valuable resources.

Few methods have been developed to address the above issues. Anaerobic methods are chosen for FW treatment due to their low carbon footprint and minimal pollutants. These approaches were based on anaerobic processes used for bioenergy generation from food waste. An illustration would be made that if all FW in Australia were converted into renewable energy sources, it could meet 3.5% of the energy requirement. A biogas production of 200 m^3/ton from FW could meet around 1% of the nation's total energy consumption in Germany. The top three nations like China, US, and India have the greatest potential for FW to create bioenergy. However, a sizable portion of FW is recovered via landfill for composting FW into bioenergy requires a greater capital investment, technological expertise, and operational costs [16]. Table 13.1 gives information on energy produced and its yield from different food wastes.

13.3 Bioenergy from Food Waste

Several studies have found that there is a huge potential for turning food waste into high value-added commodities like fuels and chemicals. However, to turn food waste into bioenergy, thermochemical or biochemical techniques can be utilized (Figure 13.1) [28]. Bioenergy is a form of renewable energy created from live organic material and used for transportation as well as the production of heat and power [29]. Plasma gasification is a cutting-edge method to boost bioenergy recovery. Biomass may be converted into biofuels (liquid fuels), which can subsequently be used to produce energy. Agricultural debris, waste from food, animal waste, waste from paper industries, waste from palm oil industries, and microalgal biomass are all examples of biomass that can be utilized to produce bioenergy. Waste produced from food has been targeted for the generation of

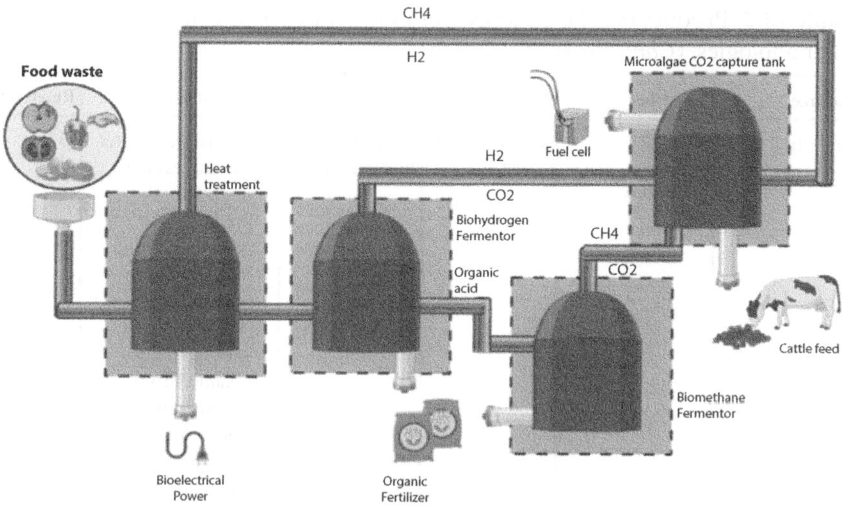

Figure 13.1 Schematic diagram of biochemical process for the production of bioenergy.

Table 13.2 Production of bioenergy from food wastes using different methodologies.

Source	Microbes	Methodology	Product	Observation	Reference
Kitchen waste water	*Synechococcus* sp. and *Chlorococcum* sp.	Microbial fuel cell	Bioelectricity	*Synechococcus* sp. and *Chlorococcum* sp. act as potential biocathode in improving power generation at microbial fuel cell. With the algae biocatholyte, it is thus possible to produce bioelectricity, sequester CO_2, and create biomass while treating kitchen wastewater in the microbial fuel cell.	[31]

(*Continued*)

Table 13.2 Production of bioenergy from food wastes using different methodologies. (*Continued*)

Source	Microbes	Methodology	Product	Observation	Reference
Industrial food waste	*Klebsiella pneumoniae*	Microbial fuel cell	Biohydrogen Bioelectricity	Industrial wastewater was successfully utilized for the production of biohydrogen and bioelectricity in microbial fuel cell. Among the various strains obtained from industrial waste water, *Klebsiella pneumonia-FA2* was found to have higher potential in bioelectricity and biohydrogen production.	[32]
Food waste	*Firmicutes, Bacteroidetes, Proteobacteria, Geobacter*	Microbial fuel cell	Bioelectricity	The microbial community isolated from canteen food waste shows the combination of fermentative and exoelectrogenic bacteria.	[33]
Rice mill waste water	-	Dual chambered microbial fuel cell	Bioelectricity	Amount of easily accessible organics for exoelectrogens increased substantially. Maximum volumetric power was greater than that of the study conducted without an acidogenic chamber. Helped in improving the bioelectricity output as well as wastewater treatment efficiency.	[34]

(*Continued*)

Table 13.2 Production of bioenergy from food wastes using different methodologies. (*Continued*)

Source	Microbes	Methodology	Product	Observation	Reference
Spent tea waste	-	Olla pot coupled Microbial fuel cell	Crop irrigation and Bioelectricity	Helped in the novel combination of biosorption, Olla-Pot irrigation, and microbial fuel cells. Integrating effective wastewater treatment with a water conservation strategy proves to be a good choice. Plant growth was not adversely affected by the combination of microbial fuel cells with Olla-pots.	[35]
Potato waste	-	Single chambered microbial fuel cell	Bioelectricity	Biodegradable organic matter in potato waste was successfully removed using MFC. Removal of organic matter was accompanied by electricity production, demonstrating that MFC technology is a promising technology for producing sustainable energy.	[36]

(*Continued*)

Table 13.2 Production of bioenergy from food wastes using different methodologies. (*Continued*)

Source	Microbes	Methodology	Product	Observation	Reference
Jackfruit waste	-	Direct combustion	Bioelectricity	The utilization of jackfruit wastes in direct burning offers a low-emission and ecologically beneficial route for creating bioenergy. The jackfruit wastes displayed appealing features for direct application in combustion. The prospective suitability for biomass boilers is supported by their combustion qualities, proving that these undesirable lignocellulosic residues can serve as an affordable source of bioenergy feedstock.	[37]
Food waste	-	Algae-assisted microbial fuel cell	Bioenergy	The developed system consisting of anodic chamber in algae assisted microbial fuel cell has been successfully used for improving bioenergy production.	[38]

(*Continued*)

Table 13.2 Production of bioenergy from food wastes using different methodologies. (*Continued*)

Source	Microbes	Methodology	Product	Observation	Reference
				Three pathways for stress alleviation may occur in the AD-AMFC system: (i) alleviating the adverse impact of VFAs by consuming them as substrates of AMFC; (ii) alkalinity generated in the catholyte facilitated pH recovery during the AD process; (iii) ammonia inhibition was relieved due to ammonium migration and algal accumulation. The novel system presents opportunities for stabilizing and enhancing the AD process in the treatment of high-strength wastes with high ammonium concentrations.	
Potato chips processing waste water	-	Dual chamber microbial fuel cell	Bioelectricity	High removal efficiency of COD up to 99% associated with a relatively significant energy generation at room temperature	[39]

bioenergy, including corn residues, cheese, wheat bran, whey, restaurant waste, soybean meal, waste generated from kitchen such vegetable and fruit seeds, pomace, peels, molasses, and sugarcane bagasse (Table 13.2) [30]. Food waste is valorized to produce biohydrogen, biogas, bioelectricity, biodiesel, biobutanol, bioethanol, as well as other valuable products like adsorbents, pigments, nutraceuticals, vermicompost, xanthan gum, and single-cell proteins. Food waste can produce chitosan, volatile fatty acids, high fructose corn syrup, citric acid, propionic acid, D-lactic acid, succinic acid, gluconic acid, and quercetin. The production of some bioenergy, including biohydrogen, bioelectricity, and biobutanol, biofuel, biohythane, and biogas, has been thoroughly covered in the section below.

13.3.1 Bioelectricity

Agricultural waste, sawmill waste, forest leftovers, food waste, rice hulls, and other lignocellulosic biomass are the primary resources of lignocellulosic feedstock that is burned to produce bioelectricity [40]. In addition to that, bioelectricity is produced by burning waste products from agriculture and forestry (biomass). However, the cutting-edge and environmentally friendly microbial fuel cell (MFC) technology can instantly transform the chemical energy contained in waste material into bio-electrical energy. Despite this, numerous techniques, including microbial fuel cells, thermochemical conversion, and combustion, may be used to produce bioelectricity. However, rapid combustion of lignocellulosic feedstock leads to reduced greenhouse gas emissions especially contrasted with traditional power generation methods. A study demonstrated that, when biomass is burned, volatile gases are produced. It also produces steam and heat when it interacts with air. In order to improve the efficacy of electricity generation, several kinds of turbine operating modes and turbine blades are employed. However, turbines that produce energy are powered by steam produced by the heat [28].

Gasification is an alternative thermochemical conversion method where leftover biomass from the field of solid waste management is gasified to generate decomposition product including chlorides, char, tar, and syngas. Apart from this, in order to save energy and save the environment, gasification is preferable over burning. For the purpose of addressing the issue of inadequate power in rural areas, gasifiers may be used [41]. Gasification is used for conversion of biomass waste produced by rural households or organizations into bioelectricity. A study also showed that an increase in power efficiency of up to 35% is obtained by integrating the combustion boilers and gasifiers. A modern way of generating electricity is the microbial fuel cell (MFC) technology. The proton exchanger membrane in MFC can separate the cathode chamber (abiotic) and anode chamber (biotic) [42]. However, in the absence of oxygen, electrogenic bacteria generate bioelectricity from organic substrates. Electrons are transmitted through external circuits, while protons are carried by proton exchange membranes. MFC can be hampered due to innumerous factors like micro-organisms, substrate loading rate, hydraulic retention rate, temperature, concentration, and pH. Additional system variables include electrodes, prices, and designs [43]. Despite this, Ti/TiO^2 electrodes have been discovered to have a greater power density than Pt electrodes. It has been shown that MFC is the most effective method for generating electricity from food waste, and MFC performance is enhanced by prior food waste hydrolysis. However, the MFC method is safe, effective, renewable, and ecologically beneficial.

Furthermore, a power density of 358.1 mV/m^2 and a current density of 847 mA/m^2 at 0.59 V were generated by bioelectricity from MFC utilizing an substrate of orange peel. The utilization of bacteria like *Pseudomonas*, *Enterococcus*, and *Paludibacter*, for generating electricity from MFC showed how much the loading rate affects MFC. Apart from this, using MFC, it is possible to produce bioelectricity from the leachate of acidogenic food waste which has produced COD elimination 90%, voltage 1.12 V, and power density 15.14 Wm^3. However, *Chlorella vulgaris* produces bioelectricity and lipids from food waste that has been treated with MFC. Some study demonstrated that, using a carbon-based electrode (especially, zeolite, graphite plate, and carbon cloth), thermomechanical pumping, and anode and microbial fuel cell, as well as metal-based electrode (tin coated copper), may generate bioelectricity from wastewater that contains xylose (TMP). Furthermore, a carbon cloth electrode was able to provide the highest power density of 333 15 mW/m^2, which also reduced 77–86% of COD. On the other hand, azadirachta indica generated a low-cost CoO-NiO cathode made of the metal oxides Ni and Co.

13.3.2 Biofuel

With the rising material consumption, food waste disposal has evolved to be a significant environmental problem. A lot of food waste is produced as a result of population increase and the food industry's quick development. A little over 1.3 T of food waste is produced annually, and as the economy and population expand, this amount is anticipated to rise [44]. Composting, landfilling, material resource optimization, and direct disposal are the primary traditional methods for treating food waste.

The egative impacts of using traditional methods are unavoidable, even though they can somewhat lessen the environmental impact of food waste. As a result, the scientific community has worked hard to enhance waste management techniques and better reduce the buildup of food waste. The negative impacts of conventional methods can be effectively mitigated by biofuel technology, which is now the most environmentally benign and promising technique. Enhanced energy molecules produced by biological processes and taken from the biomass of living creatures are referred to as "biofuels." One of the key substitutes that can meet the rising need for energy sources around the world is biofuels [45]. The different generations of biofuels are first, second, third, and fourth. The first generation of biofuel is made from edible feedstocks like wheat, potatoes, barley, rice, sugarcane, and vegetable oil, and this has attracted attention throughout the world. In an effort to avoid the controversies surrounding the first generation of biofuel, non-food items are being used as feedstock. It was suggested that a second generation of biofuel be developed to address the shortcomings of the first. Biofuels are created from non-edible feedstocks such as woody biomass and forest wastes. Since it does not require a particular type of soil, water supply, or fertilizers for growing, it does not compete with food crops. Compared to earlier generations, the second generation of biofuel emits fewer greenhouse gases. The fact that the feedstock for second-generation biofuel is waste biomass instead of being grown makes this one of its key advantages. The biomass of microalgae is used to produce third-generation biofuel. In contrast to other plants, microalgae develop more quickly and do not need a lot of land to flourish. As a result, there is no competition with the agricultural sector, residential real estate, or animal farms. The fourth generation of biofuel is concentrated on microalgal genetic engineering. Using cutting-edge technologies, it is intended to generate microalgae that can efficiently absorb huge amounts of CO_2, boost the yield of biofuel, and be more adaptable to wastewater [46].

The main categories of food wastes used to prepare biofuels include protein, starch, cellulose, fat, and sugar. Straw, bagasse, grape and apple pomace, and kitchen trash are examples of carbohydrate-rich wastes that can be transformed into biofuels through the processes of anaerobic digestion, aerobic digestion, and microbial fermentation. Agro-food wastes are utilized in the production of biofuel (Figure 13.2), a process that is extremely complex and includes a number of interconnected steps, such as prior treatment of raw material, hydrolyzing complex molecules to simpler ones using enzyme followed by fermenting monomer to biofuel, and purification [8]. Co-substrates are more frequently employed to boost the production of biofuel and address the drawbacks of a single substrate.

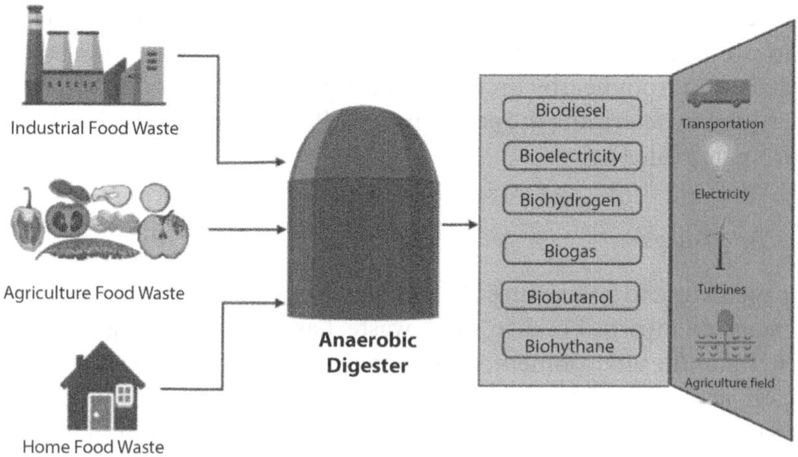

Figure 13.2 Biofuel production from industrial, agriculture and home food waste.

Production of hydrolyzed enzymes along with synthesizing biofuels is currently expected to have significant financial burdens, making it impracticable on a wide scale. The generation of microbial enzymes for catalytic processes (such as the manufacture of biofuel, hydrolytic enzymes, pigment, flavonoids, vitamins, etc.) from a variety of unused products and raw materials, which would otherwise pose as solid waste in the ecosystem, is a promising process. It is interesting to consider encouraging a method for numerous reactions catalyzed from these biomasses and then using these catalysts for synthesis in the biofuel, food, and related industries [8].

Microalgae are garnering greater attention in the third generation of biofuel because of the promising amount of biomass they produce throughout the cultivation phase, which contains substantial amounts of oil prior to the synthesis of biodiesel. Furthermore, compared to first-generation biofuels, microalgae require less space/land for growing and create considerably cleaner biodiesel when compared to petroleum fuel. They also do not compete with food crops. Microalgae cells have simpler cell walls than the second generation of biofuels, and they lack lignin and hemicellulose, making it possible to lyse them to extract fatty acids. However, choosing appropriate microalgae species, pretreatment procedures, reactor design, optimizing fermentation operating conditions, particularly in aseptic conditions where there is a significant hazard of exposure, and energy-intensive techniques of microalgae harvesting are the main obstacles to the commercialization of microalgae biofuels [47]. Food waste can be converted into biofuels, which is effective, profitable, and environmentally friendly.

To investigate efficient and effective conversion, a variety of possible waste items are transformed into biofuels, including canteen garbage and discarded grain from nearby breweries. In conclusion, further research needs to be done to determine the economic viability of using biofuel technology on a bigger scale [44].

13.3.3 Biogas

Management and processing of food waste has emerged as a critical issue in many countries, because food waste has significant negative consequences on the environment. However, the emissions of greenhouse gases (GHG) have also been impacted by processes including transportation, refrigeration, production, retail, processing, and distribution. Food waste that is dumped in landfills releases GHGs like methane and carbon dioxide into the environment. Methane has a 25-fold greater energy density than carbon dioxide. A study exhibited that 22% of GHG emissions come from agriculture, while 6.7% of GHG emissions come from food waste [48]. However, food waste may be effectively managed via anaerobic digestion. Furthermore, biogas is produced via the anaerobic digestion process and made up of different types of gas like CO_2, H_2O, and CH_4 [49]. Apart from this, in order to digest food anaerobically, biological and metabolic activities must occur without the presence of oxygen. In the ideal situation, organic matter decomposes into smaller molecules, producing end products like CO_2 and CH_4, commonly referred to as biogas. In addition to producing biogas, other gases including ammonia and hydrogen sulfides are also generated, while the phases of anaerobic digestion were studied in relation to microorganisms using 454-pyrosequencing. However, anaerobic digestion consists of four steps during the production of biogas [50]. These stages are hydrolysis, methane production, acetic acid production, and acidification. In the first stage, amylase, cellulase, lipase, protease, and xylanase are a few types of the enzymes that attach to polymeric substrates and break them down into monomers (glucose, amino acids, glycerol, and fatty acids) to start the hydrolysis process.

However, these enzymes are produced when bacteria are hydrolyzed, while during the subsequent acidogenesis step, the monomers are converted to gaseous by-products, fatty acids, and alcohols. Apart from this, both facultative and obligatory anaerobes are required for the first two phases. Anaerobes that can exist either with or without oxygen are called facultative anaerobes, whereas obligatory anaerobes organisms cannot live without oxygen. Additionally, four stages of anaerobic digestion go into producing biogas. The organic molecules produced during acetogenesis are converted to hydrogen, acetic acid, and carbon dioxide. However, the growth of bacteria in acetogenesis is rather sluggish, with a twofold period of 1.5–4 days [51]. Despite this, methanogens produce methane gas by breaking down acetate, carbon dioxide, or methyl under very anaerobic circumstances. Furthermore, hydrolysis, acetogenesis, and methanogenesis all need pH levels 6, 6–7, and 6.5–7.5, respectively. Bacteria that produce acid and methane need a pH higher than 5.0. In the 6.8–7.2 range, methanogenic bacteria function optimally.

13.3.4 Biohydrogen from Food Waste

The production of biohydrogen (H_2) from wastage has attracted the attention of experts all around the globe, since it is considered to be a sustainable green fuel. Through anaerobic fermentation, photo-fermentation, and bio-photolysis or a combination of these processes, H_2 may be produced biologically. A report showed that it is a low-cost approach to fermentation that is light-driven and recognized as the safest and greenest gaseous fuel. Numerous researchers have revealed that the synthesis of biohydrogen from the food waste (FW) can be produced from various types of culture. In addition to that, biohydrogen can be produced through various types of reactors such as fluidized reactor, photo bioreactor, membrane electrolysis cell, stirred tank reactor, anaerobic sludge blanket, and upflow anaerobic bioreactor. However, FW is a promising adsorbent since it contains a lot of readily degradable organic material, providing an economic balance and net positive energy.

Moreover, industries that process food such as potato processing, cereal, cheese, etc. produce enormous volumes of wastewater with significant quantities of sugars and carbohydrates. Subsequently, the food processing sector benefits from the usage of these wastes to produce biohydrogen, since the energy may be utilized as a source of power. Peels from fruits like bananas, pomegranates, and tangerines may be used to produce biohydrogen [52]. In addition to that, rice straw and palm oil waste may also be used in dark fermentation to produce biohydrogen [53]. Various researchers revealed that the potential biomass for the synthesis of biohydrogen include waste from pineapple and apple as well as peels from different fruits, or completely matured fruits including grapes, pears, and apples that are rich in sugar content but have no commercial value [54]. Despite this, physicochemical properties of food waste are the crucial parameters for the design, creation, and management of an anaerobic digestion system for the generation of biohydrogen. It is assumed that a number of variables, including low hydrogen partial pressure, temperature, pH, and pretreatment of food waste, are critical in influencing the synthesis and yield of biohydrogen [20]. Additionally, moisture content, particle size, nutrient content, biodegradability of food waste, and volatile solid composition are other variables that affect biohydrogen production [50]. The biohydrogen synthesis can be produced by two ways such as light dependent and light independent. A study also exhibited that light-dependent processes include biophotolysis and photo fermentation, whereas light-independent types include dark fermentation and microbial electrolytic cells. Furthermore, in photolysis, water is split in the presence of solar radiation to produce hydrogen, while in dark fermentation, bacteria consume lipids, proteins, and carbs to produce biohydrogen [55]. Biohydrogen has traditionally been produced by a process known as "dark fermentation" that employs anaerobic bacteria to break down organic molecules in the absence of light. As a result of this process, volatile fatty acids and alcohols are produced in addition to biohydrogen [56]. However, the subsequent production of biomethane uses these by-products. With the goal of simultaneously producing biohydrogen and biomethane, two stages of fermentation have been devised, the first of which is used to produce biohydrogen and the second of which is used to produce biomethane. Novel techniques for synthesizing biohydrogen include solar thermal water splitting, electro photochemical water splitting, fermentation of feedstock rich in sugar, and reforming renewable sources.

Despite this batch, continuous fermentation and semi-continuous systems have been developed for the generation of biohydrogen, and pretreatment of food waste, fermentation, vessel type, inoculum, and food waste composition are crucial requirements for the production of biohydrogen. However, one third of the food waste utilized in the manufacture of biohydrogen comes from vegetable wastes and fruit that include polysaccharides, proteins, carbohydrates, and heavy metal ions [57]. Additionally, when oxygen and light are not present, this kind of fermentation, known as dry fermentation, transforms the substrate into biohydrogen. Moreover, to prevent hydrogen emission, microbes employ hydrogen, phenol, ammonia, and heavy metal ions [58]. Some researchers suggested that, using the bacteria *Clostridium thermocellum*, date palm seeds are continuously fermented in the dark to generate biohydrogen. because they contain significant levels of lignin, hemicellulose, and cellulose [59]. However, growth rate, pH, and hydrolysis capacity are the key variables that affect this process.

13.3.5 Biohythane from Food Waste

In the USA and India, the hydrogen–methane combination known as biohythane has been used commercially as an automotive fuel. Growing interest in biohythane's due to its outstanding performance has brought attention to this technology in recent years, as a result of a number of laboratory as well as pilot-scale equipment production. Between 10% and 25% of hythane is discovered to contain hydrogen [60]. The principal energy sources emitted through landfill are natural gas as well as biogas in the form of methane [61]. Hythane synthesis requires two gas channels, one at hydrogen and a second at methane along with regulated gas flow rates. Another choice is to use substrate in a two-stage dark fermentor to create hydrogen as well as methane within the same system [62]. Hythane synthesis is influenced by the H_2/CH_4 ratio, substrate, microbial community, and reactor configuration.

Food waste and sewage sludge have been selected by the scientific community as the two waste types for the trial. Due to their extensive availability, it is anticipated that 8 MT of garbage will be processed in EU nations. They have taken 82% of waste (food waste) and 28% waste sewage sludge which is being processed by single-stage AD plants [63]. The crucial factor that influences the production of energy is pH and should be maintained in the range of 5 to 6, especially when interacting with diverse substrates like HFWs, and hydrogenase enzyme function at an appropriate level is 5.5. There are multiple approaches to adjusting pH, such as adding alkaline chemicals or using HFWs with a high protein content [64].

Although methane burns cleaner than gasoline or diesel, it does not burn as efficiently because of its slow burning rate, limited flammability, and higher ignition temperature [65]. Most of methane's limitations can be solved by adding hydrogen since hydrogen may increase methane's limited choice of flammability, flame speed, and increase its fuel efficiency. Hydrogen may be added to methane to raise its H/C ratio, which lowers GHG emissions. Additionally, the inclusion of hydrogen reduces the methane's quenching distance, making ignition easier and requiring less energy [66]. It appears costly and unsustainable to produce hydrogen as well as methane using fossil fuels, while the fermentation of these gases from biological sources using renewable resources is more environmentally friendly and less expensive. A single device may combine biohydrogen and biomethane to create biohythane, a superior fuel than those derived from fossil fuels.

In order to evaluate several reactor designs for the production of biohythane from HFWs, the Anaerobic Fixed Bed reactor (AFBR) and

Continuous Stirred Tank reactor (CSTR) for dark fermentation as well as the methanogenic level were employed for evaluation. The CSTR arrangement is the most often used reactor due to its ease of assembly and simple removal of certain AD bacteria, including methanogens during the dark fermentation process [67]. However, more delicate and sluggish methanogens cause the second stage of Alzheimer's disease [68]. A viable option would be AFBRs that can immobilize more sensitive microbes on permeable substrates with a wide base, highest absorption potential, resilience to pneumatic shocks along with organic shock, and no requirement for blending devices. These factors were taken into consideration when Mikheeva *et al.* [69] investigated a two-stage based AD using an AFBS as methanogenic fermentation and a CSTS as dark fermentation.

13.3.6 Bio-Butanol

Utilizing food waste effectively can have a favorable impact on the energy and environmental sustainability of the production of chemicals and fuels. Research on the fermentation of food waste has been conducted to produce biofuel. Butanol production is being developed as a more sophisticated biofuel than ethanol due to the growth of new biotechnologies and the increased interest in biofuel research. Numerous attempts have been made to create butanol from cellulosic biomass, including wheat straw, corn stover, barley straw, cassava bagasse, switchgrass, and miscanthus, as it is expensive to produce butanol from food grade feedstock (i.e., glucose and corn starch [70]. Wheat bran with both starch and non-starch polysaccharides was examined for the synthesis of butanol. Food waste has a significant advantage over cellulosic biomass when it comes to generating butanol. The majority of food waste contains a high amount of sugars and starch, which can readily be utilized for the butanol-producing culture (*Clostridium*), whereas cellulosic biomass needs to be pretreated under difficult circumstances, requiring a significant amount of energy consumption. Second, considerable amounts of functionalized molecules (such as proteins, fatty acids, and minerals) are present in food waste and can serve as nutrients for the growth of cultures responsible for butanol production [71]. Bio-butanol may also be made through biomass with ABE (acetone butanol ethanol) fermentation. Butanol, ethanol, and acetone are produced by *Clostridium acetobutylicum* by solventogenesis, butyric acid, and acetic acid via acidogenesis [72]. The yield of bio-butanol from ABE fermentation involves a number of steps like downstream processing, metabolic engineering for clostridium species, process monitoring, elimination of inhibitors, and pretreatment of substrates comprising starch, sugar, or lignocellulose (product recovery and purification). A variety of *Clostridium* species may produce bio-butanol from sugars, poly alcohols, organic acids, and amino acids [73].

Proliferation of bio-butanol synthesis is intended to expedite fermentation by removing the separate acidogenic phase and to prevent any link among sporulation along with solvent production [74]. Cells must be purified via membrane technology after butanol has been refined and regenerated back into fermenters [75]. The study of clostridia's metabolic processes and mechanism of biobutanol production is based on the *Clostridium* species' metabolic engineering. Research on *Clostridia* metabolism focuses on the organism's capacity to grow and tolerate oxygen, its strong butanol efficiency, strain stability, cellulose usage, as well as its maximum solvent resistance [76]. By suppressing acetoacetate decarboxylase, which reduces acetone formation, butanol production can be enhanced even further [77]. Hydrolytic enzymes produced by clostridia can degrade lignocellulosic materials.

Feedstock costs, which make about 60% of the whole operation's budget, are the biggest impediment to the manufacture of bio-butanol. Producing bio-butanol may be made economically feasible by shifting from starch-based biomass feed stocks (food additives) to the direction of agricultural leftover material, especially rice straw [78]. The synthesis of bio-butanol has made substantial use of wheat straw [79], barley [80], and maize fiber [81]. An inexpensive source of lignocellulosic biomass is rice straw, which is regarded as a waste in Asian nations. Because of its high cellulose content, this biomass can be considered as a raw material for the synthesis of bio-butanol, which would help to solve the issues with substrate economics and pollution brought on by industrial waste burning [82].

13.4 Conclusion

A lot of food waste is generated worldwide, creating pollution by releasing harmful gases and increasing the global warming effect. Disposal of these wastes in a proper manner is of utmost important for both humans and other living things. As we have discussed in this chapter, these wastes are full of organic and inorganic matter, making them suitable for energy production via different anaerobic methods and processes. They help in bio-energy production like fuels, electricity, heat, bioethanol, etc., which can be supplied to rural and other inaccessible areas. Hence, utilization of food wastes would definitely add value to the products and lead to a sustainable environment in the future.

References

1. Ferreira, J.A., Agnihotri, S., Taherzadeh, M.J., Waste biorefinery, in: *Sustainable resource recovery and zero waste approaches*, pp. 35–52, Elsevier, 2019.
2. Saqib, N.U., Sharma, H.B., Baroutian, S., Dubey, B., Sarmah, A.K., Valorisation of food waste via hydrothermal carbonisation and techno-economic feasibility assessment. *Sci. Total Environ.*, *690*, 261–276, 2019.
3. Vandermeersch, T., Alvarenga, R.A.F., Ragaert, P., Dewulf, J., Environmental sustainability assessment of food waste valorization options. *Resour. Conserv. Recycl.*, *87*, 57–64, 2014.
4. Shah, A.V., Srivastava, V.K., Mohanty, S.S., Varjani, S., Municipal solid waste as a sustainable resource for energy production: State-of-the-art review. *J. Environ. Chem. Eng.*, *9*, 4, 105717, 2021.
5. Malav, L.C., Yadav, K.K., Gupta, N., Kumar, S., Sharma, G.K., Krishnan, S., Bach, Q.V., A review on municipal solid waste as a renewable source for waste-to-energy project in India: Current practices, challenges, and future opportunities. *J. Cleaner Prod.*, *277*, 123227, 2020.
6. Hossain, S., Rahman, M.A., Chowdhury, M.A., Mohonta, S.K., Plastic pollution in Bangladesh: A review on current status emphasizing the impacts on environment and public health. *Environ. Eng. Res.*, *26*, 6, 2021.
7. Robak, K. and Balcerek, M., Review of second generation bioethanol production from residual biomass. *Food Technol. Biotechnol.*, *56*, 2, 174, 2018.
8. Arya, P.S., Yagnik, S.M., Rajput, K.N., Panchal, R.R., Raval, V.H., Valorization of agro-food wastes: ease of concomitant-enzymes production with application in food and biofuel industries. *Bioresour. Technol.*, 361, 127738, 2022.
9. Yadav, S., Singh, D., Mohanty, P., Sarangi, P.K., Biochemical and thermochemical routes of H2 production from food waste: a comparative review. *Chem. Eng. Technol.*, 46, 191–203, 2021.
10. Pour, F.H. and Makkawi, Y.T., A review of post-consumption food waste management and its potentials for biofuel production. *Energy Rep.*, *7*, 7759–7784, 2021.
11. Panigrahi, S. and Dubey, B.K., A critical review on operating parameters and strategies to improve the biogas yield from anaerobic digestion of organic fraction of municipal solid waste. *Renew. Energy*, *143*, 779–797, 2019.
12. Feiz, R., Larsson, M., Ekstrand, E.M., Hagman, L., Ometto, F., Tonderski, K., The role of biogas solutions for enhanced nutrient recovery in biobased industries—three case studies from different industrial sectors. *Resour. Conserv. Recycl.*, *175*, 105897, 2021.
13. Weide, T., Brügging, E., Wetter, C., Ierardi, A., Wichern, M., Use of organic waste for biohydrogen production and volatile fatty acids via dark fermentation and further processing to methane. *Int. J. Hydrogen Energy*, *44*, 44, 24110–24125, 2019.

14. Kumoro, A.C. and Saeed, M.T.M., Ultrasound-assisted transesterification of tropical goat fat–Palm oil blend for biodiesel synthesis. *Energy Convers. Manage.: X, 14*, 100213, 2022.
15. Gollnhofer, J.F., Weijo, H.A., Schouten, J.W., Consumer movements and value regimes: Fighting food waste in Germany by building alternative object pathways. *J. Consum. Res., 46*, 3, 460–482, 2019.
16. Dung, T.N.B., Sen, B., Chen, C.C., Kumar, G., Lin, C.Y., Food waste to bioenergy via anaerobic processes. *Energy Proc., 61*, 307–312, 2014.
17. Nanda, S., Reddy, S.N., Dalai, A.K., Kozinski, J.A., Subcritical and supercritical water gasification of lignocellulosic biomass impregnated with nickel nanocatalyst for hydrogen production. *Int. J. Hydrogen Energy, 41*, 9, 4907–4921, 2016.
18. Yang, X., Lee, S.J., Yoo, H.Y., Choi, H.S., Park, C., Kim, S.W., Biorefinery of instant noodle waste to biofuels. *Bioresour. Technol.*, 159, 17–23, 2014.
19. Arapoglou, D., Varzakas, T.H., Vlyssides, A., Israilides, C.J.W.M., Ethanol production from potato peel waste (PPW). *Waste Manage.*, 30, 10, 1898–1902, 2010.
20. Kim, J.H., Lee, J.C., Pak, D., Feasibility of producing ethanol from food waste. *Waste Manage.*, 31, 9-10, 2121–2125, 2011.
21. Matsakas, L., Kekos, D., Loizidou, M., Christakopoulos, P., Utilization of household food waste for the production of ethanol at high dry material content. *Biotechnol. Biofuels*, 7, 1, 1–9, 2014.
22. Yan, S., Chen, X., Wu, J., Wang, P., Pilot-scale production of fuel ethanol from concentrated food waste hydrolysates using Saccharomyces cerevisiae H058. *Bioprocess. Biosyst. Eng.*, 36, 937–9465, 2013.
23. Yasin, N.H.M., Man, H.C., Yusoff, M.Z.M., Hassan, M.A., Microbial characterization of hydrogen-producing bacteria in fermented food waste at different pH values. *Int. J. Hydrogen Energy*, 36, 16, 9571–9580, 2011.
24. Malaták, J. and Dlabaja, T., Hydrothermal carbonization of kitchen waste. *Res. Agric. Eng.*, 62, 2, 64–72, 2016.
25. Papanikolaou, S., Dimou, A., Fakas, S., Diamantopoulou, P., Philippoussis, A., Galiotou-Panayotou, M., Aggelis, G., Biotechnological conversion of waste cooking olive oil into lipid-rich biomass using *Aspergillus* and *Penicillium* strains. *J. Appl. Microbiol.*, 110, 5, 1138–1150, 2011.
26. Kondamudi, N., Mohapatra, S.K., Misra, M., Spent coffee grounds as a versatile source of green energy. *J. Agric. Food. Chem.*, 56, 24, 11757–11760, 2008.
27. Huang, H., Singh, V., Qureshi, N., Butanol production from food waste: a novel process for producing sustainable energy and reducing environmental pollution. *Biotechnol. Biofuels*, 8, 1–12, 2015.
28. Das, P., Chandramohan, V.P., Mathimani, T., Pugazhendhi, A., A comprehensive review on the factors affecting thermochemical conversion efficiency of algal biomass to energy. *Sci. Total Environ., 766*, 144213, 2021.

29. Varjani, S., Shah, A.V., Vyas, S., Srivastava, V.K., Processes and prospects on valorizing solid waste for the production of valuable products employing bio-routes: a systematic review. *Chemosphere, 282,* 130954, 2021.
30. Awasthi, M.K., Harirchi, S., Sar, T., Vigneswaran, V.S., Rajendran, K., Gómez-García, R., Taherzadeh, M.J., Myco-biorefinery approaches for food waste valorization: Present Status and Future Prospects. *Bioresour. Technol.*, 360, 127592, 2022.
31. Mohamed, S.N., Hiraman, P.A., Muthukumar, K., Jayabalan, T., Bioelectricity production from kitchen wastewater using microbial fuel cell with photosynthetic algal cathode. *Bioresour. Technol., 295,* 122226, 2020.
32. Ramu, S.M., Thulasinathan, B., Hari, D.G., Bora, A., Jayabalan, T., Mohammed, S.N., Alagarsamy, A., Fermentative hydrogen production and bioelectricity generation from food based industrial waste: An integrative approach. *Bioresour. Technol., 310,* 123447, 2020.
33. Asefi, B., Li, S.L., Moreno, H.A., Sanchez-Torres, V., Hu, A., Li, J., Yu, C.P., Characterization of electricity production and microbial community of food waste-fed microbial fuel cells. *Process Saf. Environ. Prot., 125,* 83–91, 2019.
34. Raychaudhuri, A. and Behera, M., Enhancement of bioelectricity generation by integrating acidogenic compartment into a dual-chambered microbial fuel cell during rice mill wastewater treatment. *Process Biochem., 105,* 19–26, 2021.
35. Kumar, K., Harindran, S.V., Gajalakshmi, S., Biosorptionally treated dye wastewater employed in "Olla-pot coupled microbial fuel cell" for tomato (Solanum lycopersicum L.) irrigation and bio-electricity production. *Bioresour. Technol. Rep., 13,* 100626, 2021.
36. Din, M., II, Iqbal, M., Hussain, Z., Khalid, R., Bioelectricity generation from waste potatoes using single chambered microbial fuel cell. *Energy Sources Part A,* 1–11, 2020.
37. Alves, J.L.F., da Silva, J.C.G., Mumbach, G.D., Di Domenico, M., da Silva Filho, V.F., de Sena, R.F., Marangoni, C., Insights into the bioenergy potential of jackfruit wastes considering their physicochemical properties, bioenergy indicators, combustion behaviors, and emission characteristics. *Renew. Energy, 155,* 1328–1338, 2020.
38. Hou, Q., Yang, Z., Chen, S., Pei, H., Using an anaerobic digestion tank as the anodic chamber of an algae-assisted microbial fuel cell to improve energy production from food waste. *Water Res., 170,* 115305, 2020.
39. Radeef, A.Y. and Ismail, Z.Z., Polarization model of microbial fuel cell for treatment of actual potato chips processing wastewater associated with power generation. *J. Electroanal. Chem., 836,* 176–181, 2019.
40. Sgarbossa, A., Boschiero, M., Pierobon, F., Cavalli, R., Zanetti, M., Comparative life cycle assessment of bioenergy production from different wood pellet supply chains. *Forests, 11,* 11, 1127, 2020.

41. Abdul Malek, A.B.M., Hasanuzzaman, M., Rahim, N.A., Prospects, progress, challenges and policies for clean power generation from biomass resources. *Clean Technol. Environ. Policy*, *22*, 6, 1229–1253, 2020.
42. Obileke, K., Onyeaka, H., Meyer, E.L., Nwokolo, N., Microbial fuel cells, a renewable energy technology for bio-electricity generation: A mini-review. *Electrochem. Commun.*, *125*, 107003, 2021.
43. Chung, T.H. and Dhar, B.R., A multi-perspective review on microbial electrochemical technologies for food waste valorization. *Bioresour. Technol.*, *342*, 125950, 2021.
44. Ananno, A.A., Masud, M.H., Chowdhury, S.A., Dabnichki, P., Ahmed, N., Arefin, A.M.E., Sustainable food waste management model for Bangladesh. *Sustain. Prod. Consumption*, *27*, 35–51, 2021.
45. Pandey, K., Yadav, A.K., Goel, C., Utilization of Food Waste for Biofuel Production, in: *Food Waste to Green Fuel: Trend & Development*, pp. 1–23, Springer, Singapore, 2022.
46. Singh, A., Prajapati, P., Vyas, S., Gaur, V.K., Sindhu, R., Binod, P., Varjani, S., A comprehensive review of feedstocks as sustainable substrates for next-generation biofuels. *Bioenergy Res.*, 16, 1–18, 2022.
47. Suparmaniam, U., Lam, M.K., Uemura, Y., Lim, J.W., Lee, K.T., Shuit, S.H., Insights into the microalgae cultivation technology and harvesting process for biofuel production: A review. *Renew. Sustain. Energy Rev.*, 115, 2019.
48. Slorach, P.C., Jeswani, H.K., Cuéllar-Franca, R., Azapagic, A., Environmental sustainability of anaerobic digestion of household food waste. *J. Environ. Manage.*, 710, 236, 798–814, 2019.
49. Vu, M.T., Noori, M.T., Min, B., Conductive magnetite nanoparticles trigger syntrophic methane production in single chamber microbial electrochemical systems. *Bioresour. Technol.*, 296, 122265, 2020.
50. Zhang, L., Loh, K.C., Lim, J.W., Zhang, J., Bioinformatics analysis of metagenomics data of biogas-producing microbial communities in anaerobic digesters: A review. *Renew. Sustain. Energy Rev.*, *100*, 110–126, 2019.
51. Koniuszewska, I., Korzeniewska, E., Harnisz, M., Czatzkowska, M., Intensification of biogas production using various technologies: A review. *Int. J. Energy Res.*, *44*, 8, 6240–6258, 2020.
52. El Barnossi, A., Moussaid, F., Housseini, A., II, Antifungal activity of Bacillussp. gn-A11-18isolated from decomposing solid green household waste in water and soil against Candida albicans and Aspergillus Niger, in: *E3S Web of Conferences*, vol. 150, EDP Sciences, p. 02003, 2020.
53. Audu, J.O., Ibrahim, N., Ibrahim, Z., Dagang, W.R.Z.W., Yahya, A., Hussin, H., Abdul-Wahab, M.F., Optimization of the operational parameters for mesophilic biohydrogen production from palm oil mill effluent using enriched mixed culture. *Biomass Convers. Biorefin.*, 13, 1–17, 2021.
54. Leong, Y.K. and Chang, J.S., Valorization of fruit wastes for circular bioeconomy: Current advances, challenges, and opportunities. *Bioresour. Technol.*, 359, 127459, 2022.

55. Ndayisenga, F., Yu, Z., Zheng, J., Wang, B., Liang, H., Phulpoto, I.A., Zhou, D., Microbial electrohydrogenesis cell and dark fermentation integrated system enhances biohydrogen production from lignocellulosic agricultural wastes: Substrate pretreatment towards optimization. *Renew. Sustain. Energy Rev.*, *145*, 111078, 2021.
56. Sarangi, P.K. and Nanda, S., Biohydrogen production through dark fermentation. *Chem. Eng. Technol.*, *43*, 4, 601–612, 2020.
57. Nabgan, W., Abdullah, T.A.T., Nabgan, B., Jalil, A.A., Nordin, A.H., Ul-Hamid, A., Ikram, M., Catalytic biohydrogen production from organic waste materials: A literature review and bibliometric analysis. *Int. J. Hydrogen Energy*, *46*, 60, 30903–30925, 2021.
58. Bhatia, S.K., Jagtap, S.S., Bedekar, A.A., Bhatia, R.K., Rajendran, K., Pugazhendhi, A., Yang, Y.H., Renewable biohydrogen production from lignocellulosic biomass using fermentation and integration of systems with other energy generation technologies. *Sci. Total Environ.*, *765*, 144429, 2021.
59. Bhatia, S.K., Kim, S.H., Yoon, J.J., Yang, Y.H., Current status and strategies for second generation biofuel production using microbial systems. *Energy Convers. Manage.*, *148*, 1142–1156, 2017.
60. Prajapati, K.B. and Singh, R., Bio-electrochemically hydrogen and methane production from co-digestion of wastes. *Energy*, *198*, 117259, 2020.
61. Zacharia, K.M.B., Yadav, S., Machhirake, N.P., Kim, S.H., Lee, B.D., Jeong, H., Kumar, R., Bio-hydrogen and bio-methane potential analysis for production of bio-hythane using various agricultural residues. *Bioresour. Technol.*, *309*, 123297, 2020.
62. Shanmugam, P., Correlation between empirical formulae based stoichiometric and experimental methane potential and calorific energy values for vegetable solid wastes. *Energy Rep.*, *7*, 19–31, 2021.
63. Micolucci, F., Gottardo, M., Cavinato, C., Pavan, P., Bolzonella, D., Mesophilic and thermophilic anaerobic digestion of the liquid fraction of pressed biowaste for high energy yields recovery. *Waste Manage.*, *48*, 227–235, 2016.
64. Lay, J.J., Fan, K.S., Ku, C.H., Influence of chemical nature of organic wastes on their conversion to hydrogen by heat-shock digested sludge. *Int. J. Hydrogen Energy*, *28*, 12, 1361–1367, 2003.
65. Wahbi, A., Tsolakis, A., Herreros, J., Emissions control technologies for natural gas engines, in: *Natural Gas Engines*, pp. 359–379, Springer, Singapore, 2019.
66. Taskiran, O.O., Improving burning speed by using hydrogen enrichment and turbulent jet ignition system in a rotary engine. *Int. J. Hydrogen Energy*, *46*, 57, 29649–29662, 2021.
67. Angeriz-Campoy, R., Álvarez-Gallego, C.J., Romero-García, L., II, Thermophilic anaerobic co-digestion of organic fraction of municipal solid waste (OFMSW) with food waste (FW): enhancement of bio-hydrogen production. *Bioresour. Technol.*, *194*, 291–296, 2015.

68. Battista, P., Salvatore, C., Berlingeri, M., Cerasa, A., Castiglioni, I., Artificial intelligence and neuropsychological measures: The case of Alzheimer's disease. *Neurosci. Biobehav.*, 114, 211–228, 2020.
69. Mikheeva, E.R., Katraeva, I.V., Vorozhtsov, D.L., Kovalev, D.A., Kovalev, A.A., Grigoriev, V.S., Litti, Y.V., Dark fermentative biohydrogen production from confectionery wastewater in continuous-flow reactors. *Int. J. Hydrogen Energy*, 47, 53, 22348–22358, 2022.
70. Lak, S.Z., Bonyadi, M., Rahimpour, M.R., Syngas from agricultural waste, in: *Advances in Synthesis Gas: Methods, Technologies and Applications*, pp. 379–409, 2023.
71. Usmani, Z., Sharma, M., Awasthi, A.K., Sharma, G.D., Cysneiros, D., Nayak, S.C., Gupta, V.K., Minimizing hazardous impact of food waste in a circular economy–Advances in resource recovery through green strategies. *J. Hazard. Mater.*, *416*, 126154, 2021.
72. Pratto, B., dos Santos-Rocha, M.S.R., Longati, A.A., de Sousa Júnior, R., Cruz, A.J.G., Experimental optimization and techno-economic analysis of bioethanol production by simultaneous saccharification and fermentation process using sugarcane straw. *Bioresour. Technol.*, *297*, 122494, 2020.
73. Chandgude, V., Välisalmi, T., Linnekoski, J., Granström, T., Pratto, B., Eerikäinen, T., Bankar, S., Reducing agents assisted fed-batch fermentation to enhance ABE yields. *Energy Convers. Manage.*, *227*, 113627, 2021.
74. Thakur, A., Kumar, A., Kaya, S., Vo, D.V.N., Sharma, A., Suppressing inhibitory compounds by nanomaterials for highly efficient biofuel production: a review. *Fuel*, *312*, 122934, 2022.
75. Bokhary, A., Leitch, M., Liao, B.Q., Liquid–liquid extraction technology for resource recovery: Applications, potential, and perspectives. *J. Water Process Eng.*, *40*, 101762, 2021.
76. Tian, L., Conway, P.M., Cervenka, N.D., Cui, J., Maloney, M., Olson, D.G., Lynd, L.R., Metabolic engineering of Clostridium thermocellum for n-butanol production from cellulose. *Biotechnol. Biofuels*, *12*, 1, 1–13, 2019.
77. Bai, S.K., Hong, Y., Wu, Y.R., Emerging technologies for genetic modification of solventogenic clostridia: From tool to strategy development. *Bioresour. Technol.*, *334*, 125222, 2021.
78. Arora, K., Kumar, P., Bose, D., Li, X., Kulshrestha, S., Potential applications of algae in biochemical and bioenergy sector. *3 Biotech.*, *11*, 6, 1–24, 2021.
79. Qureshi, N. and Maddox, I., Reduction in butanol inhibition by perstraction: Utilization of concentrated lactose/whey permeate by clostridium acetobutylicum to enhance butanol fermentation economics. *Food Bioprod. Process.*, 83, 43–52, 2005.
80. Qureshi, N., Saha, B.C., Dien, B., Hector, R.E., Cotta, M.A., Production of butanol (a biofuel) from agricultural residues: Part I–use of barley straw hydrolysate. *Biomass Bioenergy*, 34, 559–565, 2010.

81. Qureshi, N., Saha, B.C., Hector, R.E., Hughes, S.R., Cotta, M.A., Butanol production from wheat straw by simultaneous saccharification and fermentation using clostridium beijerinckii: Part I-batch fermentation. *Biomass Bioenergy*, 32, 168–175, 2008.
82. ChooChuay, C., Pongpiachan, S., Tipmanee, D., Deelaman, W., Iadtem, N., Suttinun, O., Cao, J., Effects of agricultural waste burning on PM2. 5-bound polycyclic aromatic hydrocarbons, carbonaceous compositions, and water-soluble ionic species in the ambient air of Chiang-Mai, Thailand. *Polycyclic Aromat. Compd.*, *42*, 3, 749–770, 2022.

14

Photosynthetic Microalgal Microbial Fuel Cell (PMMFC): A Novel Strategy for Wastewater Treatment and Bioenergy Generation

Jagdeep Kumar Nayak[1,2*], Rahul Gautam[2], Kundan Samal[3] and Uttam Kumar Ghosh[2]

[1]*Department of Chemical Science, Bernal Institute, University of Limerick, Limerick, Ireland*
[2]*Department of Polymer and Process Engineering, Indian Institute of Technology, Roorkee, India*
[3]*Department of Civil Engineering, KIIT University, Patia, India*

Abstract

Given the rising population and rapid industrialization, the demand for renewable energy has become an imperative to meet our energy needs. Renewable energy sources such as solar, hydro, wind, and biomass store a significant amount of energy. However, inefficient utilization of these sources prevents them from meeting the rising energy demand. Microbial fuel cell (MFC), a sustainable treatment system, could provide the energy requirement of producing electricity by utilizing organic complexes from wastewater. In addition to the conventional MFC operation, the use of algae to produce electricity in MFCs appears to be a wise approach for extracting energy from sunlight in an economical and sustainable manner. The use of microalgae in the MFC system is called photosynthetic microalgal microbial fuel cell (PMMFC) or photosynthetic microbial fuel cell (PMFC). This chapter explains the basic principle of PMFCs, its type of operation, advantages, and disadvantages. This chapter also focuses on the mechanism of supplying oxygen to the cathodic system, which can solve the reduction limitation. More studies on PMFCs could provide a sustainable solution for bioelectricity production, biomass generation, and wastewater treatment.

**Corresponding author*: jnayak@pe.iitr.ac.in

Hrudayanath Thatoi, Sonali Mohapatra, Swagat Kumar Das and Sukanta Kumar Pradhan (eds.) Applied Biotechnology and Bioinformatics: Agriculture, Pharmaceutical Research and Environment, (331–346)

Keywords: Photosynthetic microbial fuel cell, microalgae, bioelectricity, wastewater

14.1 Introduction

With the rising population and industrialization, energy demand has also risen in recent years. The conventional renewable energy sources such as solar, wind, hydro, and biomass store a significant amount of energy. However, inefficient use of these resources prevents them from meeting the rising energy demand. Renewable energy functions as a clean energy that can be produced from a variety of feedstocks to minimize the reliance on fossil fuels [1–4].

As a result of increasing population and industrialization in every part of the world, wastewaters are one of the emerging threats [5, 6]. The convergence of polluted wastewater with freshwater bodies poses a threat to society. Through the ecological life cycle, this can endanger both aquatic and human life [7]. The untreated wastewater with toxic substances requires proper wastewater treatment before they are discarded to freshwater bodies to save the environment [8]. These treatment methodologies include chemical, physical, or biological treatments in the wastewater treatment plants [9, 10]. Because of the rise in the bio-electrochemical systems (BES) applicability, microbial fuel cells (MFCs) are considered as an efficient tool to convert waste to energy. They are a sensitive and ecofriendly technology that can address both pollution removal and bioenergy generation from the waste source [11, 12]. However, MFCs have not increased productivity due to their constant need for oxygen; thus, performance is limited [13]. The aeration at the cathode accounts for nearly half of the overall cost. To address some of the drawbacks, microalgae are being considered as the source of oxygen due to their photosynthesis, improving power generation while also removing pollutants. Furthermore, the efficiency of MFC can be increased by aiming at the specific pollutant's removal. The wastewater stream containing the drugs, pesticides and herbicides, dyes, and heavy metals can be treated in MFCs [14, 15]. The addition of live microalgae to the cathode of an MFC, also termed as photosynthetic MFCs (PMFC), has proven to be more promising in terms of MFC efficiency and cost.

MFCs assisted with microalgae are regarded as an appropriate technology for simultaneous waste treatment and electricity generation, removing pollutants, and treating wastewater at the same time. All of these properties of PMFC make them an attractive candidate for potential biomass feedstock production. Microalgae integrated with MFCs has become a fascinating technology because of the fact that microalgae could act as oxygen producers and, thus, an electron acceptor at the cathode. To achieve the required efficiency of MFC, the selection of microalgae is critical, and diatoms are the most efficient of the many microalgae used to date [16]. This approach can account for approximately 25% of atmospheric CO_2 and 30% of global crude oil production [17]. Aside from that, they are abundant in a variety of important high and low value metabolites. Metal pollutants, specifically the heavy metals, are of great concern because they are not depleted over time and persist either in the environment or in living things, causing detrimental impacts. When discharged into the environment, they give rise to unpredictable alterations [18].

The latest literature showed significant research on microalgae integration in MFCs due to its advantages; however, a lot still need to be explored such as optimization of operational parameters, species selection, and design of PMFCs. This chapter focused on the basic principles and types of the PMFCs for enhanced wastewater treatment and bioelectricity generation. This chapter also covered the advantages and disadvantages of PMFCs along with the future prospect of the technology.

14.2 Microbial Fuel Cell

A "microbial fuel cell" is a device that converts biochemical energy from bacterial metabolism into bioelectric energy. The extracellular electron is produced as a result of the anaerobic oxidizing bacteria oxidizing the

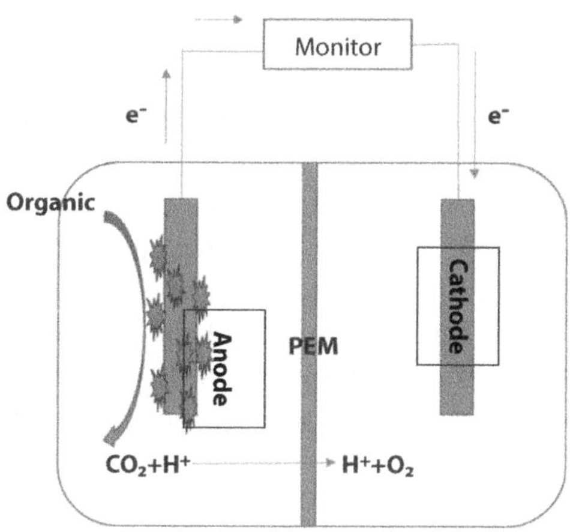

Figure 14.1 Schematic diagram of MFC.

substrate in the anodic chamber (Figure 14.1). To complete the cycle, the produced electron moves via the external circuit and into the cathodic chamber [19]. A typical "H-type" dual-chambered MFC has infected anodic and cathodic chambers on either side. Across the transport of protons (H^+) through a proton exchange membrane (PEM) made of nafion-117, oxygen is reacted with protons (H+) to complete the circuit and produce energy. The standard MFCs use a range of substrates, including organic chemicals, alcohols, acetate, glucose, and acetate. MFCs are a burgeoning technology that is drawing attention, particularly because of their potential to remediate wastewater while also acting as a source of renewable energy [20].

14.2.1 Photosynthetic Microalgal Microbial Fuel Cell (PMMFC)

Photosynthetic microbial fuel cell, a type of MFC, where photosynthetic organisms are recruited for oxidation, charge separation, and oxygen production, acts as an electron acceptor. The symbiotic relationship between algae feedstock and heterotrophic microorganisms offers significant advantages for the system. With this integration of microalgae, the traditional treatment system became photosynthetic and more environmentally and economically sustainable [21]. Thus, it is assumed that the growing cathodic algae in MFC can enhance the rate of electricity generation, by using the oxygen coming from the algal photosynthesis [22]. Thus, algae-based photosynthetic MFC represents a more sustainable advanced technique as compared to the typical MFCs for simultaneous CO_2 sequestration, biomass production, electricity generation, and wastewater treatment [23, 24].

In PMFC, the algal biocathode has been employed as an electron acceptor for conversion of CO_2 to biomass. Generally, algae are used as a mediator in the cathodic chamber (Figure 14.2). The electrons from the anodic chamber reduce the oxidized mediator if used, which is used in algal cells to release electrons and oxidation takes place. The algal cells consume the electron for the metabolic activities to grow and convert CO_2 into biomass and oxygen (Figure 14.4). The algal cells release the oxygen as the oxidized mediator and the mediator again get reduced in the catholyte by the transferred electron [25]. The following reactions are taking place in both the cathodic and anodic chamber as mentioned Equations 14.1 and 14.2, respectively [26].

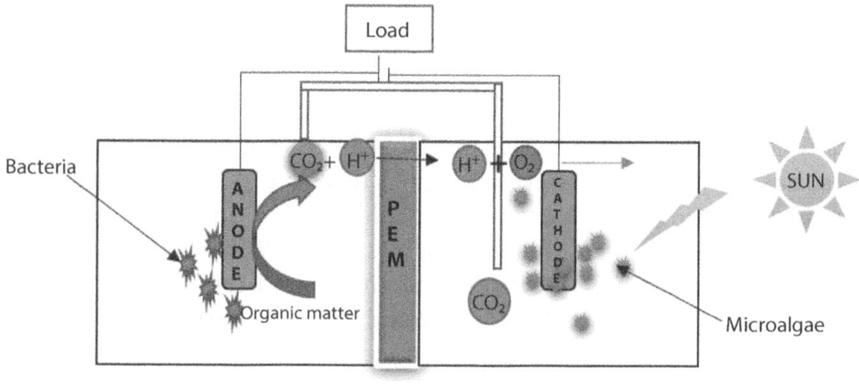

Figure 14.2 Schematic diagram of photosynthetic microbial fuel cell.

$$6CO_2 + 12H^+ + 12e^- \rightarrow C_6H_{12}O_6 \text{ (biomass)} + 3O_2 \quad (14.1)$$

$$C_6H_{12}O_6 + 6H_2O \rightarrow 6CO_2 + 24H^+ + 24e^- \quad (14.2)$$

In the presence of light, algae carry out photosynthesis with the help of CO_2 to produce biomass and organic matter, while in the dark reaction, there is no photosynthesis, and oxygen is consumed to oxidize the organic matter produced in the light to get energy (Equation 14.3) [27].

$$C_6H_{12}O_6 + 6O_2 \rightarrow 6CO_2 + 6H_2O \quad (14.3)$$

The previous studies showed that some photosynthetic cyanobacteria such as *Spirulina platensis* were used as a bio-anode catalyst, whereas the algae helped to maintain the electrochemical potential with the formation of biofilm, which can accept the electrons directly [28, 29].

Similar design and work have been done by Wang *et al.* [30], where they demonstrated the setup called microbial carbon capture cell (MCC) in which the generated CO_2 from substrate degradation was transferred from the anodic chamber to the cathodic chamber for the utilization in algal photosynthesis. This process represents a sustainable technology for both CO_2 sequestration for biomass generation and bioelectricity generation. However, this technology is still in its beginning stage. A lot of work needs to be done for its full-fledged implementation.

14.3 Types of PMFC

14.3.1 Single-Chambered PMFC

Many MFC designs have been developed using various bacterial species, but no researchers have attempted to create an algae MFC. Algae has been gaining prominence in the field of MFC in recent years. The single-chambered PMFC (SPMFC) is the same as PMFC except the proton exchange membrane is in between the anodic and cathodic chamber. In SPMFC configuration, both the anode and cathode electrodes are typically present in the same chamber and connected through an electric circuit. Furthermore, some researchers created single-chamber PFMC in which both the bacteria and algae were incubated in a chamber, and this system reported to show synergistic actions to improve the process's efficacy. The efficiency of PMFC should represent the symbiotic relationship of bacteria and microalgae. Some of the photosynthetic microbes and microalgae, i.e., *Canna indica, Chlorella vulgaris,* and *Scenedesmus abundans*, shuttle electrons to the electrode directly; thus, mediators can be omitted out. The mechanism was experimented in an SPMFC design containing the photosynthetic bioanode. Several researchers proposed this design [31], whereby a straightforward method for building an electricity generator that enables microalgae to be attached directly to the anode was proposed (Figures 14.3c,d). In addition to this, a few of the blue-green microalga can enhance the power generation acting as a biocatalyst. The requisite electro-potential may be produced by the anodic microalgal biofilm, enabling both the photosynthetic process and the respiration reaction to produce current. This design harbors both anode and cathode to a single chamber.

A synergistic combination of bacterial and microalgae cells was experimented to increase the effectiveness of the single-chamber PMFC. Bacterial cells utilize the organic substances produced by the microalgal cells, such as acetate, as a fuel source for power production. The light-dependent process of photosynthesis allows the PMFC to recharge and operate for extended periods by utilizing the feedstock produced by algae. The design can serve as a bio-battery, which is portable in nature, and can produce power in both light and dark conditions.

The photo-biological fuel cell is a unique design, having sandwiched PEM between the electrodes. Microalgae form anodic biofilm and capture atmospheric CO_2 as a carbon source, combining autotrophic and heterotrophic metabolism to sequester both atmospheric and industrial CO_2.

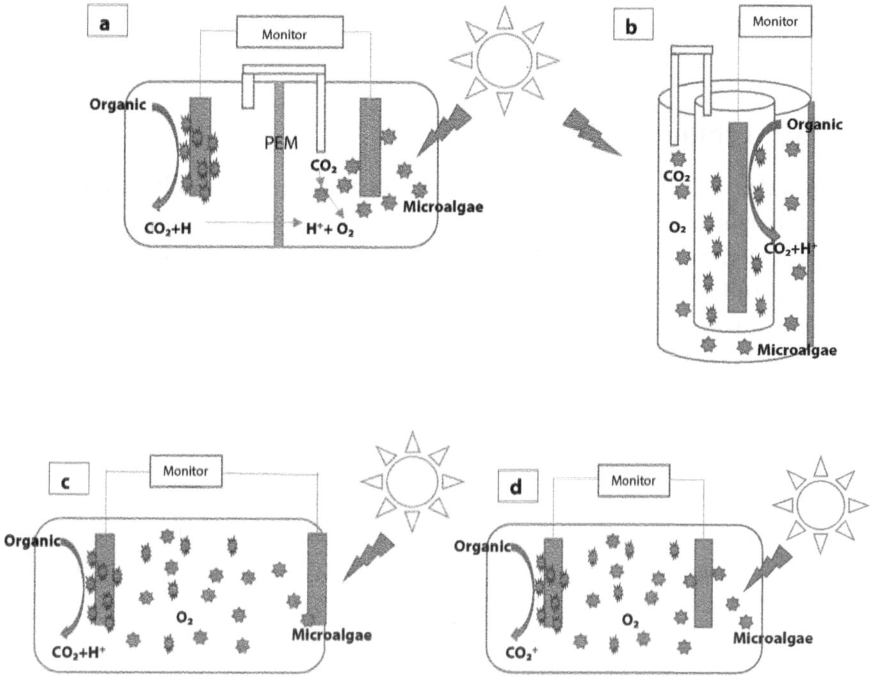

Figure 14.3 (a) Dual-chambered PMFC, (b) tubular PMFC, (c) single-chambered air cathode PMFC, (d) single-chambered PMFC.

14.3.2 Dual-Chambered PMFC

Due to the easy convertibility of dual-chambered MFC, the double-chambered PMFC is the most studied design, as the cathode algal photosynthesis serves as an oxygen source. The design of PMFC is the same as MFC, with two chambers separated by ion-permeable PEM and connected in a closed circuit [32]. The anodic chamber is filled with wastewater or sludge or any organic component to provide nutrient to the microbes and the cathodic chamber is filled with culture of microalgae with adequate supply of light, with 12–16 hours of illumination. Some studies showed the supply of carbon dioxide from the generated biogas at anode of the same PMFC to be utilized by the microalgae for photosynthesis and O_2 generation. Some studies also supplied the CO_2 from the air/tank through a pump to the cathodic chamber. The supply of the CO_2 could be through a vent connected at the top of the anode by a tube to the chamber containing the cathode electrode and for algal photosynthesis (Figure 14.3a). However, microalgae could act as a biocatalyst. The dual-chambered PMFC can solve many issues such as utilization of CO_2, can act as a biocatalyst, and can produce O_2 to substitute mechanized aeration.

14.3.3 Tubular PMFC

The tubular PMFC system is a unique configuration, which provides an equal distribution of substrate along with the microalgae. This potential device has the capability to convert organic matter into electric energy while simultaneously utilizing the gaseous output from the anode to recover nutrients through microalgal technology. A tubular photosynthetic microbial fuel cell (PMFC) system with an algae-assisted cathode is gaining attention in the scientific (Figure 14.3b) community. The extra energy needed for aeration will be spared because of the oxygen created by algae photosynthesis. In the study, a novel tubular PMFC design with algal growth and multiple electrodes in the cathode chamber was assessed under different hydraulic retention periods (HRTs). The anode could be a plexiglass cylinder with a cation exchange membrane adhered to the bottom. The anode and cathode were kept apart by the membrane to prevent any contact between them. Then, less than 1-mm-thick graphite particles were added to the cylinder. The grains of graphite included plants. The anode, which was constructed of graphite felt and put at the base of a beaker, was then inserted into the cathode compartment.

14.4 Role of Algae in PMFC

14.4.1 Algae at Anode

The use of microalgae with wastewater or sludge as substrate in the anodic chamber could be utilized in many purposes. The dead algal biomass is used to provide food for the bacteria. However, only bacteria can help in organic matter degradation and bioelectricity generation [33]. Typically, in single-chambered PMFCs, algal organisms are employed in the anode, using either live cultures or biomass from dead cells. However, the pretreated algal biomass, activated sludge, and lipid extracted algae were tested, but the low power output was noticed. To test the significance of microalgae in electron transfer, various studies have also been done. Harvesting electrons from microalgae is a unique idea. To prove that scientists utilized more than one marine microalgae in microbial fuel cell, the maximum electricity production (0.792 V) by *Tetraselmis gracilis* was noticed. Many researchers have successfully used other microalgae separately such as *Scenedesmus, Chlorella, Mougeotia*, and blue-green algae for the PMFC system for bioelectricity generation and biomass production. Apart from that, there are many studies with mixed culture and significant results. The major drawback of single-chambered PMFC or microalgae in the anodic chamber is the excessive presence of oxygen as oxygen is the inhibitor of electron transfer in anode [33]. The amount of oxygen released during photosynthesis has a significant impact on power output. By purging the nitrogen gas or adding a highly concentrated salt solution, this inhibition can be reduced.

14.4.2 Algae at Cathode

It makes sense to wonder if energy can be produced from algae in MFC since algae are eukaryotic cells that have mitochondria. Using algae to enhance the bioelectricity production would be both ecologically beneficial and economically viable since it would grow quickly, require little upkeep, and provide a variety of products that are helpful to industry, such as nutrients, biofuel, pharmaceuticals, and industrial chemicals. In case of MFC, the use of chemical catalyst in cathodic chamber is always not recommendable as it can affect the overall costing of the system. Although costly catalysts like platinum are needed to speed up the redox kinetics under ambient settings, air cathodes have also been employed, where oxygen is used as the terminal electron acceptor [34]. "Biocathode" is a novel

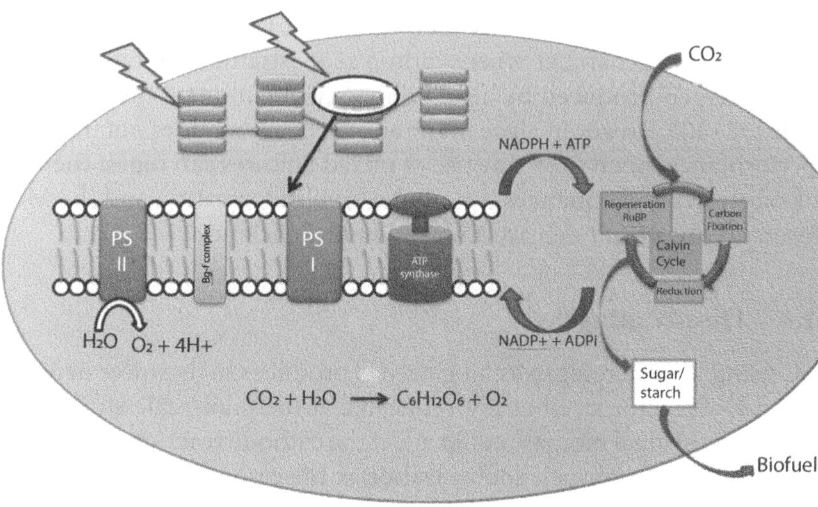

Figure 14.4 Mechanism of carbon sequestration.

term in which microorganisms act as a catalyst to speed up electrochemical reduction on the cathode surface. Biocathodes provide low-cost, environmentally friendly systems that are more effective and produce value goods [35]. It makes sense to integrate algal photosynthesis with MFCs in order to maintain or lower energy usage. In these devices, oxygen was produced on-site by the algae during photosynthesis and discharged into the cathode compartment. This will not only assist in the operation of MFCs but also enable the use of generated biomass in a variety of applications.

14.4.3 Advantages

MFCs use organic waste for energy derivation leading to waste management and simultaneous energy production. These systems use a very wide range of substrates ranging from pure substrate to waste organics that are inexpensive, creating sustainable, long-term solutions [11, 36]. If waste is not handled efficiently, it will become a burden to the environment. For instance, secondary treatment of the WTP for domestic wastewater treatment consumes 15–20% of the energy, so these systems are an efficient approach [37]. The algae not only produces oxygen but also helps in maximizing the power output. Another advantage of PMFC is the fact that algal biomass can be supplied to anode as the feed [24]. These PMFCs generally do not require any expensive cathode catalysts as the algae can do the trick also acting as biocatalysts. The algal biomass produced has different commercial aspects; the biomass could be a good source of biofuel. These systems also help in handling the eutrophication caused by N and P compounds present in the waste. The CO_2 formed at the anode can be transported to the algal chamber where carbon sequestration can be practiced. Also, the oxygen produced by algal biomass will eliminate the sparger at cathode [38–40]. However, algae in the anodic chamber does not influence the performance much [41]. The use of mixed cultures can digest the variety of substrates and enhances the anodic biofilm formation and the MFC performance for power density and coulombic efficiency.

14.4.4 Disadvantages

PMFC using the microalgae as biocatalysts produces high power densities even at a low reactor size but not replicable at the pilot-scale studies [37, 432]. The excess algal biomass could affect the cathode reaction as the cathode chamber affects oxygen concentration at the cathode, thus hampering the redox reaction efficiency. The PMFCs can deliver a voltage output of around 1 V, which requires a large stack of the MFCs to get a higher output in terms of power [36, 43, 44]. These systems face practical limitations in terms of oxygen diffusion, which affects the whole mechanism of the bio-electricity production. PEM surface area and the volume of the MFC affect the power output [24, 45]. The studies suggest that the large surface area of PEM adversely affects the performance [46, 47].

14.4.5 Challenges

The electron transfer mechanism needs to be focused to increase the energy output. The high overpotential at the cathode affects the performance of MFC; in addition, voltages of not more than 1 V are a limitation of these systems [43, 48]. If mediators are used, they can add up to the cost of the operations. These systems are pH sensitive, so the maximum achievable coulombic efficiency and current density are between pH 7 and 8. The change in pH adversely affects the biofilm and, thus, the overall performance. The high cost of electrode materials and membranes and the expensive catalysts prevent access to the technology. The membrane fouling and electrode fouling are some of the factors that increase the cost of the systems as proton accumulation at the anode turns the chamber more acidic and unfavorable for the growth of the bacteria [48, 49], while high pH at the cathode hampers the current generation. High-surface-area electrodes need to be focused on to reduce the cost and obtain the maximum benefit from the systems The optimal PEM surface area and MFC volume also need to be achieved [50–52].

14.5 Conclusion

Microbial fuel cells have received a lot of attention as a power source in biological environments because they provide clean, affordable, and long-lasting energy. Combining biomass production with power generation is appealing, but much more research and experimentation is required to overcome the barriers to efficient and economically viable PAMFCs. This chapter discussed all the possible aspects of PMFCs, including their advantages and disadvantages. The mechanism of inhibition of oxygen limitation in cathode has been discussed. The future aspects of PMFCs and the drawbacks of MFCs and PMFCs have been explained.

References

1. Samal, K., Roshan, R., Bhunia, P., Science of the Total Environment Design and development of a hybrid macrophyte assisted vermi fi lter for the treatment of dairy wastewater : A statistical and kinetic modelling approach. *Sci. Total Environ.*, 645, 156–169, 2018, https://doi.org/10.1016/j.scitotenv.2018.07.118.
2. Gautam, R., Nayak, J.K., Daverey, A., Ghosh, U.K., Emerging sustainable opportunities for waste to bioenergy: an overview, in: *Waste-to-Energy Approaches Towards Zero Waste*, pp. 1–55, 2022, https://doi.org/10.1016/B978-0-323-85387-3.00001-X.
3. Amit, J.K.N. and Ghosh, U.K., An innovative mixotrophic approach of distillery spent wash with sewage wastewater for biodegradation and bioelectricity generation using microbial fuel cell. *J. Water Process Eng.*, 23, 306–313, 2018, https://doi.org/10.1016/j.jwpe.2018.04.003.
4. Mogili, N.V., Murugesan, N., Ayothiraman, S., Gautam, R., Deshavath, N.N., Reddy Erva, R., Biohydrogen production from wastewater and organic solid wastes, in: *Waste-to-Energy Approaches Towards Zero Waste*, pp. 165–195, 2022, https://doi.org/10.1016/B978-0-323-85387-3.00009-4.
5. Anwar, H., Shahid, M., Natasha, N.K., Khalid, S., Tariq, T.Z., Ahmad, S., Nadeem, M., Abbas, G., Risk assessment of potentially toxic metal(loid)s in Vigna radiata L. under wastewater and freshwater irrigation. *Chemosphere*, 265, 129124, 2021, https://doi.org/10.1016/J.CHEMOSPHERE.2020.129124.
6. Rathi, B.S., Kumar, P.S., Vo, D.V.N., Critical review on hazardous pollutants in water environment: Occurrence, monitoring, fate, removal technologies and risk assessment. *Sci. Total Environ.*, 797, 149134, 2021, https://doi.org/10.1016/J.SCITOTENV.2021.149134.
7. Alvarado, A., West, S., Abbt-Braun, G., Horn, H., Hydrolysis of particulate organic matter from municipal wastewater under aerobic treatment. *Chemosphere*, 263, 128329, 2021, https://doi.org/10.1016/J.CHEMOSPHERE.2020.128329.
8. Sridharan, R., Krishnaswamy, V., Kumar, P.S., Vidhya, T.A., Sivamurugan, V., Kumar, D.T., Doss, C.G.P., Vo, D.V.N., Analysis and effective separation of toxic pollutants from water resources using MBBR: Pathway prediction using alkaliphilic P. mendocina. *Sci. Total Environ.*, 797, 149135, 2021, https://doi.org/10.1016/J.SCITOTENV.2021.149135.
9. Hulshoff Pol, L.W., Lens, P.N.L., Stams, A.J.M., Lettinga, G., Anaerobic treatment of sulphate-rich wastewaters. *Biodegradation*, 9, 213–224, 1998, https://doi.org/10.1023/a:1008307929134.
10. Das, M.P., Bashwant, M., Kumar, K., Das, J., Control of pharmaceutical effluent parameters through bioremediation. *J. Chem. Pharm. Res.*, 4, 1061–1065, 2012.
11. Nayak, J.K. and Ghosh, U.K., Post treatment of microalgae treated pharmaceutical wastewater in photosynthetic microbial fuel cell (PMFC) and

biodiesel production. *Biomass Bioenergy*, 131, 1–10, 2019, https://doi.org/10.1016/j.biombioe.2019.105415.

12. Gautam, R., Nayak, J.K., Talapatra, K.N., Amit, Ghosh, U.K., Assessment of different organic substrates for Bio-Electricity and Bio-Hydrogen generation in an Integrated Bio-Electrochemical System. *Mater. Today Proc.*, 80, 3, 6–10, 2021, https://doi.org/10.1016/j.matpr.2021.06.223.
13. Do, M.H., Ngo, H.H., Guo, W., Chang, S.W., Nguyen, D.D., Liu, Y., Varjani, S., Kumar, M., Microbial fuel cell-based biosensor for online monitoring wastewater quality: A critical review. *Sci. Total Environ.*, 712, 135612, 2020, https://doi.org/10.1016/J.SCITOTENV.2019.135612.
14. Peng, Y.Z., Li, Y.Z., Peng, C.Y., Wang, S.Y., Nitrogen removal from pharmaceutical manufacturing wastewater with high concentration of ammonia and free ammonia via partial nitrification and denitrification. *Water Sci. Technol.*, 50, 31–36, 2004, https://doi.org/10.2166/wst.2004.0356.
15. Pandey, A.K., Gaur, V.K., Udayan, A., Varjani, S., Kim, S.H., Wong, J.W.C., Biocatalytic remediation of industrial pollutants for environmental sustainability: Research needs and opportunities. *Chemosphere*, 272, 129936, 2021, https://doi.org/10.1016/J.CHEMOSPHERE.2021.129936.
16. Khan, M.J., Harish, A.A., Schoefs, B., Pugazhendhi, A., Varjani, S., Rajendran, K., Bhatia, S.K., Saratale, G.D., Saratale, R.G., Vinayak, V., Insights into diatom microalgal farming for treatment of wastewater and pretreatment of algal cells by ultrasonication for value creation. *Environ. Res.*, 201, 111550, 2021, https://doi.org/10.1016/J.ENVRES.2021.111550.
17. Ramachandra, T.V. and Hebbale, D., Bioethanol from macroalgae: Prospects and challenges. *Renew. Sustain. Energy Rev.*, 117, 109479, 2020, https://doi.org/10.1016/j.rser.2019.109479.
18. Varjani, S., Rakholiya, P., Shindhal, T., Shah, A.V., Ngo, H.H., Trends in dye industry effluent treatment and recovery of value added products. *J. Water Process Eng.*, 39, 101734, 2021, https://doi.org/10.1016/J.JWPE.2020.101734.
19. Nayak, J.K., Mishra, S., Amit *et al.* Simultaneous Production of Bio-energy and Bio-treatment of Wastewater Using Photosynthetic Microbial Fuel Cell: Optimization and Kinetic Modeling Approach. *Waste Biomass Valor,* 14, 69–84, 2023. https://doi.org/10.1007/s12649-022-01836-4.
20. Kumar Nayak, J., Gautam, R., Ghosh, U.K., Bioremediation potential of bacterial consortium on different wastewaters for electricity and biomass feedstock generation. *Biomass Convers. Biorefin.*, 1, 1–14, 2022, https://doi.org/10.1007/S13399-022-02992-2.
21. Amit, N.K., Verma, S., Park, J., Jaiswal, A.K., Ghosh, U.K., Gautam, R., Utilization of nano-sized waste lime sludge particles in harvesting marine microalgae for biodiesel feedstock production. *Nanotechnol. Environ. Eng.*, 7, 99–107, 2022, https://doi.org/10.1007/S41204-021-00195-0.
22. Juang, D.F., Lee, C.H., Hsueh, S.C., Comparison of electrogenic capabilities of microbial fuel cell with different light power on algae grown

cathode. *Bioresour. Technol.*, 123, 23–29, 2012, https://doi.org/10.1016/j.biortech.2012.07.041.

23. Rosenbaum, M., He, Z., Angenent, L.T., Light energy to bioelectricity: Photosynthetic microbial fuel cells. *Curr. Opin. Biotechnol.*, 21, 259–264, 2010, https://doi.org/10.1016/j.copbio.2010.03.010.
24. Gajda, I., Greenman, J., Melhuish, C., Ieropoulos, I., Self-sustainable electricity production from algae grown in a microbial fuel cell system. *Biomass Bioenergy*, 82, 87–93, 2015, https://doi.org/10.1016/j.biombioe.2015.05.017.
25. Zhou, M., He, H., Jin, T., Wang, H., Power generation enhancement in novel microbial carbon capture cells with immobilized Chlorella vulgaris. *J. Power Sources*, 214, 216–219, 2012, https://doi.org/10.1016/j.jpowsour.2012.04.043.
26. Powell, E.E., Mapiour, M.L., Evitts, R.W., Hill, G.A., Growth kinetics of Chlorella vulgaris and its use as a cathodic half cell. *Bioresour. Technol.*, 100, 269–274, 2009, https://doi.org/10.1016/j.biortech.2008.05.032.
27. González Del Campo, A., Cañizares, P., Rodrigo, M.A., Fernández, F.J., Lobato, J., Microbial fuel cell with an algae-assisted cathode: A preliminary assessment. *J. Power Sources*, 242, 638–645, 2013, https://doi.org/10.1016/j.jpowsour.2013.05.110.
28. Lin, C.C., Wei, C.H., Chen, C.I., Shieh, C.J., Liu, Y.C., Characteristics of the photosynthesis microbial fuel cell with a Spirulina platensis biofilm. *Bioresour. Technol.*, 135, 640–643, 2013, https://doi.org/10.1016/j.biortech.2012.09.138.
29. Fu, C.C., Hung, T.C., Wu, W.T., Wen, T.C., Su, C.H., Current and voltage responses in instant photosynthetic microbial cells with Spirulina platensis. *Biochem. Eng. J.*, 52, 175–180, 2010, https://doi.org/10.1016/j.bej.2010.08.004.
30. Wang, X., Feng, Y., Liu, J., Lee, H., Li, C., Li, N., Ren, N., Sequestration of CO2 discharged from anode by algal cathode in microbial carbon capture cells (MCCs). *Biosens. Bioelectron.*, 25, 2639–2643, 2010, https://doi.org/10.1016/j.bios.2010.04.036.
31. Yongjin, Z., Pisciotta, J., Billmyre, R.B., Baskakov, I.V., Photosynthetic microbial fuel cells with positive light response. *Biotechnol. Bioeng.*, 104, 939–946, 2009, https://doi.org/10.1002/BIT.22466.
32. Angioni, S., Millia, L., Mustarelli, P., Doria, E., Temporiti, M.E., Mannucci, B., Corana, F., Quartarone, E., Photosynthetic microbial fuel cell with polybenzimidazole membrane: synergy between bacteria and algae for wastewater removal and biorefinery. *Heliyon*, 4, 1–26, 2018, https://doi.org/10.1016/j.heliyon.2018.e00560.
33. Xu, C., Poon, K., Choi, M.M.F., Wang, R., Using live algae at the anode of a microbial fuel cell to generate electricity. *Environ. Sci. Pollut. Res.*, 22, 15621–15635, 2015, https://doi.org/10.1007/S11356-015-4744-8/TABLES/2.
34. Gude, V.G., Microb, J., Technol, B., Kokabian, B., Gadhamshetty, V., Beneficial Bioelectrochemical Systems for production of energy, water, and Biomass production Microbial & Biochemical Technology an open access journal Biofuel Cells and Bioelectrochemical systems Beneficial Bioelectrochemical

Systems for Energy, Water. *J. Microbial. Biochem. Technol.*, 6, 2013, https://doi.org/10.4172/1948-5948.S6-005.
35. Sun, Y., Wei, J., Liang, P., Huang, X., Microbial community analysis in biocathode microbial fuel cells packed with different materials. *AMB Express*, 2, 1–8, 2012, https://doi.org/10.1186/2191-0855-2-21/FIGURES/6.
36. Logan, B.E. and Regan, J.M., Microbial fuel cells - Challenges and applications. *Environ. Sci. Technol.*, 40, 5172–5180, 2006, https://doi.org/10.1021/es0627592.
37. Gu, Y., Li, Y., Li, X., Luo, P., Wang, H., Wang, X., Wu, J., Li, F., Energy Self-sufficient Wastewater Treatment Plants: Feasibilities and Challenges. *Energy Proc.*, 105, 3741–3751, 2017, https://doi.org/10.1016/J.EGYPRO.2017.03.868.
38. Nguyen, H.T.H. and Min, B., Leachate treatment and electricity generation using an algae-cathode microbial fuel cell with continuous flow through the chambers in series. *Sci. Total Environ.*, 723, 138054, 2020, https://doi.org/10.1016/J.SCITOTENV.2020.138054.
39. Velvizhi, G. and Venkata Mohan, S., Electrogenic activity and electron losses under increasing organic load of recalcitrant pharmaceutical wastewater. *Int. J. Hydrogen Energy*, 37, 5969–5978, 2012, https://doi.org/10.1016/j.ijhydene.2011.12.112.
40. Venkata Mohan, S., Velvizhi, G., Annie Modestra, J., Srikanth, S., Microbial fuel cell: Critical factors regulating bio-catalyzed electrochemical process and recent advancements. *Renew. Sustain. Energy Rev.*, 40, 779–797, 2014, https://doi.org/10.1016/j.rser.2014.07.109.
41. Arun, S., Sinharoy, A., Pakshirajan, K., Lens, P.N.L., Algae based microbial fuel cells for wastewater treatment and recovery of value-added products. *Renew. Sustain. Energy Rev.*, 132, 2020, https://doi.org/10.1016/j.rser.2020.110041.
42. Gupta, S., Nayak, A., Roy, C., Yadav, A.K., An algal assisted constructed wetland-microbial fuel cell integrated with sand filter for efficient wastewater treatment and electricity production. *Chemosphere*, 263, 128132, 1–10, 2021, https://doi.org/10.1016/j.chemosphere.2020.128132.
43. Venkata Mohan, S., Srikanth, S., Chiranjeevi, P., Arora, S., Chandra, R., Algal biocathode for in situ terminal electron acceptor (TEA) production: Synergetic association of bacteria–microalgae metabolism for the functioning of biofuel cell. *Bioresour. Technol.*, 166, 566–574, 2014, https://doi.org/10.1016/J.BIORTECH.2014.05.081.
44. Logroño, W., Pérez, M., Urquizo, G., Kadier, A., Echeverría, M., Recalde, C., Rákhely, G., Single chamber microbial fuel cell (SCMFC) with a cathodic microalgal biofilm: A preliminary assessment of the generation of bioelectricity and biodegradation of real dye textile wastewater. *Chemosphere*, 176, 378–388, 2017, https://doi.org/10.1016/J.CHEMOSPHERE.2017.02.099.
45. Cusick, R.D., Kiely, P.D., Logan, B.E., A monetary comparison of energy recovered from microbial fuel cells and microbial electrolysis cells fed

winery or domestic wastewaters. *Int. J. Hydrogen Energy*, 35, 8855–8861, 2010, https://doi.org/10.1016/j.ijhydene.2010.06.077.

46. Kannan, N. and Donnellan, P., Algae-assisted microbial fuel cells: A practical overview. *Bioresour. Technol. Rep.*, 15, 100747, 2021, https://doi.org/10.1016/J.BITEB.2021.100747.
47. Nayak, J.K. and Ghosh, U.K., Microalgae Cultivation for Pretreatment of Pharmaceutical Wastewater Associated with Microbial Fuel Cell and Biomass Feed Stock Production. *2nd WaterEnergyNEXUS – International Conference Microalgae cultivation for pretreatemnt of pharmaceutical wastewate*, pp. 3–6, 2020, https://doi.org/10.1007/978-3-030-13068-8.
48. Jung, R.K., Cheng, S., Oh, S.E., Logan, B.E., Power generation using different cation, anion, and ultrafiltration membranes in microbial fuel cells. *Environ. Sci. Technol.*, 41, 1004–1009, 2007, https://doi.org/10.1021/es062202m.
49. Pant, D., Van Bogaert, G., Porto-Carrero, C., Diels, L., Vanbroekhoven, K., Anode and cathode materials characterization for a microbial fuel cell in half cell configuration. *Water Sci. Technol.*, 63, 2457–2461, 2011, https://doi.org/10.2166/wst.2011.217.
50. Greenman, J., Gajda, I., You, J., Mendis, B.A., Obata, O., Pasternak, G., Ieropoulos, I., Microbial fuel cells and their electrified biofilms. *Biofilm*, 3, 100057, 2021, https://doi.org/10.1016/J.BIOFLM.2021.100057.
51. Gajda, I., Greenman, J., Ieropoulos, I.A., Recent advancements in real-world microbial fuel cell applications. *Curr. Opin. Electrochem.*, 11, 78–83, 2018, https://doi.org/10.1016/j.coelec.2018.09.006.
52. Liu, H., Cheng, S., Logan, B.E., Production of electricity from acetate or butyrate using a single-chamber microbial fuel cell. *Environ. Sci. Technol.*, 39, 658–662, 2005, https://doi.org/10.1021/es048927c.

15

Self-Cleaning Aquarium: The Microbial Biofilm Approach for Ammonia Bioremediation

Shaon Ray Chaudhuri[1]*, Tethi Biswas[2] and Indranil Mukherjee[2,3]

[1]*Department of Microbiology, Tripura University, Suryamaninagar, Tripura West, Tripura, India*
[2]*Centre of Excellence in Environmental Technology and Management, Maulana Abul Kalam Azad University of Technology, West Bengal, Haringhata, Nadia, West Bengal, India*
[3]*School of Management Studies, Maulana Abul Kalam Azad University of Technology, West Bengal, Haringhata, Nadia, West Bengal, India*

Abstract

Fresh water scarcity is a global issue that impacts our day-to-day activity, threatening our food security. Aquaculture is a major consumer of water and a bulk wastewater generator. There is about 16.9 m^3 of fresh water consumption for every kilogram of fish production. In order to sustain aquaculture production, adequate fish feed is required. Some portion of it remains unutilized while the rest is consumed by the fish and, in turn, results in the generation of fish excreta. The unutilized fish feed and the fish excreta add to the ammonia concentration of the aquaculture wastewater, which, in turn, results in toxicity. In case of an aquarium, the water has to be replaced every 10 to 15 days, resulting in the consumption of a large volume of water with discharge of untreated water into the environment. This study, conducted on a 255- and an 840-L aquarium, reports a stable, sludge-free, bacterial biofilm-based system that could treat the aquarium water continuously during operation, eliminating the need for replacement of the water for 711 days. Through this biotechnological approach, the consumption of fresh water for non-potable application could be reduced to a major extent, while keeping the environment safe.

**Corresponding author*: shaon.raychaudhuri@gmail.com

Hrudayanath Thatoi, Sonali Mohapatra, Swagat Kumar Das and Sukanta Kumar Pradhan (eds.) Applied Biotechnology and Bioinformatics: Agriculture, Pharmaceutical Research and Environment, (347–368) © 2025 Scrivener Publishing LLC

Keywords: Aquaculture, ammonia, bacteria, biofilm, environmental sustenance, bioremediation

15.1 Current Scenario of Fresh Water Scarcity and Impact of Aquaculture

The development of a nation is coupled to its industrialization, which, in turn, drains its natural reserves and adversely impacts the environment, unless the pollutants are adequately treated. The Environmental Protection Agency in India reports an annual release of 3.22×10^9 m^3 of wastewater. The global scenario as per the National Research Development Corporation is equally bleak with 80% of the total produced wastewater being discharged untreated. The discharge seeps into the surrounding fresh water bodies, polluting the miniscule quantity of the reserves. Indiscriminate use of the fresh water for non-portable applications deplete the existing reserves, making day-to-day activity difficult and maintenance of livelihood a challenge due to water scarcity. Only 0.01% of the daily water consumption (3.49 m^3) is for drinking purposes. The majority (3.6×10^{12} m^3/year) accounts for the virtual water used for agriculture, animal raring, and aquaculture, releasing effluent that needs to be adequately treated to prevent environmental pollution. With increasing population, the virtual water demand increases, expecting to reach 2.7×10^{13} m^3/year by 2050 [1, 2]. As per the World Resources Institute, 167 countries are water stressed, with a prediction of 33 countries losing access to drinking water by 2040 [3]. Aquaculture is one of the major industries of the food sector that consumes a large volume of water and generates effluent that needs to be diluted substantially before it could be reused for the purpose of aquaculture. The Chinese blue revolution has favorably impacted the aquaculture industry in India, making it the second largest fish producer (14.16 Mt in 2019-2020) in the world [1, 4]. Water consumption during aquaculture is for both raising and sustaining growth of the fish (20 L of water being necessary for each kg of fish production during direct use) in addition to that required for fish feed production (indirect use requiring 1.7 m^3 per kilogram of fish in fresh water while 2.9 m^3/kg in brackish water). Literature suggests around 16.9 m^3 of fresh water consumption per kilogram of fish produced [1, 5]. The aquaculture industry contributes about 47% of the global fish production, which accounts to a daily production of 80 million tons. The World Bank estimates a requirement of 62% enhancement of fish production by 2030 to meet the increasing demand.

The demand for fish production increases and so does the volume of effluent generated that needs to be treated before discharge or reuse while fresh water is continuously required to sustain aquaculture production. The seepage, evaporation, and dilution/replacement of used water from fish tanks also account for a large volume of fresh water consumption [6]. In case of ornamental fish cultivation in aquarium, the water has to be replaced every 10 to 15 days to ensure fish survivability. Whatever the point of generation, it is the uneaten fish feed and fish excreta that are responsible for the bulk of pollution in aquaculture [7, 8].

Lime and salts are primarily used to maintain the aquaculture environment in this industry. Lime is used to maintain the pH at the pond bottom (usually acidic in nature) during pond treatment. Chemicals like lime, sodium chloride, sodium bicarbonate, and calcium sulfate are added during aquaculture, which helps in fish cultivation but adversely impacts the environment if the effluent is discharged without proper treatment [9–11]. The major pollutant is nitrogen (0.8–66 kg of N/t fish produced/day) as ammonia [12–15] and nitrite to a lesser extent. Aquaculture effluent has been found to contain between 0.5 and 4 mg/L of ammonia and 0–0.91 mg/L of nitrite. Ammonia above 0.5 mg/L results in toxicity of fish due to inhibition of excretion while nitrite above 0.02 mg/L causes brown blood disease of fish, necessitating simultaneous removal of ammonia and nitrite from the effluent [16–19]. Phosphorus as phosphate is also present as a pollutant to a certain extent, and depends on the cultivated fish variety [20–22]. These pollutants lead to eutrophication [23, 24] in the receiving bodies followed by generation of anoxic condition due to decomposing algal biomass. The environmental protection agency mandates the treatment of the effluent before discharging into the environment [25, 26], a point of constant concern for the aquaculture industry. Hence, a rapid, economical, and eco-friendly treatment strategy is required [27].

15.2 Existing Technologies for Aquaculture Effluent Treatment for Environmental Sustenance

Attempts to reduce effluent generation also included feeding as per demand through automated systems; re-collecting the unutilized fish feed for reducing the need for fish feed production as well as injecting of oxygen for enhancing the natural bioremediation of the fish feed and fish excreta by the natural microflora in the fish ponds [28]; and restricting cultivation in smaller area through cage culture and polyvinyl chloride bag cultures [29, 30]. These methods are often not adoptable at the industrial scale due to the expenditure involved [31]. The concentrated effluent generated through these methods has to go through conventional effluent treatment involving solid removal and ammonia removal through oxidation, enhancing the dissolved oxygen concentration through aeration and enabling better activity of the native microbial community in the effluent treatment unit, clarification of the treated effluent, and finally disinfection before discharge [13, 27]. The entire process is time- and energy-intensive, making their adoption at a large scale crippling for smaller establishments.

Different microbes are known to be involved in effluent treatment. *Klebsiella pneumoniae*, *Rhodococcus* sp., *Bacillus* sp., *B. anammoxidans*, *Bacillus subtilis*, *Pseudomonas alcaligenes*, *Pseudomonas stutzeri*, *Pseudomonas aeruginosa*, *Nocardioides albus*, *Stigeoclonium nanum*, *Nitrosomonas* sp., *Nitrospira* sp., and *Thiobacillus* sp. [13, 32–42] are the reported microbes known to be involved in bioremediation of ammonia, nitrate, and chemical oxygen demand under suspended and immobilized conditions during treatment. The time required for bioremediation varies between 24 hours and 336 hours in suspended systems, membrane bioreactors, and sequencing batch reactors. *Lemna minor* is also used for phytoremediation of ammonia, nitrate, and chemical oxygen demand [13] from aquaculture effluent. These systems were scaled up from 0.5 L to 11.81 m^3 (11,810 L) capacity with different bioremedial efficiency. Combinations of biological, chemical, and physical systems have also been tested at different scales for aquaculture effluent treatment. Microbes (*Nitrosomonas* sp., *Nitrospira* sp., and *Thiobacillus* sp.) embedded in zeolite and pyrite as biofilters have been found to be effective at a scale of 15 L to remove approximately 97% ammonia, 92% total inorganic nitrogen, and 42% phosphate within 24 hours of hydraulic retention time [13]. *Stigeoclonium nanum* immobilized on alginate beads could reduce 90% ammonia within 5 days at a 1.2 L capacity [42], showing a removal efficiency amounting to 19.54 mg of ammonia for each kilogram of immobilized matrix per day. Yet, another microbial system with *Pseudomonas stutzeri*, *Pseudomonas aeruginosa*, and *Nocardioides albus* showed around 96% ammonia removal within 14 days [39] with sodium alginate as immobilization matrix at a scale of 0.5 L. Indigenous microbes from the aquaculture sludge and the aquaculture effluent demonstrated 96% ammonia reduction in 4 days at a scale of 500 L in a sequencing batch reactor [38], and 96% ammonia removal and nearly 100% removal of total suspended solid in less than 41 hours from a membrane reactor treating recirculating aquaculture effluent at a scale of 11.81 m^3 [40]. The microbes immobilized on a filter bed convert the suspended/dissolved pollutants into a non-toxic end-product or biomass [43].

All these systems show a limited shelf life of operation under immobilized conditions and the need for replacement in case of membrane reactors with inoculum charging at regular intervals for other systems. The desired alternative would be a one-time charging system with sustained performance, higher than its suspended counterpart [44, 45]. The improved performance is associated with higher cell density upon immobilization in biofilm along with extracellular polymeric substance production, which provides resistance to external perturbation [46–48].

Ammonia removal takes place by converting it into nitrate through two steps. The first step involves aerobic nitrification wherein ammonium ion gets converted to nitrite by ammonia-oxidizing bacteria. The nitrite-oxidizing bacteria converts the nitrite to nitrate in the second step. This is a slow process that can be made faster through bioaugmentation. The other biological method of anaerobic ammonia oxidation is anammox. Hydroxylamine is produced by reduction of the electron acceptor, i.e., nitrite. Hydroxylamine then reacts with the electron donor ammonium, leading to the production of dinitrogen gas. The intermediate product generated is hydrazine, which, during oxidation to dinitrogen gas, generates electron, that in turn is used for the initial step of formation of hydroxylamine from nitrite. A concentration of 2 μM oxygen inhibits anammox reversibly. *Nitrosomonas eutropha* carry out denitrification under oxygen-limiting conditions. The nitrifiers are simultaneously present in the anammox system under limiting oxygen concentration. In such a setup, the nitrifiers oxidize ammonia to nitrite (reducing the available oxygen concentration) while the nitrite and the remaining ammonia are converted by anammox into gaseous nitrogen through gradual supply of air. Bacteria like *N. eutropha* ensure reduced level of oxygen while *B. anammoxidans* maintains a low concentration of nitrite in the system. This symbiotic functioning of the aerobic and the anaerobic system is essential for maintaining ecological balance and is referred to as "Completely autotrophic nitrogen removal over nitrite" or CANON. Adoption of anammox reduces the operation and maintenance cost by about 90% as aeration cost and sludge generation is reduced. SHARON or the Single reactor system for High-rate Ammonia Removal over Nitrite, reduced ammonia by 50% while producing 50% nitrite in the presence of aeration, When the effluent from this system was coupled with anammox, its combination with anammox resulted in the conversion of the nitrogen source into 95% nitrogen gas and 5% nitrate nitrogen in absence of external aeration. This process could be successfully scaled up from an operating volume of 2 L to a pilot scale of 1800 m^3.The advantages of anammox is that it is an elegant, sustainable, and economical way of removing ammonium (along with nitrite into nitrogen gas) from wastewater and gas using less oxygen, reducing both CO_2 equivalent gas emission and running cost (up to 60%). Anammox being carried out by autotrophs needs no additional organic carbon, reducing chemical input and sludge generation [49–51].

15.3 The Novel Rapid Biofilm Reactor-Based Ammonia Removing System

As reported earlier by Ray Chaudhuri [2, 52], the attempt for screening efficient microbial candidate for the development of tailor-made wastewater-specific microbial formulation of environmental origin was carried out in enriched medium. Different microbial isolates were screened for their ammonia-removing ability. Their combinations were developed and the most suitable candidate was used for the study [2]. The isolate from the activated sludge of the artificial shrimp farm in the state of Odisha in India (20°58'43"N 86°52'01"E) was found to be the most efficient for bioremediation of ammonia. One percent (v/v) inoculation, corresponding to 1.97×10^5 colony-forming unit/ml, showed 34.51% and 63.24% ammonia removal in 24 hours under suspended and immobilized condition. However, the optimum inoculum for bacterial performance was 4% (v/v) [53]. The bacterial strain *Bacillus albus* ASSF01(GenBank MK318642) was 99.89% identical to *Bacillus albus* FORCN041 at the partial 16S rDNA level. This gram-positive diplobacillus produced extracellular protease, amylase, DNase, and catalase while lacking the ability to produce oxidase and lipase. Carbohydrates like maltose, dextrose, trehalose, salicin, esculin, cellobiose, and sodium gluconate could be utilized by the isolate for growth. The isolate under suspended condition has a lag phase of 2 hours and log phase of 9 hours (3rd–11th hour) that reached the stationary phase at the 12th hour. The doubling time of the isolate was 67 min 12 s under suspended condition, while the same upon immobilization was 15 min 36 s, indicating the formulation to be more suitable for biofilm-based system development. The formulation showed growth at pH 5–10 with 8 being the optimum in a temperature ranging from 27 to 45°C with 37°C being the optimum. The isolate was found to be sensitive towards amikacin (30 micrograms), azithromycin (30 micrograms), cefadroxil (30 micrograms), imipenem (10 micrograms), levofloxacin (5 micrograms), meropenem (10 micrograms), and ofloxacin (2 micrograms), while showing resistance against ampicillin, amoxicillin, cefepime, cefoperazone, ceftriaxone, cefuroxime, colistin, ertapenem, gentamycin, linezolid, metronidazole, moxifloxacin, neomycin, nitrofurantoin, rifampicin, and trimethoprim.

The isolate was found to be a structured biofilm former that depicted the start of formation of the biofilm within 2 hours that reached saturation by the 6th hour showing extracellular polymeric substance deposition in between the cells under a scanning electron microscope. The biofilm saturation was most rapid with 4% inoculum.

The optimum inoculum for ammonia reduction under suspended state was found to be 4% (v/v), which corresponded to a viable cell count of 7.88×10^5 CFU/ml. When the biofilm was allowed to form and stabilize over a period of couple of days, it was found that the initial inoculum did not impact the extent of ammonia reduction, provided the reactor was allowed to stabilize for 4 days. However, the extent of ammonia reduction was highest with an inoculum of 4% (v/v) and above, if the saturation of ammonia reduction had to be achieved within 24 hours. This formulation formed as biofilm on a raschig ring could remain stable during effluent treatment for 35 months before its efficiency dropped. Hence, it is an excellent example of a stable sludge-free system using tailor-made microbial formulation for wastewater treatment.

This formulation showed inhibition in performance under enriched conditions (in Luria Bertani Broth) with the highest performance under minimal medium growth conditions. This property could ensure a major boost for reducing the cost of inoculum development for commercial application. The cost of setting up a 1-m^3 biofilm reactor for aquaculture wastewater treatment could be reduced to USD 0.085 from USD 15.27.

The immobilized formulation could reduce ammonia from different kinds of wastewater with different hydraulic retention times (HRTs). The HRT was dependent on the initial ammonia concentration and the other pollutants in the wastewater. Aquaculture effluent and dairy effluent, with and without additional ammonia seeding, were treated efficiently with this formulation. There was a reduction in the concentration of ammonia in wastewater to 0.5 mg/L or less within 12, 14, 15, 16, 42, 48, and 98 hours starting from the initial concentrations of 1.7 mg/L, 3.2 mg/L, 3.4 mg/L, 4.7 mg/L, 27 mg/L, 47 mg/L, and 100 mg/L, respectively. Ammonia removal was at a rate of 1.69×10^{-9} M/s under ambient conditions during treatment of aquaculture effluent. There was simultaneous reduction of nitrite and phosphate from the wastewater by this formulation. This formulation could remove 41.6 mg of ammonia per kilogram of immobilization matrix per day while the only other literature [42] showing rapid ammonia removal could remediate 19.54 mg of ammonia per kilogram of the immobilization matrix per day. That makes this formulation the fastest in terms of ammonia removal from wastewater and has been reported in different literatures by the group [53–55].

The formulation was used for treating tougher effluent like the refinery wastewater. Three different types of wastewater streams from the refinery were treated using the formulation. The streams were subjected to different extents of pre-treatment and hence had different loads of pollutants. The sample of each of the stream was treated in 50-ml biofilm reactors with immobilized microbial formulation in nine replicates in order to ensure proper statistical validation of the finding. The degree of bioremediation was determined in a 24-hour cycle up to a time of 120 hours by measurement of ammonia through usage of Nessler's reagent method [56] (Table 15.1).

For waste stream 1, the *p* value was 1.13E-29, for waste stream 2, the *p* value was 4.74E-30, while for waste stream 3, the *p* value was 5.6E-28. In each case, there was significant variation in the extent of ammonia reduction at different time points for all three samples. A detailed analysis between every 2 time points was also conducted. The results were statistically analyzed using ANOVA at 95% confidence level as shown in Table 15.2. It revealed a

Table 15.1 Average concentration of ammonia in the bioreactors in mg/L.

	Waste stream 1	Waste stream 2	Waste stream 3
0 day	113.03	104.91	94.72
1 days	94.4	90.63	82.81
2 days	83.76	77.71	72.21
3 days	80.12	71.44	67.08
4 days	74.84	67.85	64.49
5 days	70.75	64.81	61.77
Percentage reduction of ammonia with time in the bioreactors			
At 24 hours	16.5%	13.6%	12.6%
At 48 hours	25.9%	25.9%	23.8%
At 72 hours	29.1%	31.9%	27.6%
At 96 hours	33.8%	35.3%	31.9%
At 120 hours	37.4%	38.2%	34.8%

Table 15.2 Statistical validation of the data (p value) showing significant variation with time.

	Days 0 and 1	Days 0 and 2	Days 0 and 3	Days 0 and 4	Days 0 and 5
Waste stream 1	5.08E-07	1.54E-10	9.01E-11	1.45E-12	3.47E-13
Waste stream 2	5.6E-28	5.3E-15	8.05E-15	4.28E-13	4.64E-15
Waste stream 3	9.57E-07	3.83E-13	1.13E-10	2.27E-14	3.99E-16

significant reduction in ammonia concentration with an increase in HRT as shown above with crude samples.

The significant increase in percentage reduction with time proves that the microbial formulation was efficient in treating refinery effluent. A second set of treatment was done with samples diluted two times (50% dilution) to check for the presence of inhibitors in the effluent. The extent of ammonia removal was measured as per Gautam [56], and the chemical oxygen demand (COD) assessment of samples from these sets was carried out as per Merck's method. The detailed procedure of COD assessment is provided in Table 15.3.

The data presented in Table 15.4 are in terms of ammonia concentration (mg/L) at different time points and the percentage of reduction.

For waste stream 1, the p value was 5.6E-46, for waste stream 2, the p value was 1.43E-44, while for waste stream 3, the p value was 1.96E-55. In each case, there was significant variation in the extent of ammonia reduction at different times point for all three samples. A detailed analysis between every 2 time points was conducted (Table 15.5). It revealed a significant ammonia reduction with time as shown below with diluted samples.

The measurement of COD and the extent of reduction in COD at different time points are presented in Tables 15.6 and 15.7.

From the results, it is clear that this process could reduce the ammonia concentration from the provided wastewater streams efficiently. There were inhibitors in samples obtained from waste stream 2 and waste stream 3. However, such inhibitory effect was not seen in the case of waste stream 1. The lower the concentration of the pollutant, the better the performance of the system. From this study, it was clear that the developed process worked

Table 15.3 Protocol for COD measurement.

The following method is used for COD assessment of the samples. Three replicates of each sample were pooled and one COD measurement reaction was set up for each sample.
The measurement of the COD of the sample was done by using COD digester (Merck Spectroquant TR 320) followed by the use of a spectrophotometric technique. In a COD vial, 3 ml of sample was taken; then, 0.3 ml of solution A {Mercury (II) sulfate (Merck Spectroquant COD solution A with measuring range of 4–1500 mg/L} and 2.3 ml of solution B {Sulfuric acid, and Potassium di-chromate (Merck Spectroquant COD solution B with measuring range of 100–1500 mg/L)} were added to it and mixed thoroughly. On placing the COD vial inside the COD digester (Spectroquant TR 320), the solution was processed for 2 hours at 148°C. Three milliliters of distilled water (instead of the sample) was added to the reactant solutions mentioned previously to prepare the blank solution. A spectrophotometer (Biobase BK-UV1000) was used to measure the absorbance at 605 nm. The following formula was used to compute the COD concentration of the sample:
COD concentration (mg/L) = (K* Absorbance+β), where K = 2249.00 and β (interface) = 0.00. The samples (both influent and effluent) were diluted to the desired degree for conducting the analysis.

Table 15.4 Average concentration of ammonia in the bioreactors with 2-fold diluted effluent in mg/L.

	Waste stream 1	**Waste stream 2**	**Waste stream 3**
0 day	70.82	62.16	57.08
1 days	59.18	51.31	46.47
2 days	54.04	46.68	40.73
3 days	48.69	40.13	34.04
4 days	42.94	36.25	30.08
5 days	36.74	31.91	25.43
Percentage reduction of ammonia with time in the bioreactors			
In 24 hours	16.4%	17.5%	18.6%
In 48 hours	23.7%	24.9%	28.6%
In 72 hours	31.2%	35.4%	40.3%
In 96 hours	39.4%	41.7%	48.8%
In 120 hours	48.1%	48.7%	57.5%

Table 15.5 Statistical validation of the data showing significant variation with time.

	Days 0 and 1	Days 0 and 2	Days 0 and 3	Days 0 and 4	Days 0 and 5
Waste stream 1	1.91E-11	1.11E-13	9.85E-17	2.14E-18	8.49E-20
Waste stream 2	5.25E-12	1.97E-13	9.52E-17	1.24E-16	4.24E-19
Waste stream 3	3.77E-14	8.4E-19	1.75E-21	6.85E-23	2.56E-23

Table 15.6 Chemical oxygen demand (COD) in mg/L at different time point in the three different samples.

Time	Waste stream 1	Waste stream 2	Waste stream 3
0 hours	629.72	449.80	1200.97
24 hours	382.33	337.35	1106.51
48 hours	292.37	224.90	985.06
72 hours	269.88	179.92	809.64
96 hours	202.41	157.43	600.48
120 hours	179.92	134.94	519.52

Table 15.7 Percentage reduction of COD(%) at different time point in the three different samples.

Time	Waste stream 1	Waste stream 2	Waste stream 3
24 hours	39.29	25.00	7.87
48 hours	53.57	50.00	17.98
72 hours	57.14	60.00	32.58
96 hours	67.86	65.00	50.00
120 hours	71.43	70.00	56.74

for the tested effluent. The efficiency/performance can be optimized further for the current sample for enhanced performance.

15.4 The Case Study of the Self-Cleaning Aquarium

It is expected that the Global Ornamental Fish Market will expand from USD 6800 million in 2019 to USD 11,336.47 million by 2025, with this growth taking place at 10% compounded annual rate through 2025. This envisaged change is thought to be associated with the change in the lifestyle. Scientific studies have demonstrated the ornamental fishes in domestic aquariums to have a positive psychological impact on home dwellers. This has made aquarium a home décor item for millennials. A portion of the population has financial surplus to be invested on such items and they are willing to spend. Nowadays, loneliness is a major crisis for this section of the population. A pet can often come as a stress buster, if one is willing to keep them. Dwelling in high-rise buildings with certain restrictions, a busy work schedule, and a lot of traveling make it tough to maintain most other forms of pets, except for fishes in aquariums. Aquariums with colorful fish were known to help unwind, reduce stress, and calm down household members. The growing trend towards luxury lifestyle with increased affordability is a driving force towards the growth in the ornamental fish and aquarium market. This has promoted breeding of exotic varieties of fishes.

In the Global Ornamental Fish Market, there exists segmentation based on the type of fish (tropical fresh water, temperate fresh water, and marine ornamental), the application domain (residential or commercial), the geographical region (North America, South America, Europe, Asia-Pacific, Middle East, and Africa), and the point of sale (dedicated stores, multispecialty stores, online purchase, and others). In 2019, the tropical freshwater fish species had the highest share of 58.27%, and this is projected to experience the maximum growth owing to the ease of operations. In terms of application, the residential segment is currently the dominant player in the global market, but it is predicted that by 2025, the commercial segment will overshoot the residential segment, due to the increasing trends of adoption of aquariums at complexes, shopping malls, airports, restaurants, and hotels.Companies like Aqua Nautic Specialist (ANS) Pte Ltd, Tropical Fish International Pte Ltd, Sunbean Aquarium Pte Ltd, Sanyo Aquarium (Pte) Ltd, Qian Hu Fish Farm Trading, AlgaeBarn LLC, BioAquatix.com, Imperial Tropical, Inc., Oasis Fish Farm Co., and Sagar Fish Aquarium are some of the prominent organizations in the hugely competitive global ornamental fish market. The leading organizations also invest in fish farming to further boost the market.

However, a major drawback in maintaining these aquariums is the need to clean the units every 10 to 15 days while avoiding the uncontrolled growth of algae on the walls and in the water inside the aquarium. This growth is mainly due to the presence of ammonia, nitrate, nitrite, and phosphate in the water from the uneaten fish feed as well as the fish excreta. This cleaning involves consumption of fresh water and release of polluted water into the environment. On an average, the volume of fresh water consumed annually by a 255-L aquarium will be between 6225 and 9490 L. It is necessary to examine the total water consumption of all the aquariums taken collectively. The existing treatment technologies for aquaculture effluent require between 24 hours and 14 days for the treatment. In case of aquariums, the continuous use of filters and oxygen purgers help in maintaining aquatic life for 10 to 15 days with optimum fish feeding. What is required is a continuous self-cleaning system for the aquarium.

The microbial formulation mentioned in the above section was used as stabilized biofilms (17 months old) on raschig rings that were made to float on the surface of the aquarium water (Figure 15.1a), ensuring sufficient interaction with the pollutants in the water while not hampering the aesthetics of the aquatic life (Figure 15.1b). The material of the ring should be such that the biofilm should form stably on them and the rings should float on the upper surface (Figures 15.1b and c) and not within the water column as shown in Figure 15.1a, which, in turn, will disturb the fish population. The normal feeding and aeration practice was continued and the water quality, in terms of change in color and algal growth on the walls, was monitored routinely (Figures 15.1c and d). Two such aquariums were used for the trial, one in a domestic (255 L) and another in a commercial (840 L) setup (University Guest House). The water quality was maintained beyond 365 days and continued up to 711 days. Hence, the exchange of water in an aquarium could be delayed to a major extent, ensuring preservation of fresh water for potable applications.

The water quality was tested at day 0 (before the biofilm containing raschig rings were introduced) and after 3 months of operation from both sites to understand (at a quantitative level) the extent of purification. This was compared with the water quality in the flat bottom, shallow fish ponds (called *Bheri* in local language), which are known to produce 1/3 of the fish demand of the metropolitan city of Kolkata utilizing 1/3 of the liquid waste of the city [57–59] as the source of fish feed production. These water bodies have a definite architecture that results in the purification of the water with enhanced fish production. The water quality was measured

Figure 15.1 Pictures of aquarium with raschig rings. Pictures in the two rows from left to right. (a) Raschig rings of heavier material in the 255-L aquarium. Rings floating in the water column disturb the aquatic life. (b) 840-L aquarium with biofilm containing raschig rings on the water surface (front view) on day zero; floating rings do not disturb the aesthetics. (c) Front view of the 840-L aquarium on day 366. (d) 711th day of operation. The water quality was suitable for continuing aquaculture, sustaining fish survival.

using Palintest Photometer 7100 as per the manufacturer's protocol. The pH (PCS testr 35), biochemical oxygen demand [56], and COD (Merk Spectroquant TR 320) were measured as per standard procedure. The data generated from the study are provided in Table 15.8.

The input COD of the aquarium water is higher than the COD of the Bheri water due to the different chemicals that are added to the aquarium water during recharging. Those are being removed by the floating biofilms. The quality is better or at par with that of *Bheri* water. Hence, this is a successful demonstration of a self-cleaning aquarium that will be ecofriendly and ecosafe. The process has been granted as an Indian Patent vide number 424161 on 6 March 2023 [60].

Table 15.8 Water quality before and after 3 months of treatment in the commercial and domestic aquarium with a comparison with *Bheri* water.

Parameters	**Input water of Guest House Aquarium on day zero**	**Output water of Guest House Aquarium after 3 months of operation**	**Input water of the domestic aquarium**	**Output water of domestic aquarium after 3 months of operation**	***Bheri* water**
Calcium hardness (mg/l)	101	68	88	51	74
Total hardness (mg/l)	118	77	98	62	76
Alkalinity(mg/l)	200	40	185	30	96
Fluoride (mg/l)	1	0.03	0.8	0.02	0.3
Nitrate (mg/l)	1.28	6.61	0.57	1.38	2.4
Nitrite (mg/l)	0.008	0.041	0	0.002	0.68
Ammonia (mg/l)	0	0.01	0	0	0.03
Ammonium (mg/l)	0.11	0.02	0.08	0	0.12
Phosphate (mg/l)	4.8	3.9	2.1	1.3	7.1
Sulfate (mg/l)	0	0	0	0	0
Sulfite (mg/l)	0	0	0	0	0
Sulfide (mg/l)	0	0	0	0	0
Aluminum (mg/l)	0	0	0	0	0
Zinc (mg/l)	0	0	0	0	0
Iron (mg/l)	0	0	0	0	0.15

(*Continued*)

Table 15.8 Water quality before and after 3 months of treatment in the commercial and domestic aquarium with a comparison with *Bheri* water. (*Continued*)

Parameters	Input water of Guest House Aquarium on day zero	Output water of Guest House Aquarium after 3 months of operation	Input water of the domestic aquarium	Output water of domestic aquarium after 3 months of operation	*Bheri* water
Magnesium (mg/l)	0	0	0	0	0
Magnesium hardness (mg/l)	17	2	9	0	5
Potassium (mg/l)	2	2.02	0	0	3.1
Free chlorine (mg/l)	0.04	0.02	0.03	0.02	0.03
pH	6.7	6.85	6.75	6.85	6.85
BOD (mg/l)	4.2	3.6	3.6	2.4	10.2
COD (mg/l)	247.4	170.9	232.4	159.7	194.9

15.5 Conclusion and Future Application

In this chapter, the authors report an ecofriendly and economical process for microbial treatment of ammonia containing wastewater, some of them being aquaculture, food, or refinery wastewater. This could also be adopted for other kinds of ammonia-containing wastewater like the coke oven wastewater.

Through the adoption of this patented technology, the process of ammonia removal can be made faster with minimal energy expenditure. Hence, the mis-utilization of fresh water for aquaculture (for dilution or replacement) could also be reduced to a major extent with a drastic decrease in the amount of wastewater generation from the above sector. In the case of wastewater treatment for recirculation [53, 55], it has been scaled up to

16.29 L per day. In the case of the self-cleaning aquarium, it has been scaled up to 840 L. However, there is future scope of improving the efficiency of the process further, before transitioning into full-scale commercial application. The steps to be optimized for continuous cleaning of aquaculture water are as follows:

1. Appropriate selection of material and arrangement for matrix for the development of the self-cleaning aquarium system with the highest surface area. The matrix should be such that it floats on the water while being suitable for adhesion of the microbial biofilm on the surface.
2. Optimization of inoculum for biofilm formation on selected matrix in terms of adequate inoculum percentage and incubation condition (temperature and time) for stabilized biofilm development.
3. Optimization of the number of biofilm containing raschig rings as well as the ratio of BOD:N:P to be maintained per unit volume for sustained rapid performance.
4. Performance evaluation for shelf life determination of the bioactive component (biofilm containing raschig rings).
5. Impact assessment of the system on brackish aquaculture.
6. Exploring commercialization of the developed process through industry partners for aquarium.

This microbial approach adoption will make ammonia-rich effluent treatment economical, adoptable, and ecofriendly while avoiding misuse of fresh water for non-potable applications. Hence, adequate biotechnological intervention can lead to appropriate, sludge-free bioremedial process development for environmental sustenance.

Acknowledgments

The authors acknowledge the Ministry of Environment, Government of India (under the FAST Scheme) as well as University Grants Commission-Department of Atomic Energy (under the CRS scheme) for financial assistance. They acknowledge Tripura University and Maulana Abul Kalam Azad University of Technology West Bengal for providing the laboratory facility, the computational facility, and the aquarium for the trial.

References

1. Hussain, M.I., Muscolo, A., Farooq, M., Ahmad, W., Sustainable use and management of non-conventional water resources for rehabilitation of marginal lands in arid and semiarid environments. *Agric. Water Manage.*, 221, 462, 2019.
2. Chaudhuri, Microbial Biofilm Reactor for Sustained Waste Water Treatment and Reuse. S.R., Karchiyappan, Karri, Dehghani (Eds.), pp. 345–368 Springer, Cham, Switzerland 2022.
3. Ranking the World's Most Water-Stressed Countries in 2040. *World Resour. Institute (WRI)*, https://www.wri.org/insights/ranking-worlds-most-water-stressed-countries-2040, 2015, accessed on 17/03/2023.
4. Aquaculture, Food and Agriculture Organization of the United Nations (FAO), https://www.fao.org/fishery/en/aquaculture, 2021, accessed on 17/03/2023.
5. Verdegem, M.C.J. and Bosma, R.H., Water withdrawal for brackish and inland aquaculture, and options to produce more fish in ponds with present water use. *Water Policy*, 11, 52, 2009.
6. Sharma, K.K., Mohapatra, B.C., Das, P.C., Sarkar, B., Chand, S., Water budgets for freshwater aquaculture ponds with reference to effluent volume. *Agric. Sci.*, 4, 353, 2013.
7. Cho, C.Y. and Bureau., D.P., Reduction of waste output from salmonid aquaculture through feeds and feeding. *Progress. Fish Cult.*, 59, 155, 1997.
8. Axler, R., Larsen, C., Tikkanen, C., McDonald, M., Yokom, S., Water quality issues associated with aquaculture: A case study in mine pit lakes. *Water Environ. Res.*, 68, 995, 1996.
9. Dauda, A.B., Ajadi, A., Tola-Fabunmi, A.S., Akinwole, A.O., Waste production in aquaculture: Sources, components and managements in different culture systems. *Aquacult. Fish.*, 4, 81, 2019.
10. Boyd, C.E. and McNevin, A.A., *Chemical in aquaculture. Aquaculture, resource use, and the environment*, 1st ed, pp. 172–210, John Wiley & Sons, Inc, New Jersey, 2014.
11. Goldburg, R. and Triplett, T., Murky waters: Environmental effects of aquaculture in the United States, in: *Waste Treatment Methods in Aquaculture*, Environmental Defense Fund Publication, Washington, District of Columbia USA, 1997.
12. Randall, D.J. and Wright, P.A., Ammonia distribution and excretion in fish. *Fish Physiol. Biochem.*, 3, 107, 1987.
13. Fei, X., Sun, S., He, S., Huang, J., Zhou, W., Application of a novel two-stage biofiltration system for simulated brackish aquaculture waste-water treatment. *Environ. Sci. Pollut. Res. Int.*, 27, 636, 2019.
14. Crab, R., Avnimelech, Y., Defoirdt, T., Bossier, P., Verstraete, W., Nitrogen removal techniques in aquaculture for a sustainable production. *Aquaculture*, 270, 1, 2007.

15. Knoll, S., Zwisler, W., Simon, M., Bacterial colonization of early stages of limnetic diatom micro aggregates. *Aquat. Microbiol. Ecol.*, 25, 141, 2001.
16. *Managing Ammonia in Fish Ponds*, Southern Regional Aquaculture Centre Publication No. 4603, https://fisheries.tamu.edu/files/2013/09/SRAC-Publication-No.-4603-Managing-Ammonia-in-Fish-Ponds.pdf. 2004. Accessed 14 July 2021.
17. Boyd, C.E. and Gautier, D., Effluent composition and water quality standards. *Glob. Aquacult. Advocate*, 3, 61, 2000.
18. Hopkins, T.E., Wood, C.M., Walsh, P.J., Interactions of cortisol and nitrogen metabolism in the ureogenic gulf toadfish Opsanus beta. *J. Exp. Biol.*, 198, 2229, 1995.
19. Porrello, S., Ferrari, G., Lenzi, M., Persia, E., Ammonia variations in phytotreatment ponds of land based fish farm wastewater. *Aquacult. Eng.*, 219, 485, 2003.
20. Bureau, D. and Cho, C., Phosphorus utilization by rainbow trout (*Oncorhynchus mykiss*): estimation of dissolved phosphorus waste output. *Aquaculture*, 179, 127, 1999.
21. Sugiura, S.H., Dong, F.M., Hardy, R.W., A new approach to estimating the minimum dietary requirement of phosphorus for large rainbow trout based on nonfecal excretions of phosphorus and nitrogen. *J. Nutri.*, 130, 865, 2000.
22. Piedrahita, R., Reducing the potential environmental impact of tank aquaculture effluents through intensification and recirculation. *Aquaculture*, 226, 35, 2003.
23. Doupe, R.G., Alder, J., Lymberyl, A.J., Environmental and product quality in finfish aquaculture development: An example from inland Western Australia. *Aqua. Res.*, 30, 595, 1999.
24. Boyd, D., Wilson, M., Howell, T., Recommendations for operational water quality monitoring at cage culture aquaculture operations, in: *Environment Monitoring and Reporting Branch*, Ministry of the Environment, Canada, 2001.
25. Moccia, R.D., Naylor, S., Reid, G., An overview of aquaculture in Ontario, in: *University of Guelph Extension Centre Fact Sheet*, Publ. No 96-003, Canada, 1997.
26. Fisheries and Aquaculture, National Aquaculture Legislation Overview. Food and Agriculture Organization of the United Nations (FAO), 2020, http://www.fao.org/fishery/legalframework/nalo_india/en, accessed on 18/03/2023.
27. Shan, H. and Obbard, J., Ammonia removal from prawn aquaculture water using immobilized nitrifying bacteria. *Appl. Microbiol. Biotechnol.*, 57, 791, 2002.
28. Mayer, I. and McLean, E., Bioengineering and biotechnological strategies for reduced waste aquaculture. *Water Sci. Technol.*, 31, 85, 1995.
29. Czysz, W., Denne, A. *et al.*, Waste Water Treatment Technology, in: *Origin, Collection, Treatment, and Analysis of Waste Water*, Springer- Verlag, Berlin, Germany, 1989.

30. Tlusty, M.F., Snook, K., Pepper, V.A., Anderson, M.R., The potential for soluble and transport loss of particulate aquaculture waste. *Aqua. Res.*, 31, 745, 2001.
31. Pillay, T.V.R., Aquaculture and the environment, in: *Fishing News Books*, Cambridge, Massachusetts USA, 1992.
32. Su, J.J., Yeh, K.S., Tseng, P.W., A strain of *Pseudomonas* sp. isolated from piggery wastewater treatment systems with heterotrophic nitrification capability in Taiwan. *Curr. Microbiol.*, 53, 77, 2006.
33. Padhi, S.K., Tripathy, S., Sen, R., Sinha Mahapatra, A., Mohanty, S., Maiti, N.K., Characterisation of heterotrophic nitrifying and aerobic denitrifying *Klebsiella pneumonia* CF-S9 strain for bioremediation of wastewater. *Int. Biodeter. Biodeg.*, 78, 67, 2013.
34. Zhao, B., He, Y.L., Hughes, J., Zhang, X.F., Nitrogen removal capability through simultaneous heterotrophic nitrification and aerobic denitrification by *Bacillus* sp. LY. *Environ. Technol.*, 31, 409, 2010.
35. Barman, P., Kati, A., Mandal, A.K., Bandyopadhyay, P.P., Mohapatra, P.K., Biopotentiality of *Bacillus cereus* PB45 for nitrogenous waste detoxification in *ex situ* model. *Aquacult. Int.*, 25, 1167, 2016.
36. Chen, P., Li, J., Li, Q.X., Wang, Y., Li, S., Ren, T., Wang, L., Simultaneous heterotrophic nitrification and aerobic denitrification by bacterium *Rhodococcus* sp. CPZ24. *Biores. Technol.*, 116, 266, 2012.
37. Yang, X.P., Wang, S.M., De-W, Z., Zhou, L.X., Isolation and nitrogen removal characteristics of an aerobic heterotrophic nitrifying–denitrifying bacterium, *Bacillus subtilis* A1. *Biores. Technol.*, 102, 854, 2011.
38. Lyles, C., Boopathy, R., Fontenot, Q., Kilgen, M., Biological treatment of shrimp aquaculture wastewater using a sequencing batch reactor. *Appl. Biochem. Biotechnol.*, 151, 474, 2008.
39. Erna, N.M., Banerjee, S., Shariff, M., Yusoff, F.M., Screening, Identification and Immobilization of Ammonia Oxidizing Bacterial Consortium Collected from Mangrove Areas and Shrimp Farms. *Asian J. Anim. Vet. Adv.*, 8, 73, 2013.
40. Sharrer, M.J., Tal, Y., Ferrier, D., Hankins, J.A., Summerfelt, S.T., Membrane biological reactor treatment of a saline backwash flow from a recirculating aquaculture system. *Aquacult. Eng.*, 36, 159, 2007.
41. Omitoyin, B.O., Ajani, E.K., Okeleye, O.I., Akpoilih, B.U., Ogunjobi, A.A., Biological treatments of fish farm effluent and its reuse in the culture of Nile Tilapia (*Oreochromis niloticus*). *J. Aquac. Res. Dev.*, 8, 1000469, 2017.
42. Ramli, N.M., Verdegem, M.C.J., Yusoff, F.M., Zulkifely, M.K., Verreth, J.A.J., Removal of ammonia and nitrate in recirculating aquaculture systems by the epiphyte *Stigeoclonium nanum* immobilized in alginate beads. *Aquacult. Environ. Interact.*, 9, 213, 2017.
43. Geoffrey, B., Literature review of phosphorus removal technologies, and their feasibility at recirculation fish farms in Ontario, in: Final Project Report, University of Guelph, Canada, 2000.

44. Asri, M., Elabed, S. *et al.*, Biofilm based systems for industrial wastewater treatment, in: *Handbook of environmental materials management*, C. Hussain (Ed.), pp. 1–21, Springer, Switzerland, 2018.
45. Qureshi, N., Annous, B.A., Ezeji, T.C., Karcher, P., Maddox, I.S., Biofilm reactors for industrial bioconversion processes: employing potential of enhanced reaction rates. *Microb. Cell Fact.*, 4, 24, 2005.
46. Ray Chaudhuri, S., Mukherjee, I. *et al.*, Developing tailor made microbial consortium for effluent remediation, in: *Nuclear Material Performance*, R.O.A. Rahman and H. E.-D. M. Saleh (Eds.), pp. 17–35, Intech, United Kindgom, 2016.
47. Azizi, S., Valipour, A., Sithebe, T., Evaluation of different wastewater treatment processes and development of a modified attached growth bioreactor as a decentralized approach for small communities. *Sci. World J.*, 1, 156870, 2013.
48. Ray Chaudhuri, S., Sharmin, J. *et al.*, Novel microbial system developed from low level radioactive waste treatment plant for environmental sustenance, in: *Management of Hazardous Wastes*, H. E.-D. M. Saleh and R.O.A. Rahman (Eds.), pp. 121–154, Intech, Croatia, 2016.
49. A. Mulder, Anoxic ammonium oxidation. US patent 5078884, 1992, assigned to DSM IP Assets BV.
50. G. Peng, Method for removing the contamination of C, N utilizing heterotrophic ammonia oxidizing bacteria. US patent 8394272 B2, 2007, assigned to G Peng.
51. M.B. Vanotti, A.A. Szogi, M.J. Rothrock Jr., Anammox bacterium isolate. US patent 8574885 B2, 2010, assigned to The United States of America, as represented by the Secretary of Agriculture.
52. Ray Chaudhuri, S., Tailor-made microbial wastewater treatment, in: *Bio Nano Filtration in Industrial Effluent Treatment-Advanced and Innovative Approaches*, M. Shah (Ed.), pp. 69–83, CRC Press. Taylor & Francis Group, LLC, Boca Raton, USA, 2023.
53. Gogoi, M., Bhattacharya, P., Sen, S.K., Mukherjee, I., Bhushan, S., Ray Chaudhuri, S., Aquaculture effluent treatment with ammonia remover *Bacillus albus* (ASSF01). *J. Environ. Chem. Eng.*, 9, 105697, 2021.
54. Gogoi, M., *Development of Microbial Prototype for Effluent Treatment and Reuse in Dairy and Aquaculture Industry. PhD Thesis*, Tripura University, 2021.
55. Gogoi, M., Mukherjee, I., Ray Chaudhuri, S., Characterization of ammonia remover *Bacillus albus* (ASSF01) in terms of biofilm formation ability with application in aquaculture effluent treatment. *Environ. Sci. Pollut. Res.*, 29, 61838, 2022.
56. Gautam, S.P., Guide manual: water and waste water analysis, in: Central Pollution Control Board, Government of India, https://www.indiawaterportal.org/articles/water-and-wastewater-analysis-guidemanual-central-pollution-control-board⟩, 2011. accessed 27 January 2020.

57. Ray Chaudhuri, S., Mishra, M., Nandy, P., Thakur, A.R., Waste management: A case study of ongoing traditional practices at East Calcutta Wetland. *Am. J. Agric. Biol. Sci.*, 3, 315, 2008.
58. Ray Chaudhuri, S., Mishra, M., Salodkar, S., Sudarshan, M., Thakur, A.R., Traditional Aquaculture Practice at East Calcutta Wetland: The Safety Assessment. *Am. J. Environ. Sci.*, 4, 140, 2008.
59. Ray Chaudhuri, S., Mukherjee, I., Ghosh, D., Thakur, A.R., East Kolkata Wetland: a multifunctional niche of international importance. *Online J. Biol. Sci.*, 12, 80, 2012.
60. S. Ray Chaudhuri, A process and system for ammonia removal from wastewater. Indian Patent 424161, 2023, assigned to S Ray Chaudhuri.

16

Metagenomics Unveiled: Deciphering Microbial Responses to Climate Change

Megha Kaviraj[1,2,3*], Manojit Singh[4], Soumendranath Chatterjee[1] and Upendra Kumar[2]

[1]The University of Burdwan, Burdwan, West Bengal, India
[2]ICAR-National Rice Research Institute, Cuttack, Odisha, India
[3]University of Birmingham, Edgbaston, Birmingham, England, United Kingdom
[4]Triveni Devi Bhalotia College, Raniganj, West Bengal, India

Abstract

Metagenomic-based studies have revealed a vast variety of microorganisms involved in nutrient cycling in both terrestrial and aquatic systems. These microorganisms mediate the transformation of carbon compounds and other biogeochemical processes, such as the nitrogen and sulfur cycle, which can impact groundwater, surface water, and atmospheric trace gas concentrations. This new information reshapes our understanding of microbial communities and provides critical inputs for advanced climate models, enabling accurate predictions of feedback in biogeochemical functioning and their impact on climate. It offers insights into microbial interactions, ecological relationships, and conservation efforts, aiding in the identification of keystone species and ecosystem health indicators for effective restoration strategies. Therefore, in this chapter, we discussed (1) climate change and its impact on the environment and microbiome, (2) the latest metagenomic tools and their implications on some case studies, (3) microbial adaptation and mitigation strategies to escape environmental stressors, and (4) a conceptual framework for a climate model and future research directions that can reduce the detrimental impacts of climate change on microbial communities and the surrounding environment.

Keywords**:** Metagenomics, climate change, greenhouse gas emission, ecosystem health, environment

**Corresponding author*: meghakaviraj31@gmail.com

Hrudayanath Thatoi, Sonali Mohapatra, Swagat Kumar Das and Sukanta Kumar Pradhan (eds.) Applied Biotechnology and Bioinformatics: Agriculture, Pharmaceutical Research and Environment, (369–394)

16.1 Introduction

Microorganisms have a role in the cycling of carbon and nutrients, which has an impact on human health, animals, and plants, as well as on agriculture and the world food system. Microbes first appeared at least 3.8 billion years before life on Earth and are likely to continue to exist for a very long time after any potential extinction events [1]. However, determining their function in the ecosystem is challenging due to their enormous diversity and diverse responses to environmental changes. The study of genetic material (DNA or RNA) obtained directly from environmental samples is known as metagenomics, a specialized branch of genomics [2, 3]. Metagenomics investigates the complete genetic content of a mixed community of microorganisms, often known as a microbial community or microbiome, in contrast to typical genomics, which focuses on the examination of the genetic material of individual organisms [4]. The importance of metagenomics in analyzing environmental samples is based on its capacity to offer a thorough understanding of the microbial diversity and functional potential within an ecosystem in particular [5, 27].

Additionally, metagenomics enables the identification and characterization of numerous unculturable microorganisms, such as bacteria, archaea, viruses, fungi, and other types of microorganisms [1, 5]. This sheds light on the wide range of life forms that coexist in a particular ecosystem. A tiny percentage of the microorganisms found in environmental samples can be grown using conventional microbiological methods. By directly examining the genetic code, metagenomics breaks over this limitation and makes it possible to investigate the vast majority of microbes that are difficult or impossible to grow in a laboratory [6]. Now, researchers can comprehend how these microorganisms contribute to the nutrient cycle, energy production, and other crucial ecosystem activities by identifying particular genes and their metabolic pathways [7]. The stability and health of diverse ecosystems, including soil, water bodies, the ocean, and even human microbiomes, are vitally dependent on metagenomics. New enzymes, bioactive substances, and other important biotechnological resources can be found by analyzing the genetic diversity of environmental samples [8]. Moreover, metagenomics offers insights into microbial interactions, ecological relationships, and conservation efforts, aiding in the identification of keystone species and ecosystem health indicators for effective restoration strategies [6, 9].

Overall, metagenomics has transformed the study of microbial communities and ecosystems by providing an effective set of tools for digging into the unexplored realm of microbes and their effects on the environment and human health [10]. The study of microbial metagenomics is essential for comprehending and managing climate change in a range of ways (Figure 16.1). In both terrestrial and marine ecosystems, microorganisms play a critical role in the cycling of carbon and other nutrients. Predicting how climate change will affect carbon storage and greenhouse gas emissions, such as carbon dioxide (CO_2) and methane (CH_4), requires a deep understanding of microbially driven carbon cycle processes [11, 12]. Other biogeochemical cycles, including the nitrogen, sulfur, and phosphorus cycles, are also governed by microbes. The supply of vital nutrients in ecosystems and their reaction to environmental changes are influenced by the diversity and roles of the microbes involved in these cycles, which are revealed through metagenomics. Methane is a strong greenhouse gas, and microbial activity is primarily responsible for both its synthesis and consumption [13, 14]. Methanogenic and methanotrophic bacteria, which make and consume methane, respectively, in a variety of ecosystems including wetlands, permafrost zones, and the oceans, can be better understood by metagenomic research [15–18].

Particularly vulnerable to climate change are the Arctic and Antarctic polar regions. Climate change-related shifts in microbial ecosystems may also have an impact on infectious diseases, *viz.*, Cholera, Giardiasis, Salmonellosis, Lyme disease, and many respiratory infections [1, 10]. Therefore, metagenomics helps to understand these vector–microbe interactions and predicts climate change effects. It guides restoration and

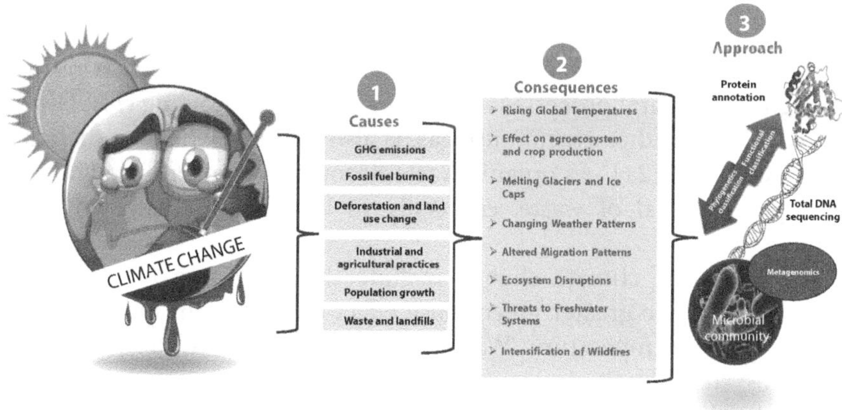

Figure 16.1 Pictorial representations of causes for climate change and its consequence events in environments, where the metagenomic approach gives an idea of phylogenetic and functional details of the entire microbial makeup.

conservation efforts, and discovers novel microorganisms for bioenergy production, providing sustainable alternatives and reducing greenhouse gas emissions. Overall, in this chapter, we tried to shed light on how recent metagenomics techniques help to understand complex relationships between microbial communities and processes causing climate change. With the use of this knowledge, climate models can be improved, specific conservation plans can be created, and effective mitigation methods may be put into effect to mitigate the adverse impacts of climate change on microbial ecosystems and the wider environment.

16.2 Climate Change and Its Impact on the Environment and Microbiome

Long-term alterations to global or regional temperature patterns, precipitation patterns, and other aspects of the climate are referred to as "climate change." It is mainly caused by the buildup of greenhouse gases (GHGs), which trap heat and cause the Earth to warm [19, 20]. The major human activities that emit these GHGs are discussed below-

The main reason for greenhouse gas emissions caused by humans is the burning of fossil fuels like coal, oil, and natural gas for energy generation. Carbon dioxide (CO_2) and other pollutants are released into the atmosphere during this process. The ability of the Earth to absorb CO_2 from the atmosphere is diminished by the clearing of forests for logging, agriculture, and other uses [19]. By storing carbon in the soil and trees, forests serve as carbon sinks. Some industrial processes directly release GHGs into the atmosphere. Manufacturing, producing cement, and chemical production are some of these. Methane (CH_4), a strong greenhouse gas, is released through agricultural processes like raising cattle and growing rice. Nitrous oxide (N_2O), another greenhouse gas, is also released when synthetic fertilizers are used [21]. As organic waste decomposes anaerobically during improper disposal of waste, especially in landfills, methane emissions result.

The effects of climate change include the following: (1) Increasing temperatures: The greater greenhouse effect has led to an increase in the average global temperature. Warmer temperatures have the potential to alter ecosystems, melt glaciers and ice sheets, and affect weather patterns. (2) Sea-level rise: Sea levels are increasing as ice sheets and glaciers melt and ocean waters warm up from global warming. This may lead to coastal erosion, floods in low-lying areas, and dangers to coastal infrastructure and communities. (3) Extreme weather events: Hurricanes, heat waves, droughts, and heavy rains are only a few of the instances of extreme weather events that are becoming more frequent and intense as a result of climate change. Ecosystems, economies, and human populations may sustain extensive harm as a result of these occurrences. (4) Ecosystem disruptions: Climate and precipitation changes can have a significant impact on ecosystems. Climate change can affect biodiversity and ecosystem services by altering the distribution of plant and animal species, interfering with migratory patterns, and causing habitat loss. (5) Ocean acidification: Ocean acidification is caused by increased CO_2 levels, which raise oceanic carbonic acid concentrations. Marine life is negatively impacted by this, especially organisms that produce calcium carbonate skeletons and shells. (6) Health effects: The deterioration of air quality, the expansion of infectious diseases, and the escalation of heat-related disorders are all ways that climate change may negatively impact the well-being of individuals. Particularly at risk are vulnerable populations. (7) Social and economic disruptions: As a result of sea level rise and extreme weather, climate change may cause communities to be evacuated. It may also affect agriculture and food security, leading to economic instability.

The influence of climate change on microbial communities and their ecological roles is one of the study areas that are receiving greater interest [22–24]. Climate change is having significant effects on ecosystems all over the world. Microbes, such as bacteria, archaea, fungi, and viruses, are essential to many ecological processes, and they can have a cascade impact on entire ecosystems when they respond to changing environmental conditions [24]. Here are some ways that microbial populations and their ecological roles may be impacted by climate change:

Microbial community composition can change as a result of variations in temperature, precipitation patterns, and other climatic conditions [25]. There may be changes in the dynamics of ecosystems when some microbial species become more prevalent while others become extinct. The carbon (C), nitrogen (N), and phosphorus (P) cycles, among others, are mostly driven by microbes [6]. Through their effects on microbial activity, changes in precipitation and temperature can affect the pace at which these cycles occur. For example, higher decomposition rates may result from warmer temperatures, which may result in the release of more CO_2 into the atmosphere. Climate changes have a significant impact on soil bacteria [26]. Warmer temperatures could accelerate the decomposition of organic matter, which will impact the quantity and quality of soil nutrients. Changes in precipitation patterns can also affect the amount of soil moisture, which, in turn, affects microbial activity.

The distribution and prevalence of pathogenic microorganisms can also alter as a result of climate change. The prevalence of diseases carried by bacteria may change as a result of warming temperatures and altering rainfall patterns [23]. For instance, alterations in the temperature and quality of the water can have an impact on the spread of waterborne diseases. Animals and plants frequently [26] coexist well with microbes. To help plants absorb nutrients, mycorrhizal fungi are one example. These interactions can be upset by changes in precipitation and temperature, which may have an impact on the health of the plants and the environment. For ocean productivity and health, marine microbial populations are essential [2, 24]. Ocean acidification and warming can change the distribution and activity of marine microorganisms, which, in turn, can have an impact on nutrient cycling, C sequestration, and the entire marine food web. Extreme weather events including droughts, floods, and heatwaves could become more frequent and intense as a result of climate change [22]. These occurrences could affect microbial communities and their functions in maintaining ecosystem health, resulting in both short- and long-term ecological consequences. Therefore, researchers use field observations, laboratory experiments, and molecular techniques to study climate change's interactions with microbial communities, a crucial aspect for understanding ecosystem responses, conservation strategies, and biodiversity mitigation.

16.3 Metagenomics as a Tool for Climate Change Research

Metagenomics helps to assess changes in microbial biodiversity and functionality in response to climate-related stressors [1]. By comparing metagenomic data from different time points or locations, one can identify shifts in microbial diversity and abundance under various ranges of temperature and precipitation which might favor specific groups of taxa over others [3]. It can reveal the changes in microbial diversity in terms of species richness and evenness in response to different climate-related stressors. Climate change can impact the direct functional capabilities of microbial communities, affecting critical ecosystem processes. Researchers can predict the functional annotation of microbial communities by identifying genes and their metabolism pathways from the metagenomic data [5]. Metagenomics also allows scientists to identify specific genetic adaptations in microbial communities in response to climate change, revealing genes more prevalent in response to specific stressors [6, 27]. Long-term monitoring of microbial communities in different ecosystems can track shifts in biodiversity and functionality as climate change progresses. This knowledge deepens our understanding of ecosystem responses to climate change and informs strategies for mitigating its effects and preserving biodiversity and ecosystem services. Here are a few of the latest metagenomic tools that are widely used: (1) MetaPhlAn—This method has been shown useful in better understanding microbial community structures and is used for taxonomic profiling of metagenomic samples. (2) MG-RAST—Metagenomics Rapid Annotation using Subsystem Technology or MG-RAST, an automated workflow pipeline for the analysis of metagenomic data that offers insights into functional and taxonomic characteristics. (3) Kraken—Kraken is a taxonomy sequence classifier that quickly and effectively classifies metagenomic sequences. (4) MEGAN—The MetaGenome ANalyzer, or MEGAN, is a powerful tool for interactively exploring huge metagenomic datasets. (5) QIIME (Quantitative Insights Into Microbial Ecology)—Originally developed to analyze 16S rRNA, QIIME now handles metagenomic data and provides a comprehensive set of tools for microbiome study.

The latest interest metagenomics projects were as follows: The Tara Oceans Expedition is an international research that investigated marine microbial diversity using metagenomics, offering insights into the function of oceans in the management of global climate [28]. The Earth Microbiome Project has made a substantial contribution to the global microbiome and its possible influence on climate change [29]. The study reconstructs bacterial and archaeal genomes from shotgun metagenomes, capturing the ecology and evolution of host-associated and environmental microbiomes. Over 10,000 metagenomes from various habitats covering every continent and ocean on Earth were collected, resulting in 52,515 genomes representing 12,556 unique species-level operational taxonomic units across 135 phyla [29]. The Human Microbiome Project (HMP) analyzed human microbial communities in healthy and disease states, generating over 48TB of data. The HMP Data Coordination Center (HMPDACC) provides a unified repository, search functionality, analysis pipelines, and standardized protocols for community use [30]. Therefore, climate change prediction using metagenomic technologies is becoming more and more promising. Complex microbial interactions and their effects on climate dynamics can be uncovered by integrating machine learning and big data analytics. To create reliable models that improve our capacity to predict and adapt to climatic changes, metagenomic researchers and climate scientists will need to keep working together. This will lead to a deeper comprehension of the complex interplay between microbial communities and patterns of global climate.

16.4 Microbial Adaptation to Climate Change

Over time, microbial communities have shown an amazing capacity for adjusting to shifting environmental conditions, particularly those brought on by climate change. Microbes can flourish in new or altered habitats because of adaptations that can take place at both the genetic and physiological levels [31]. Here are a few examples of how microorganisms have evolved to cope with shifting climatic conditions: Thermophilic bacteria have evolved to flourish in hot springs and other environments with high temperatures. These bacteria have developed protein and enzyme structures that are heat-resistant, allowing them to function well in high temperatures [32]. Certain microbial communities could benefit from these adaptations and boost their presence in warmer habitats as global temperatures rise. Cold-adapted microorganisms live in polar regions and other cold environments, as compared to thermophiles. They have developed strategies to keep their cell membranes fluid and create antifreeze proteins that stop the development of ice crystals inside cells. They can maintain metabolic processes despite intense cold due to these adaptations [33]. Psychrophiles can produce unique peptides and proteins known as ice-binding proteins (IBPs), which regulate the formation of ice crystals. Among these are antifreeze proteins (AFPs) that prevent big ice crystals formation inside cells, which might cause damage or potentially death of cells. These organisms can produce and use energy efficiently in cold conditions because they frequently have specific routes for energy metabolism. Improved healing mechanisms in cells aid in preventing possible harm from the cold and maintain the integrity of cellular processes. Desiccation-resistant traits have been developed by several bacteria as a means of adaptation to arid environments. These adaptations may include defense mechanisms like spores, which have a strong resistance to drying and can hibernate for a long time until conditions are favorable.

Methane is a potent greenhouse gas that is produced by some methanogenic bacteria and some archaea. Some methanogens have evolved to alter carbon and nutrition availability in response to the changing environment [14]. Some methanogens have additionally developed metabolic pathways that enable them to utilize different carbon sources, minimizing their reliance on methane synthesizing [22]. Primary producers known as cyanobacteria are involved in carbon fixation and photosynthesis. The effectiveness of these organisms' carbon fixation mechanisms can be affected by variations in temperature and nutrient availability. Some cyanobacteria have evolved by changing their metabolic and photosynthetic machinery to increase their capacity for carbon fixation under various climatic circumstances.

The availability of carbonate ions that marine organisms utilize to construct their shells and skeletons is impacted by ocean acidification, which is caused by rising carbon dioxide levels. To adjust to shifting carbonate chemistry, some marine microorganisms, including coccolithophores and foraminifera, have demonstrated modifications in their calcification processes. Pathogen prevalence and distribution may change as a result of climate change. Some pathogens possess modifications that improve their survival and transmission in climatically diverse environments. Changes in gene expression, regulation, and the acquisition of additional functional genes via horizontal gene transfer are frequently involved in microbial adaptations at the genomic level [15, 34]. Microbes may utilize the advantage of new niches, use alternate energy sources, and adapt to shifting nutritional availability due to these adaptations. Although exciting, microbial responses to climate change also have effects on ecosystem dynamics and human health. Understanding these adaptations is essential for anticipating how microbial communities will adapt to ongoing climate change and for developing strategies for managing the effects that these changes will have on the environment and human health. Marine microorganisms have a difficult time surviving in contaminated water bodies, which are frequently the result of human activity. These bacteria may experience stress due to exposure to contaminants, such as heavy metals, hydrocarbons, and agricultural runoff, which may affect their metabolic and ecological processes [35]. Certain marine microorganisms have evolved complex detoxification pathways to eliminate toxic chemicals, including metal-binding proteins, enzymatic degradation, and efflux pumps [36]. Furthermore, to more effectively withstand contaminated conditions, microbial communities may display adaptive changes in their gene expression and composition. Determining the general health of marine ecosystems and creating efficient environmental remediation plans require an understanding of the molecular mechanisms of marine microorganisms to water pollution. Research in this area advances our understanding of microbe adaptability and resistance in the face of anthropogenic disturbances, offering insights that may inform sustainable practices for safeguarding marine biodiversity.

16.5 Feedback Loops and Climate Change

Feedback loops are mechanisms where a change in one variable causes inhibition or enhancement of a shift in another variable. Feedback loops can either accelerate (positive feedback) or reduce (negative feedback) the consequences of warming in the context of climate change [37]. The atmosphere, oceans, glaciers, and ecosystems are among the many examples of the different parts of the Earth's systems that these loops could include [38]. In feedback loops associated with climate change, microbial communities are another important component. For example, we consider the potent greenhouse gas methane (CH_4), which could be available in a range of habitats, including wetlands, rice paddies, and the intestinal tracts of some animals, and is produced by microbes, particularly methanogenic archaea. These microorganisms decompose organic matter in oxygen-deprived environments, releasing methane as a by-product. This microbial activity can both amplify and reduce climate change.

Positive Feedback Loop—Amplifying Effects: Microbial methane production may occasionally contribute to a positive feedback loop that amplifies climate change. The increased microbial activity could result in increased methane emissions as temperatures rise. Methane is a strong greenhouse gas, so when it is released into the atmosphere, it warms the environment even more, perhaps causing a faster rise in global temperature. Additionally, the thawing of permafrost in northern locations can result in the release of significant amounts of organic material, which serves as a substrate for bacteria that produce methane and increase greenhouse gas emissions [34, 39].

Negative Feedback Loop—Mitigating Effects: However, there are also circumstances in which microbial activity could contribute to establishing detrimental feedback loops that reduce the impact of climate change. For instance, certain microbial activities can store carbon in the soil, lowering the quantities of carbon dioxide in the atmosphere. Furthermore, methanotrophic bacteria, which consume methane, are the equivalents of methane-producing microbes in some wetlands. By acting as a methane sink, these bacteria limit the amount of methane released into the atmosphere [33].

Besides, human management, human actions have the potential to affect these feedback loops [40, 41]. As an illustration, alterations in land use, such as drained wetlands for farming or construction, may interfere with microbiological processes naturally occurring and increase methane emissions [42]. On the other hand, conservation initiatives that support wetland restoration could help in re-establishing the equilibrium between methane production and consumption. For effective global warming projections, it is crucial to comprehend the complex relationships between microbial communities, greenhouse gas emissions, and climate change. Researchers and policymakers could find strategies to mitigate the effects of climate change and preserve a more stable and balanced climate system by regulating and harnessing microbial activities.

16.6 Metagenomics in Climate Change Mitigation

Insights into the roles played by microbial communities in the cycling of carbon, nutrients, and ecosystem functioning are provided by metagenomics, which can be extremely useful in the development of effective climate change mitigation methods [43]. Knowing how microbial populations react to shifting environmental factors can guide targeted treatments that improve ecosystem resilience and encourage carbon storage, ultimately aiding in greenhouse gas mitigation efforts (Figure 16.2). Microbial communities play a key role in the ecosystems' ability to cycle carbon [44]. Researchers can recognize and investigate important microbial species and functional genes involved in carbon sequestration mechanisms by using metagenomics. Here are several ways metagenomics can help mitigate climate change: Mycorrhizal fungi and bacteria, as well as other soil microorganisms, are essential for carbon sequestration. Metagenomics can assist in identifying microbial communities that increase the stability and accumulation of soil organic matter, increasing the amount of carbon stored in the soil [45]. By identifying microbial communities that support carbon sequestration, metagenomics can direct efforts to restore wetland ecosystems. The re-establishment of microbial activities that collect and store carbon can result from the restoration of wetlands. The effectiveness of Carbon Capture and Storage (CCS) technologies can be improved by microbes. Microbial communities that improve CO_2 uptake and mineralization in geological storage sites can be optimized via metagenomics. Microbial communities utilized in bioenergy production systems, such as biogas digesters and biofuel production, can be engineered with the help of metagenomics. Methane emissions from waste and agricultural systems can be decreased through the efficient microbial breakdown of organic matter [44]. Methane and nitrous oxide, two potent greenhouse gases, can be mitigated by microbial communities, which can then be used to reduce greenhouse gas emissions: Methane-consuming microorganisms can be uncovered via metagenomics. Nitrous oxide (N_2O) is produced by specific microbial communities, such as denitrifying bacteria. Insights from metagenomics into the factors influencing N_2O emissions can help establish more practical methods for lowering N_2O release from agricultural soils and other sources [46].

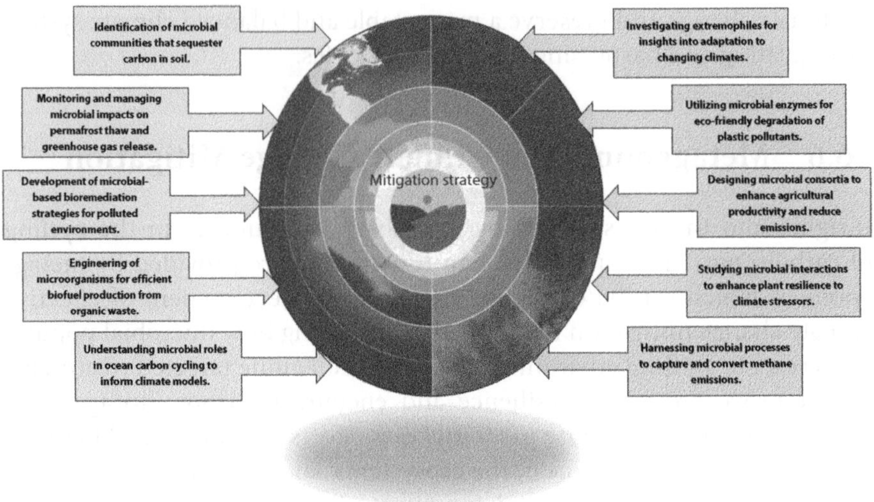

Figure 16.2 Schematic diagram of climate change mitigation strategies through metagenomic microbial interventions.

Metagenomics, despite its potential for mitigating climate change, faces challenges due to the complexity of microbial communities and the need for interdisciplinary cooperation (Figure 16.2). However, it can improve carbon sequestration, reduce greenhouse gas emissions, and advance sustainable land and resource management techniques by shedding light on their roles in carbon cycling, greenhouse gas emissions, and ecosystem processes.

16.7 Case Studies and Research Findings

Here are a few scenarios of empirical research that employed metagenomics to look into matters related to climate change:

16.7.1 Arctic Soil Microbial Communities

An investigation was conducted in the Arctic tundra to study more about the responses of microbial populations to warming temperatures and permafrost thawing. Metagenomic analysis was used to examine soil microbial diversity, carbon turnover, and functional genes involved in nutrient cycling [18]. With rising temperatures and melting permafrost, the study found changes in the composition of the microbial population. There may have been a rise in the amount of carbon released from permafrost that had been frozen as a result of some microbial groups connected to organic matter decomposition becoming more prevalent [13, 15]. A recent metagenomic study in glacial forefields uncovered microbial communities at the phylum level, which were Proteobacteria, Actinobacteria, and Bacteroidota (up to 77.15% of total diversity), followed by Acidobacteriota, Bdellovibrionota, Firmicutes, Myxococcota, and so on. At the order level, 10 samples shared most of the orders, including some abundant ones, namely, Burkholderiales, Micrococcales, Cytophagales, and Flavobacteriales.

16.7.2 Amazon Rainforest Microbial Biodiversity

Metagenomics was used to analyze microbial communities in the soils of the Amazon rainforest to look into how deforestation and changes in land use affect microbial biodiversity and function. According to the study, microbial communities observed major changes as a result of deforestation and land conversion for agriculture, which reduced diversity and affected the soil microbiome's ability for function. This has effects on the cycle of nutrients and the adaptability of ecosystems to climate change [47, 48]. Based on the analyses, the bacterial communities comprised the following phyla: Acidobacteria, Firmicutes, Verrucomicrobia, Actinobacteria, Proteobacteria, Planctomycetes, and Cyanobacteria [49].

16.7.3 Methane-Emitting Wetlands

To better understand the variables influencing production and consumption, microbial communities in methane-emitting wetlands were examined using metagenomics [44]. The study identified certain microbial species that are involved in the oxidation and generation of methane. This knowledge can inform methods for controlling microbial communities and lowering wetlands' methane emissions, which enhance greenhouse gas concentrations. Metagenomic evidence of suppressed methanogenic pathways along soil profile after wetland conversion to cropland reported that eight orders of methanogens, including Methanococcales, Methanobacteriales, Methanomassiliicoccales, Methanocellales, Methanomicrobiales, Methanopyrales, Methanosarcinales, and one of unclassified order in Methanomicrobia class, significantly declined after cultivation [50].

16.7.4 Coral Microbiomes and Climate Resilience

Metagenomics was also employed to investigate coral microbiomes in response to increasing ocean temperatures and acidity. They aimed for insight into corals' ability for climatic adaptation. According to the study, some corals contain microbial populations that are more tolerant of temperature stress, which could increase their resilience to climate change. However, the combined effects of warmth, acidity, and other stressors still pose a threat to the general health of corals [51–53]. About 25% of all the bacterial species producing sulfur oxidation genes in outer coral belonged to *Rhodopseudomonas*, whereas in inner corals, the highest abundances were more evenly distributed among *Rhodopseudomonas*, *Ruegeria*, and *Roseobacter* [54].

16.7.5 Permafrost Microbial Responses

In order to understand how thawing permafrost soils affect the carbon cycle and greenhouse gas emissions, metagenomics was utilized to examine the microbial communities there. The study showed that thawing permafrost changed the makeup of the microbial population, favoring bacteria that can decompose organic materials. Due to the increased carbon release caused by permafrost thawing, climate change may be amplified [13, 15, 18]. Proteobacteria and Actinobacteria were prominent members of the permafrost community, but Firmicutes, Bacteroidetes, and Chloroflexi can change significantly with age and other physicochemical limitations [55].

16.7.6 Industrial Waste Contaminated Microbe Response

A range of pollutants, including organic compounds, heavy metals, nutrients, and other contaminants, are released into water bodies as a result of industrial activity. Runoff from industrial sites, inappropriate chemical disposal, and the discharge of effluents from manufacturing operations are some of the sources of pollutants that end up in waterways. Some microorganisms are susceptible to modifications in their surroundings. Many studies reported those functional genes involved in processes like stress response, detoxification, and biodegradation through looking at the metagenome, for example, in pulp and paper industry wastewater, revealing 25356 OTUs. Major phyla identified included Proteobacteria, Bacteroidetes, Firmicutes, Chloroflexi, Actinobacteria, Spirochetes, Patesibacteria, Acidobacteria, and unknown microbes [56]. Interestingly natural selection may cause bacteria to evolve genetically as a response to contaminant exposure over time. Microbial community disruption can have a domino impact on the ecosystem as a whole, influencing aquatic organism health, water quality, and nutrient cycling. In a nutshell, the presence of industrial wastes can modify the composition and characteristics of aquatic microbial communities.

Table 16.1 Glimpses of some recently studied metagenomics sequencing approaches in an array of different habitats.

Location	Studied habitat	Techniques/platforms used	Sources
Italy	Human gut	Shotgun metagenomics	[59]
China	Soil	16S rRNA sequencing	[60]
Australia	Marine sediments	Metagenomic assembly	[61]
Canada	Freshwater lake	Metatranscriptomics	[62]
Brazil	Oral cavity	Shotgun metagenomics	[63]
Germany	Skin and ear canal microbiota	16S rRNA sequencing	[64]
USA	Wastewater treatment plant	Shotgun metagenomics	[65]
Norway	Arctic permafrost	Metagenomic assembly	[66]
Japan	Deep sea hydrothermal vent	Metatranscriptomics	[67]
France	Plant rhizosphere	16S rRNA sequencing	[68]
India	Cow rumen	Shotgun metagenomics	[69]
Iceland	Hot springs	Metagenomic assembly	[70]
USA	Human vaginal microbiome	Metatranscriptomics	[71]
UK	Deep subsurface aquifer	Shotgun metagenomics	[72]
Netherlands	Air	16S rRNA sequencing	[73]
Australia	Termite gut	Shotgun metagenomics	[74]
Thailand	Coral reef	Metagenomic assembly	[75]
Mexico	Hydrocarbon-contaminated soil	Metatranscriptomics	[76]
South Korea	Human fecal samples	Shotgun metagenomics	[77]
South Africa	Acid mine drainage	16S rRNA sequencing	[78]

These case studies demonstrate how metagenomics has shed light on how microbial populations respond to elements associated with climate change (Table 16.1). Understanding ecosystem dynamics, carbon cycling, and greenhouse gas emissions in a changing climate depends on these insights. Researchers can create more informed methods for climate change adaptation and mitigation by revealing the functions of microbial populations in varied habitats.

16.8 Metagenomic Climate Model Frame

Metagenomics can significantly improve climate models, conservation plans, and mitigation methods by unraveling the intricate web of microbial life [57]. By integrating metagenomic data into climate models, researchers gain unprecedented insights into microbial communities' role in shaping ecosystems and influencing global climate patterns. This allows for a more accurate representation of carbon cycling, nutrient dynamics, and greenhouse gas emissions. Metagenomics aids in conservation by understanding microbial biodiversity, identifying key species, and ensuring ecosystem stability. It aids in monitoring health, detecting ecological stress, and implementing targeted interventions. It also aids in developing effective mitigation methods, such as bioremediation strategies for pollution control, promoting sustainable and ecologically friendly solutions. Recently, one study reported on a new term, i.e., "Metaphenomics", which is a meta-analysis of soil, plant, or organism phenomes to establish dose–response relationships across phenotypic characteristics and environmental factors [58]. It helps understand microbiome functioning and can overcome challenges in agriculture and ecological sustainability. A conceptual model frame is illustrated in Figure 16.3 and described as follows: (1) Metagenomic data acquisition—collect metagenomic sequences from a repository database ranging from a variety of ecosystems, such as aquatic, marine, and terrestrial environments. (2) Climate model integration—to improve knowledge of microbial contributions to climate processes, and incorporate metagenomic data into current climate models. For more precise forecasts, include microbial feedback processes in climate simulations, such as greenhouse gas emissions and carbon cycling. (3) Identification of keystone species: utilize metagenomic data analysis to identify keystone microbial species crucial for ecosystem stability and resilience, and incorporate their information in conservation planning. (4) Prioritize conservation planning and mitigation methods: develop conservation plans for keystone species and implement strategies to protect and restore microbial diversity, acknowledging their role in ecosystem health and climate regulation. Identify microbial pathways influencing greenhouse gas production and consumption. Develop targeted mitigation methods that leverage microbial activities to reduce emissions or enhance carbon sequestration. (5) Ecosystem monitoring: create systems of continuous monitoring to evaluate changes in microbial populations over time utilizing metagenomic technologies. Adapt conservation and mitigation plans in real time based on available data. (6) Community engagement and collaboration: Engage stakeholders, legislators, and local communities about findings and actions. Encourage coordination between the public, scientists, and conservationists to adopt and promote sustainable practices. (7) Iterative feedback loop: develop an iterative feedback loop wherein ongoing research on metagenomics informs new climate models, mitigation initiatives, and conservation programs. This conceptual framework highlights how metagenomic data should be integrated at different phases, emphasizing how it can help improve climate models, create conservation strategies, and direct efficient mitigation techniques. Every stage advances a comprehensive strategy by highlighting the relationship between microbial populations, climate dynamics, and ecosystem health.

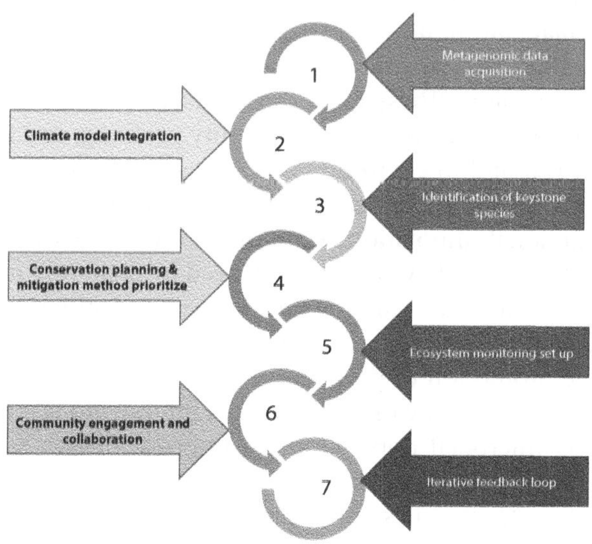

Figure 16.3 A conceptual model frame to understand microbiome functioning and climatic challenges in ecosystems.

16.9 Challenges and Future Directions

Metagenomics provides insightful information about microbial communities and how they respond to climate change. There are several limitations and boundaries to take into account: (1) Metagenomic data are complicated and large, necessitating sophisticated computer resources and bioinformatics knowledge for analysis. (2) The variety and dynamics of microbial communities can be difficult to correctly record since environmental samples might be highly changeable. (3) Functional inference—While metagenomics can reveal the existence of a gene, it cannot always reveal its expression or metabolic activity. (4) Limited ecological context—Metagenomics frequently offers a brief overview of microbial communities but lacks the temporal and spatial context required to fully comprehend ecosystem dynamics. (5) Reference databases—Metagenomic analysis accuracy depends on reference databases, which might not fully cover all microbial species.

The integration of metagenomics with other omics approaches, such as metatranscriptomics, metaproteomics, and metabolomics, can give a more complete picture of the structure and function of microbial communities. Metagenomics-based long-term monitoring can help us understand how microbial populations change in response to climate change throughout time, allowing us to identify trends and patterns. Investigators can examine the microbial communities in particular areas of environmental samples using methods like spatial metagenomics, which sheds light on the microscale interactions. These tools can be used to handle the complexity of metagenomic data, enhance community profiling, and forecast how microbes will respond to environmental changes. Metagenomic data can be used to predict microbial metabolic capacities more precisely using improved functional annotation techniques. Predictions of how microbial communities will behave can be improved by including metagenomic data in ecological models. Large-scale collaborative initiatives can aid in the development of thorough databases, the standardization of procedures, and the promotion of data sharing, which results in more reliable and broadly applicable conclusions. Targeted interventions could be made by engineering microbial populations to improve carbon sequestration, methane oxidation, and other climate-related actions.

The use of metagenomics in climate change research will probably improve in accuracy and usefulness as techniques and technology advance. A deeper comprehension of how microbes are responding to climate change and the development of successful mitigation and adaptation measures will be aided by the integration of multi-omics techniques, improvements in computational tools, and across-discipline research collaboration.

16.10 Conclusion

Metagenomic research is a transformative tool in climate change science, enabling us to understand the complex interactions between changing environmental conditions and microbial communities. These communities, often hidden from the naked eye, have immense influence over ecosystem dynamics, carbon cycling, and greenhouse gas emissions. By unraveling the genomic tapestry of microbial communities, we can understand how shifts in temperature, precipitation, and land use reverberate through these communities, predicting how ecosystems will adapt or thrive in the face of environmental upheaval. Metagenomics' significance extends beyond academia, shaping sustainable practices and policies aimed at mitigating climate change impacts. It equips us with insights to design interventions that promote ecological resilience. As we face the crossroads of climate change, metagenomics invites a transdisciplinary approach, uniting researchers from diverse fields to tackle a global challenge. With advancements in technology, metagenomics charts a course toward deeper insights and more effective strategies. With metagenomics guiding us, we can steer the trajectory of climate change, making informed decisions that foster a harmonious coexistence between humanity and the natural world.

Acknowledgments

The authors are grateful to the Director, ICAR-NRRI, Cuttack, Odisha, The University of Burdwan, Burdwan, West Bengal, India and The University of Birmingham, Edgbaston, Birmingham, United Kingdom for providing moral support for accomplishing this study.

Author Contributions

MK was involved in the full frame preparation of this chapter. MK, MS, SNC, and UK gathered the sources and contributed to the interpretation of the information and preparation of the manuscript. Everyone who has contributed thoroughly reviewed and edited the final manuscript. All the authors have read and approved the final version for publication.

Conflict of Interest

The authors declare that they have no conflict of interest.

References

1. Kačergius, A., Sivojienė, D., Gudiukaitė, R., Bakšienė, E., Masevičienė, A., Žičkienė, L., Comparison of the Structure of Soil Microbial Communities of Different Ecosystems Using the Microbiome Sequencing Approach. *Soil Syst.*, 7, 3, 70, 2023.
2. Castelle, C.J., Wrighton, K.C., Thomas, B.C., Hug, L.A., Brown, C.T., Wilkins, M.J., Banfield, J.F., Genomic expansion of domain archaea highlights roles for organisms from new phyla in anaerobic carbon cycling. *Curr. Biol.*, 25, 6, 690–701, 2015.
3. Hug, L.A., Thomas, B.C., Sharon, I., Brown, C.T., Sharma, R., Hettich, R.L., Banfield, J.F., Critical biogeochemical functions in the subsurface are associated with bacteria from new phyla and little studied lineages. *Environ. Microbiol.*, 18, 1, 159–173, 2016.
4. Arora, B., Spycher, N.F., Steefel, C., II, Molins, S., Bill, M., Conrad, M.E., Yabusaki, S.B., Influence of hydrological, biogeochemical and temperature transients on subsurface carbon fluxes in a flood plain environment. *Biogeochemistry*, 127, 367–396, 2016.
5. Stefani, F.O., Bell, T.H., Marchand, C., De La Providencia, I.E., El Yassimi, A., St-Arnaud, M., Hijri, M., Culture-dependent and-independent methods capture different microbial community fractions in hydrocarbon-contaminated soils. *PLoS One*, 10, 6, e0128272, 2015.
6. Baker, B.J., Lazar, C.S., Teske, A.P., Dick, G.J., Genomic resolution of linkages in carbon, nitrogen, and sulfur cycling among widespread estuary sediment bacteria. *Microbiome*, 3, 1–12, 2015.
7. Feng, Y., Zhang, Y., Ying, C., Wang, D., Du, C., Nanopore-based fourth-generation DNA sequencing technology. *Genomics Proteomics Bioinf.*, 13, 1, 4–16, 2015.
8. Gasc, C., Ribière, C., Parisot, N., Beugnot, R., Defois, C., Petit-Biderre, C., Peyret, P., Capturing prokaryotic dark matter genomes. *Res. J. Microbiol.*, 166, 10, 814–830, 2015.
9. Worden, A.Z., Follows, M.J., Giovannoni, S.J., Wilken, S., Zimmerman, A.E., Keeling, P.J., Rethinking the marine carbon cycle: factoring in the multifarious lifestyles of microbes. *Science*, 347, 6223, 1257594, 2015.
10. Baveye, P.C., To sequence or not to sequence the whole-soil metagenome? *Nat. Rev. Microbiol.*, 7, 10, 756–756, 2009.

11. Tringe, S.G., Von Mering, C., Kobayashi, A., Salamov, A.A., Chen, K., Chang, H.W., Rubin, E.M., Comparative metagenomics of microbial communities. *Science*, 308, 5721, 554–557, 2005.
12. Tyson, G.W., Chapman, J., Hugenholtz, P., Allen, E.E., Ram, R.J., Richardson, P.M., Banfield, J.F., Community structure and metabolism through reconstruction of microbial genomes from the environment. *Nature*, 428, 6978, 37–43, 2004.
13. Singleton, C.M., McCalley, C.K., Woodcroft, B.J., Boyd, J.A., Evans, P.N., Hodgkins, S.B., Tyson, G.W., Methanotrophy across a natural permafrost thaw environment. *ISME J.*, 12, 10, 2544–2558, 2018.
14. Wik, M., Crill, P.M., Varner, R.K., & Bastviken, D., Multiyear measurements of ebullitive methane flux from three subarctic lakes. *J. Geophys. Res. Biogeosci.*, 118, 3, 1307–1321, 2013.
15. Jansson, J.K. and Taş, N., The microbial ecology of permafrost. *Nat. Rev. Microbiol.*, 12, 6, 414–425, 2014.
16. Vogel, T.M., Simonet, P., Jansson, J.K., Hirsch, P.R., Tiedje, J.M., Van Elsas, J.D., Philippot, L., TerraGenome: a consortium for the sequencing of a soil metagenome. *Nat. Rev. Microbiol.*, 7, 4, 252–252, 2009.
17. Nealson, K.H. and Venter, J.C., Metagenomics and the global ocean survey: what's in it for us, and why should we care? *ISME J.*, 1, 3, 185–187, 2007.
18. Woodcroft, B.J., Singleton, C.M., Boyd, J.A., Evans, P.N., Emerson, J.B., Zayed, A.A., Tyson, G.W., Genome-centric., View of carbon processing in thawing permafrost. *Nature*, 560, 7716, 49–54, 2018.
19. Abeydeera, L.H.U.W., Mesthrige, J.W., Samarasinghalage, T., II, Global research on carbon emissions: a scientometric review. *Sustainability*, 11, 3972, 2019.
20. Betts, R.A., Alfieri, L., Bradshaw, C., Caesar, J., Feyen, L., Friedlingstein, P., Wyser, K., Changes in climate extremes, fresh water availability and vulnerability to food insecurity projected at 1.5 C and 2 C global warming with a higher-resolution global climate model. *Philos. Trans. R. Soc. London, Ser. A*, 376, 2119, 20160452, 2018.
21. Wei, T., Yang, S., Moore, J.C., Shi, P., Cui, X., Duan, Q., Dong, W., Developed and developing world responsibilities for historical climate change and CO2 mitigation. *PNAS*, 109, 32, 12911–12915, 2012.
22. Meena, M., Yadav, G., Sonigra, P., Nagda, A., Mehta, T., Swapnil, P., Kumar, S., Multifarious responses of forest soil microbial community toward climate change. *Microb. Ecol.*, 86, 1, 49–74, 2023.
23. Bissett, A., Brown, M.V., Siciliano, S.D., Thrall, P.H., Microbial community responses to anthropogenically induced environmental change: towards a systems approach. *Ecol. Lett.*, 16, 128–139, 2013.
24. Hutchins, D.A., Jansson, J.K., Remais, J.V., Rich, V., II, Singh, B.K., Trivedi, P., Climate change microbiology—problems and perspectives. *Nat. Rev. Microbiol.*, 17, 6, 391–396, 2019.

25. Malhi, G.S., Kaur, M., Kaushik, P., Alyemeni, M.N., Alsahli, A.A., Ahmad, P., Arbuscular mycorrhiza in combating abiotic stresses in vegetables: An eco-friendly approach. *Saudi J. Biol. Sci.*, 28, 2, 1465–1476, 2021.
26. Compant, S., Van Der Heijden, M.G., Sessitsch, A., Climate change effects on beneficial plant–microorganism interactions. *FEMS Microbiol. Ecol.*, 73, 2, 197–214, 2010.
27. Garza, D.R. and Dutilh, B.E., From cultured to uncultured genome sequences: metagenomics and modeling microbial ecosystems. *Cell. Mol. Life Sci.*, 72, 4287–4308, 2015.
28. Zhang, H. and Ning, K., The Tara Oceans project: new opportunities and greater challenges ahead. *GPB*, 13, 5, 275, 2015.
29. Nayfach, S., Roux, S., Seshadri, R., Udwary, D., Varghese, N., Schulz, F., Eloe-Fadrosh, E.A., A genomic catalog of Earth's microbiomes. *Nat. Biotechnol.*, 39, 4, 499–509, 2021.
30. Creasy, H.H., Felix, V., Aluvathingal, J., Crabtree, J., Ifeonu, O., Matsumura, J., White, O., HMPDACC: a Human Microbiome Project Multi-omic data resource. *Nucleic Acids Res.*, 49, D1, D734–D742, 2021.
31. Wani, A.K., Akhtar, N., Sher, F., Navarrete, A.A., Américo-Pinheiro, J.H.P., Microbial adaptation to different environmental conditions: molecular perspective of evolved genetic and cellular systems. *Arch. Microbiol.*, 204, 2, 144, 2022.
32. Hussain, M.M., Farooqi, Z.U.R., Rasheed, F., Din, W.M.U., Role of microorganisms as climate engineers: Mitigation and adaptations to climate change, in: *Climate change and microbes*, pp. 1–24, Apple Academic Press, New York, 2022.
33. Malik, A.A., Martiny, J.B., Brodie, E.L., Martiny, A.C., Treseder, K.K., Allison, S.D., Defining trait-based microbial strategies with consequences for soil carbon cycling under climate change. *ISME J.*, 14, 1, 1–9, 2020.
34. Guo, X., Gao, Q., Yuan, M., Wang, G., Zhou, X., Feng, J., Zhou, J., Gene-informed decomposition model predicts lower soil carbon loss due to persistent microbial adaptation to warming. *Nat. Commun.*, 11, 1, 4897, 2020.
35. Fulke, A.B., Kotian, A., Giripunje, M.D., Marine microbial response to heavy metals: mechanism, implications and future prospect. *Bull. Environ. Contam. Toxicol.*, 105, 2, 182–197, 2020.
36. Madhavi, A., Srinivasulu, M., Chandra, M.S., Rangaswamy, V., Detoxification of Heavy Metals Using Marine Metal Resistant Bacteria: A New Method for the Bioremediation of Contaminated Alkaline Environments, in: *Innovations in Biotechnology for a Sustainable Future*, pp. 297–332, 2021.
37. Wheeler, S.A., Nauges, C., Zuo, A., How stable are Australian farmers' climate change risk perceptions? New evidence of the feedback loop between risk perceptions and behaviour. *Glob. Environ. Change*, 68, 102274, 2021.
38. Asaka, J.O., Climate change-terrorism nexus? a preliminary review/analysis of the literature. *Perspect. Terror.*, 15, 1, 81–92, 2021.

39. Hui-Min, L., II, Xue-Chun, W.A.N.G., Xiao-Fan, Z.H.A.O., Ye, Q., II, Understanding systemic risk induced by climate change. *Adv. Clim. Change Res.*, 12, 3, 384–394, 2021.
40. Williams, P.C., Bartlett, A.W., Howard-Jones, A., McMullan, B., Khatami, A., Britton, P.N., Marais, B.J., Impact of climate change and biodiversity collapse on the global emergence and spread of infectious diseases. *J. Paediatr. Child Health*, 57, 11, 1811–1818, 2021.
41. Richards, C., *Reframing global warming for better risk management: climate change, food insecurity and societal collapse (Doctoral dissertation, University of Cambridge)*, 2022.
42. Brosch, T., Affect and emotions as drivers of climate change perception and action: a review. *Curr. Opin. Behav. Sci.*, 42, 15–21, 2021.
43. Asselstine, V., Lam, S., Miglior, F., Brito, L.F., Sweett, H., Guan, L., Cánovas, A., The potential for mitigation of methane emissions in ruminants through the application of metagenomics, metabolomics, and other-OMICS technologies. *J. Anim. Res.*, 99, 10, skab193, 2021.
44. Padhy, S.R., Bhattacharyya, P., Nayak, S.K., Dash, P.K., Mohapatra, T., A unique bacterial and archaeal diversity make mangrove a green production system compared to rice in wetland ecology: A metagenomic approach. *Sci. Total Environ.*, 781, 146713, 2021.
45. Liu, C., Liu, H., Liu, X., Li, G., Zhang, Y., Zhang, M., Li, Z., Metagenomic analysis insights into the influence of 3, 4-dimethylpyrazole phosphate application on nitrous oxide mitigation efficiency across different climate zones in Eastern China. *Environ. Res.*, 236, 116761, 2023.
46. Mendoza, O., Pretell, K., Ramirez, B., Sandoval, J., Caballero, J.L., Dorado, G., Metagenomic Analyses of Biofilms on Whiteleg Shrimp (Litopenaeus vannamei) Effluents: Implications for Worldwide Aquaculture Bioremediation and Environmental Sustainability in the Current Trend of Climate Change and Global Warming–State of the Art and Experimental Proof of Concept. *Rev. Fish. Sci. Aquac.*, 29, 4, 431–444, 2021.
47. Venturini, A.M., Gontijo, J.B., Mandro, J.A., Paula, F.S., Yoshiura, C.A., da França, A.G., Tsai, S.M., Genome-resolved metagenomics reveals novel archaeal and bacterial genomes from Amazonian Forest and pasture soils. *Microb. Genom.*, 8, 7, 000853, 2022.
48. Mandro, J.A., Nakamura, F.M., Gontijo, J.B., Tsai, S.M., Venturini, A.M., Metagenome-Assembled Genomes from Amazonian Soil Microbial Consortia. *Microbiol. Resour. Announc.*, 11, 11, e00804–22, 2022.
49. Melo, V.F., Barros, L.S., Silva, M.C., Veloso, T.G., Senwo, Z.N., Matos, K.S., Nunes, T.K., Soil bacterial diversities and response to deforestation, land use and burning in North Amazon, Brazil. *Appl. Soil Ecol.*, 158, 103775, 2021.
50. Wang, N., Zhu, X., Zuo, Y., Liu, J., Yuan, F., Guo, Z., Xu, X., Metagenomic evidence of suppressed methanogenic pathways along soil profile after wetland conversion to cropland. *Front. Microbiol.*, 13, 930694, 2022.

51. Palladino, G., Caroselli, E., Tavella, T., D'Amico, F., Prada, F., Mancuso, A., Biagi, E., Metagenomic shifts in mucus, tissue and skeleton of the coral *Balanophyllia europaea* living along a natural CO2 gradient. *ISME Commun.*, 2, 1, 65, 2022.
52. Keller-Costa, T., Lago-Lestón, A., Saraiva, J.P., Toscan, R., Silva, S.G., Gonçalves, J., Costa, R., Metagenomic insights into the taxonomy, function, and dysbiosis of prokaryotic communities in octocorals. *Microbiome*, 9, 1, 1–21, 2021.
53. Meenatchi, R., Thinesh, T., Brindangnanam, P., Hassan, S., Kiran, G.S., Selvin, J., Revealing the impact of global mass bleaching on coral microbiome through 16S rRNA gene-based metagenomic analysis. *Microbiol. Res.*, 233, 126408, 2020.
54. Lima, L.F., Alker, A.T., Papudeshi, B., Morris, M.M., Edwards, R.A., De Putron, S.J., Dinsdale, E.A., Coral and seawater metagenomes reveal key microbial functions to coral health and ecosystem functioning shaped at reef scale. *Microb. Ecol.*, 86, 1, 392–407, 2023.
55. Waldrop, M.P., Chabot, C.L., Liebner, S., Holm, S., Snyder, M.W., Dillon, M., Mackelprang, R., Permafrost microbial communities and functional genes are structured by latitudinal and soil geochemical gradients. *IMSE J.*, 17, 8, 1–12, 2023.
56. Sharma, P., Tripathi, S., Chandra, R., Metagenomic analysis for profiling of microbial communities and tolerance in metal-polluted pulp and paper industry wastewater. *Bioresour. Technol.*, 324, 124681, 2021.
57. Ko, K.K., Chng, K.R., Nagarajan, N., Metagenomics-enabled microbial surveillance. *Nat. Microbiol.*, 7, 4, 486–496, 2022.
58. Azeem, M., Soundari, P.G., Ali, A., Tahir, M., II, Imran, M., Bashir, S., Zhang, Z., Soil metaphenomics: a step forward in metagenomics. *Arch. Agron. Soil Sci.*, 68, 12, 1645–1663, 2022.
59. Rampelli, S., Soverini, M., D'Amico, F., Barone, M., Tavella, T., Monti, D., Candela, M., Shotgun metagenomics of gut microbiota in humans with up to extreme longevity and the increasing role of xenobiotic degradation. *Msystems*, 5, 2, e00124–20, 2020.
60. Wang, C.Y., Zhou, X., Guo, D., Zhao, J.H., Yan, L., Feng, G.Z., Zhao, L.P., Soil pH is the primary factor driving the distribution and function of microorganisms in farmland soils in northeastern China. *Ann. Microbiol.*, 69, 13, 1461–1473, 2019.
61. Sierra, M.A., Ryon, K.A., Tierney, B.T., Foox, J., Bhattacharya, C., Afshin, E., Tighe, S.W., Microbiome and metagenomic analysis of Lake Hillier Australia reveals pigment-rich polyextremophiles and wide-ranging metabolic adaptations. *Environ. Microbiome*, 17, 1, 1–20, 2022.
62. Sadeghi, J., Shahraki, A.H., Chaganti, S.R., Heath, D., Functional gene transcription variation in bacterial metatranscriptomes in large freshwater Lake Ecosystems: Implications for ecosystem and human health. *Environ. Res.*, 231, 116298, 2023.

63. Oliveira, S.G., Nishiyama, R.R., Trigo, C.A., Mattos-Guaraldi, A.L., Dávila, A.M., Jardim, R., Aguiar, F.H., Core of the saliva microbiome: an analysis of the MG-RAST data. *BMC Oral. Health*, 21, 1–10, 2021.
64. Apostolopoulos, N., Glaeser, S.P., Bagwe, R., Janssen, S., Mayer, U., Ewers, C., Thom, N., Description and comparison of the skin and ear canal microbiota of non-allergic and allergic German shepherd dogs using next generation sequencing. *PLoS One*, 16, 5, e0250695, 2021.
65. Lindner, B.G., Suttner, B., Zhu, K.J., Conrad, R.E., Rodriguez-R, L.M., Hatt, J.K., Konstantinidis, K.T., Toward shotgun metagenomic approaches for microbial source tracking sewage spills based on laboratory mesocosms. *Water Res.*, 210, 117993, 2022.
66. Xue, Y., Jonassen, I., Øvreås, L., Taş, N., Metagenome-assembled genome distribution and key functionality highlight importance of aerobic metabolism in Svalbard permafrost. *FEMS Microbiol. Ecol.*, 96, 5, fiaa057, 2020.
67. Motoki, K., Watsuji, T.O., Takaki, Y., Takai, K., Iwasaki, W., Meta transcriptomics by *in situ* RNA stabilization directly and comprehensively revealed episymbiotic microbial communities of deep-sea squat lobsters. *Msystems*, 5, 5, e00551–20, 2020.
68. Pandey, S. and Gupta, S., Diversity analysis of ACC deaminase producing bacteria associated with rhizosphere of coconut tree (Cocos nucifera L.) grown in Lakshadweep islands of India and their ability to promote plant growth under saline conditions. *J. Biotechnol.*, 324, 183–197, 2020.
69. Khatoon, M., Patel, S.H., Pandit, R.J., Jakhesara, S.J., Rank, D.N., Joshi, C.G., Kunjadiya, A.P., Rumen and fecal microbial profiles in cattle fed high lignin diets using metagenome analysis. *Anaerobe*, 73, 102508, 2022.
70. Allioux, M., Yvenou, S., Merkel, A., Cozannet, M., Aubé, J., Pommellec, J., Alain, K., A metagenomic insight into the microbiomes of geothermal springs in the Subantarctic Kerguelen Islands. *Sci. Rep.*, 12, 1, 22243, 2022.
71. Ojala, T., Kankuri, E., Kankainen, M., Understanding human health through metatranscriptomics. *Trends Mol. Med.*, 29, 5, 376–389, 2023.
72. Soares, A., Edwards, A., An, D., Bagnoud, A., Bradley, J., Barnhart, E., Mitchell, A.C., A global perspective on bacterial diversity in the terrestrial deep subsurface. *Microbiology*, 169, 1, 001172, 2023.
73. Van Kersen, W., Bossers, A., de Steenhuijsen Piters, W.A., de Rooij, M.M., Bonten, M., Fluit, A.C., Smit, L.A., Air pollution from livestock farms and the oropharyngeal microbiome of COPD patients and controls. *Environ. Int.*, 169, 107497, 2022.
74. Chiri, E., Nauer, P.A., Lappan, R., Jirapanjawat, T., Waite, D.W., Handley, K.M., Greening, C., Termite gas emissions select for hydrogenotrophic microbial communities in termite mounds. *PNAS*, 118, 30, e2102625118, 2021.
75. Titioatchasai, J., Surachat, K., Kim, J.H., Mayakun, J., Diversity of Microbial Communities Associated with Epilithic Macroalgae in Different Coral Reef

Conditions and Damselfish Territories of the Gulf of Thailand. *J. Mar. Sci. Eng.*, 11, 3, 514, 2023.

76. Howe, K.L., Seitz, K.W., Campbell, L.G., Baker, B.J., Thrash, J.C., Rabalais, N.N., Mason, O.U., Metagenomics and metatranscriptomics reveal broadly distributed, active, novel methanotrophs in the Gulf of Mexico hypoxic zone and in the marine water column. *FEMS Microbiol. Ecol.*, 99, 2, fiac153, 2023.
77. Manoharan, R.K., Srinivasan, S., Shanmugam, G., Ahn, Y.H., Shotgun metagenomic analysis reveals the prevalence of antibiotic resistance genes and mobile genetic elements in full scale hospital wastewater treatment plants. *J. Environ. Manage.*, 296, 113270, 2021.
78. Lukhele, T., Selvarajan, R., Nyoni, H., Mamba, B.B., Msagati, T.A.M., Diversity and functional profile of bacterial communities at Lancaster acid mine drainage dam, South Africa as revealed by 16S rRNA gene high-throughput sequencing analysis. *Extremophiles*, 23, 719–734, 2019.

17

Biosensor: A Tool for Assessment of Soil Pollutants

Saheed Garnaik[1]* and Jagamohan Nayak[2]

[1]Department of Soil Science, Odisha University of Agriculture and Technology, Bhubaneswar, India
[2]Department of Agronomy, Bidhan Chandra Krishi Viswavidyalaya, Mohanpur, India

Abstract

The present food demand needs a wide application of fertilizers and agrochemicals to crops for higher production. This leads to soil quality deterioration in line with toxic chemicals' deposition in the soil environment. Then, it indirectly and directly affects human and plant life. Along with agrochemicals, many household and medical wastes add contaminant to the soil. The monitoring of the soil environment is not an easy task. Hence, new technologies need to be explored to tackle the unprecedented soil conditions. Such technologies may help to know the soil constituents along with the toxic elements in the soil environment. Recently, the sensor technology has emerged as a promising tool for the characterization of the soil environment. Nano-biosensors can predict the soil constituents and toxic elements accurately. This will help in nutrient management in crop production for overcoming the quality constraints in food grains and vegetables. Biosensors can be a component of precision farming with a low-cost investment. Nano-biosensors combined with microbial biosensors can improve the sensitivity of the detection in a stipulated time. This chapter will discuss various biosensors and their application in detecting soil pollutants.

***Keywords*:** Nano-biosensors, soil environment, detectors, monitoring, nano-technology

**Corresponding author*: shibajisaheed@gmail.com

Hrudayanath Thatoi, Sonali Mohapatra, Swagat Kumar Das and Sukanta Kumar Pradhan (eds.) Applied Biotechnology and Bioinformatics: Agriculture, Pharmaceutical Research and Environment, (395–406) © 2025 Scrivener Publishing LLC

17.1 Introduction

Soil pollution is the physical, chemical, and biological degradation of soil. This will render the soil quality and soil health. The better the soil health, the better is the crop sustainability. The sustainability of agricultural production originates from the soil. The soil is the source of nutrients and water for plant growth. Also, human establishment is directly and indirectly dependent on the soil conditions. However, the soil pollutants can interfere with the life processes of plants, animals, and humans. Hence, soil phenomenon assessment is a tricky task. Nowadays, the increasing demand for food production has intensified the use of inorganic fertilizers and agrochemicals. Along with pesticides, cosmetics, personal care items, household cleaners, industrial effluents, and medications eventually add pollutants to the soil environment [1]. Additionally, due to human activities, water supplies have been contaminated with biological micropollutants like bacteria and viruses. These agents are considered as emerging or re-emerging pathogens because of their potential pathogenicity, which has caused new awareness. Various waterborne diseases are induced by biological micropollutants, e.g., enteric bacteria, viruses, mycoplasmas, and protozoa, which are the leading causes of fatalities in the world [2].

Soil pollution monitoring followed by management can reduce the ill effects of toxic elements. Hence, the monitoring of soil pollutants may help in improving the soil health. As a result, this can sustain the crop production and make the water sources (surface and ground water) toxin free. The traditional analysis methods are laborious, time-consuming, and tedious. Also, the chemical used for estimation of toxic elements in soil is ultimately disposed to soil, which later remains as a pollutant and contaminates the soil and water. Typically, for agricultural reasons, water and soil parameters are tracked and documented using conventional analytical procedures that demand preparation and work for sample. The fundamental reason for employing nano-biosensors is the requirement for more straightforward, sensitive methods. Nano-biosensors can be built with recycled nanomaterials such that different nano-biosensors can achieve the same objectives with great sensitivity and no sample preparation [3]. The quantity of environmental pollutants as a whole is frequently overestimated by conventional methods, and the bioavailability of the contaminants individually is frequently impossible to quantify. Saini *et al.* [4] reported the framework of a whole cell-based biosensor for naphthalene detection. Since then, these whole cell-based biosensors have been widely used for detection of the bioavailability of a variety of environmental pollutants. These biosensors have been shown to be affordable, straightforward, and effective [5].

The development of more environment friendly technologies to identify dangerous chemicals in soil and water is a global concern. Hence, the current chapter focuses on the biosensors that are promising for detection of soil conditions solely or combined with other biosensors for precise results. Also, this chapter discuss their application and future prospectives.

17.2 Working Principles

The biologically sensitive element and a transducer combination converts biological material into an electrical response in the form of a signal. Based on the materials used, the output of transducer can be either current or voltage. The amplitude of the output voltage signal is very low, and it is overlaid on a high-frequency signal (noise). As a result, the signal gets amplified (via an Op Amp-based current-to-voltage converter) before being routed via a Low-Pass RC Filter. This process of signal amplification followed by filtering is performed by the signal processing unit. The analog signal is the output of the signal processing unit. This result represents the biological quantity under consideration. The analog signal may be immediately shown on an LCD monitor. This analog signal is digitally transformed. This is performed due to ease in analyzing, processing, and storing a digital signal. Biosensors are made up of sensors/detectors, transducers, and an electrical circuit. The working principle is presented graphically in Figure 17.1 (adapted and modified from [6, 7]).

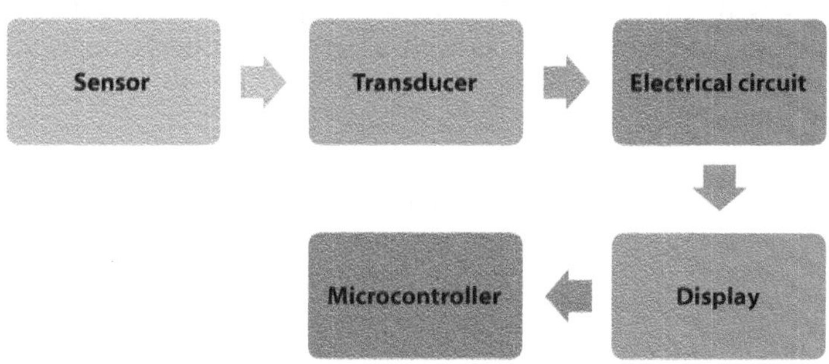

Figure 17.1 Flowchart showing the mechanism of the working principle of biosensors.

17.3 Types of Biosensors

Numerous biosensors are in use for the detection of contaminants in soil as well as water in recent times. The most important biosensors will be discussed below (Figure 17.2).

17.3.1 Nanomaterial-Based Biosensors

Nano-biosensors are simple to use, inexpensive to run, and effective at a variety of detection scales. Nano-biosensors can measure physicochemical parameters, monitor atmospheric conditions, and identify microbes in remote areas. Real-time nano-biosensors can increase agricultural output by measuring temperature and humidity and regulating fertilizer and pesticide application at a specific time and location [3].

Various nanomaterials (e.g., carbon-based, metal-based, composites, etc.) have been developed for agricultural applications, with the goal of reducing crop loss, monitoring climatic conditions, diagnosing plant diseases, and detecting plant metabolites and toxins inside the soil. Traditional pesticide detection methods, i.e., gas chromatography (GC), high-performance liquid chromatography (HPLC), and inductively coupled plasma-mass spectrometry (ICP-MS), are being superseded by nano sensor technologies and the use of bio components. The association of these nanomaterials with bio components increases the sensitivity, selectivity, and speed of identification of targeted soil contaminants, i.e., pesticides.

Carbon nanotubes, quantum dots, and nanoparticles are used in biosensor designs to increase sensitivity and signal transduction. Biosensors are a very sensitive and specific technology for detecting environmental pollutants. As a result, several different types of biosensors have been created [8]. Additionally, sensors that are low cost, compact, and environmentally benign are required. Leading biosensors are provided by nanotechnology through new nanofabrication as well as green synthesis methodologies.

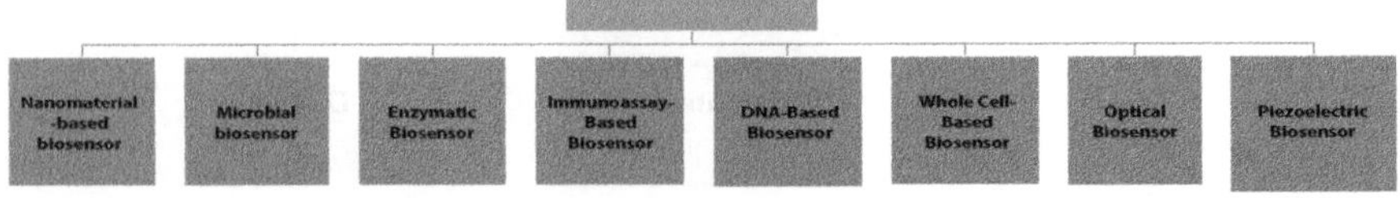

Figure 17.2 Type of biosensors commonly used to detect contaminants in soil environment.

Nano-biosensors used for pollution detection have a high sensitivity and a short detection time. Furthermore, the literature reports the detection limits ranging from nanomolar to picomolar [9]. Aguilar-Pérez *et al.* [10] examined nanotechnology in green chemical sensors for ecological applications, with a focus on pollution detection in water and soil.

17.3.2 Microbial Biosensors

Microorganisms can be genetically altered or selected to create a detectable signal in response to specific contaminants. Various microbial biosensors have been created for detecting heavy metals, herbicides, organic pollutants, and other environmental toxins [11]. Recent studies used microbial biosensors for estimating soil pollutants, i.e., phenols, pesticides, and heavy metals (Cu, Pb, As, Hg, and Cd), along with other hazardous substances using land and aquatic biota [12]. When microbial biosensors are combined with nano-biosensors, their sensitivity is increased [13].

17.3.3 Enzymatic Biosensors

An enzyme-based biosensor implements a particular enzyme as a biological sensing element, together with a transducer. This setup converts the signal produced during enzymatic reaction into a measurable response proportional to the analyte concentration [12]. The enzymes operate as highly selective catalysts for reactions with contaminants, resulting in signal generation [14].

17.3.4 Immunoassay-Based Biosensors

Antibodies or antigens are used as recognition elements. Environmental pollutants, hormones, and pharmaceutical residues have all been detected using immunoassay-based biosensors. The SERS-LFIA biosensor combines with SiO_2-Au nanoparticles (NPs) for the specific and sensitive detection of ricin, staphylococcal enterotoxin B (SEB), and botulinum neurotoxin type A (BoNT/A). SiO_2-Au NPs were used to construct SERS tags with desirable properties such as light weight, homogeneous particle size, excellent dispersion, and high SERS performance [15].

17.3.5 DNA-Based Biosensors

Nucleic acids serve as recognizable components in DNA-based biosensors. Their operation is based on two mechanisms: (a) the hybridization process between the target DNA and its complementary strand immobilized on a sensing area through spontaneous H-bonding between adenine-thymine (A-T) and cytosine-guanine (C-G) pairs; and (b) the modification of the single stranded-DNA structure by the molecules of the target analyte [12].

17.3.6 Optical Biosensors

Optical biosensors work by monitoring the changes in light intensity, wavelength, or surface plasmon resonance as a result of pollutant contact. These sensors have excellent sensitivity and can monitor in real time. To address the increased need for legislative acts on environmental pollution management and early warning, the development of revolutionary analytical equipment that are exact, sensitive, specific, quick, and easy-to-use is required. As a strong alternative to traditional analytical procedures, optical biosensors provide very sensitive, real-time, and high-frequency monitoring of contaminants without considerable sample preparation [16].

17.3.7 Piezoelectric Biosensors

Piezoelectric biosensors rely on piezoelectric crystals, which distort mechanically when contaminants connect to their recognition element. This deformation provides a measurable electrical signal. Because of the large number of pesticides now in use, there is a growing interest in creating biosensors for their detection. Piezoelectric devices have emerged as the most appealing among all physical transducers due to their simplicity, cheap instrumentation costs, ability to detect in real time and label-free, and overall high sensitivity. Marrazza [17] provided an overview of biosensors for organophosphate and carbamate pesticide detection based on quartz crystal microbalances that have been published in the literature.

17.4 Application of Biosensors

17.4.1 Heavy Metal Detection

Pollutants such as biocides and heavy metals are often found in regions of fast industrial expansion and are seen as a worldwide threat. Toxins in the environment are commonly connected to degradation of health and the loss of biodiversity. Pollution detection must be quick and precise in order to reduce these dangers. Although traditional wet analysis methods for soil pollutants have sensitivity and specificity, there are still many

Table 17.1 Commonly used biosensors for heavy metal detection.

Pollutant	Type of biosensor	Reference
Heavy metal (Cd, Pb, As, Hg, etc.)	Green fluorescent protein-based whole cell bacterial biosensor	[19]
	Urease–glutamic dehydrogenase biosensor	[20]
	Optical detection module-based biosensor using fortified bacterial beads	[21]
	Nano-biosensors	[22]
	Genetic circuits in microbial biosensors	[23]

barriers to their widespread use, e.g., the need for expensive equipment, labor-intensive procedures, and long detection times [18] (Table 17.1).

Direct determination of pollutants as well as monitoring of soil and water impose significant challenges in environmental and analytical chemistry. Existing tactics for real-time analyses and monitoring of contaminated specimens are restricted due to a lack of non-destructive procedures and broad detection devices.

17.4.2 Organic Pollutant Monitoring

Biosensors improve the accuracy and sensitivity of technologies like GC-MS and HPLC in the detection of organic pollutants such as pesticides, herbicides, polycyclic aromatic hydrocarbons (PAHs), and volatile organic compounds (VOCs). Biosensors have been used to detect pesticide residues in agricultural settings, hydrocarbon pollutants in soil or water, and volatile organic compounds (VOCs) in interior spaces [24]. The importance of biosensor data in improved pollution management and control, as well as their potential utility in precision agriculture and industrial emissions monitoring, cannot be overstated.

17.4.3 Emerging Pollutants

Pharmaceuticals, personal care items, microplastics, and nanomaterials may all be detected with biosensors (enzyme-based and nano-based). As a result, the problems in monitoring these pollutants, as well as the potential role of biosensors in increased detection and risk assessment, are readily possible. However, the efficacy of specific biosensor platforms designed to identify these new threats is based on their sensitivity, specificity, and durability. This finding might usher in a new age of identifying emerging toxins in the environment, such as those that have an impact on aquatic ecosystems and human health.

17.5 Advantages, Disadvantages, and Adoption of Biosensors

17.5.1 Advantages and Disadvantages of Biosensors

Biosensors are analytical devices that combine a biological component with a physicochemical detector to identify and measure various substances. These devices have revolutionized many fields, including medical diagnostics, environmental monitoring, food safety, and biotechnology. By leveraging the specificity of biological molecules such as enzymes, antibodies, or nucleic acids, biosensors offer precise and rapid detection of target analytes. Their integration with modern technology has enabled the development of compact, portable, and user-friendly devices that provide real-time data, making them invaluable tools in various applications. However, despite their numerous advantages, biosensors also face several challenges that must be addressed to maximize their potential and reliability. Table 17.2 explains important advantages and disadvantages of the use of biosensor.

Table 17.2 Advantages and disadvantages of biosensors while detecting soil pollutants.

Advantages	Disadvantages
Rapid response times	Calibration every time
Real-time monitoring capabilities	Low reproducibility
Reduce the need for laboratory analysis	Issues related to stability
Cost-effectiveness	Impractical portability
Reduced sample volume requirements	Poor selectivity

17.5.2 Strategies Explored to Overcome These Limitations and Improve the Overall Performance of Biosensors

- Biosensors are well-suited for on-site, in-field applications, making them ideal for distant and difficult environmental conditions requiring continuous monitoring.
- Nano-biosensors show excellent performance and reliability for pollutant recognition. However, policymakers and research groups should plan public and occupational policies to address environmental and external factors that affect the materials' properties.
- Any misgivings or worries that stakeholders may have about biosensor technology must be addressed. Public outreach, education, and awareness initiatives should be used to overcome opposition to change and promote the benefits of biosensor adoption.
- The potential for bias or misunderstanding of biosensor data, as well as measures for mitigating such risks, should be investigated further.

17.5.3 Integration and Adoption

The potential advantages of embedding biosensors into current soil environment monitoring networks include the real-time data supplied by biosensors for identifying abrupt pollution incidents, analyzing pollution dynamics, and guiding prompt reaction measures. It can improve the efficiency and efficacy of biosensors in environmental monitoring programs. Biosensor networks' scalability and capacity to cover broad geographical areas can be strategically deployed to fill monitoring gaps and offer extensive spatial coverage, particularly in resource-constrained countries. Biosensors have the potential to transform how environmental data are gathered and handled. Most importantly, biosensor networks have the potential to minimize the strain on traditional laboratory-based approaches while also streamlining data collecting and interpretation.

17.6 Ethical Considerations and Future Challenges

To utilize biosensors ethically, a full investigation of the ethical aspects connected to biosensor use in environmental monitoring must be implemented. To prevent data piracy, data ownership and its consequences in biosensor-generated data should be thoroughly checked. Individuals, communities, companies, and governments, among others, should share responsibility for data ownership, access, and usage. Data accuracy, quality assurance, and data validation are key difficulties in biosensor networks. Calibration, sensor maintenance, and frequent performance tests to assure trustworthy data output are all uncertain. To guarantee scientific integrity and public trust, the importance of transparency and openness in sharing data and results should be examined. Environmental justice ideas must be incorporated into biosensor deployment and monitoring initiatives.

17.7 Conclusion

The use of biosensors in environmental pollution detection has various advantages, including quick reaction times, cost-effectiveness, and the opportunity for on-site, in-field monitoring. Biosensors are good instruments for environmental monitoring since they are sensitive, selective, simple, and quick. These technologies are still being actively studied and developed to meet diverse environmental concerns and contribute to environmental conservation and sustainability. Biosensor technologies have been shown to be a faster expanding study subject during the last decades. Market trends show a 10.4% rise in biosensor development for a variety of applications, including biopharma, food and beverages, biodefense, and environmental monitoring. However, in comparison to biomedical applications, biosensor technologies for soil environment applications are still in their early stages and face several challenges due to intrinsic environmental analysis characteristics. Nanotechnology was developed in order to achieve the goal of environmental stability. Nanomaterial fabrication, design, and functionalization decrease waste formation while preserving environmentally favorable synthesis procedures. The use of nano-biosensors to detect pollutants has shown excellent performance and a variety of desirable qualities like small size, sensitivity, low cost, and mobility. Furthermore, the green synthesis of nanoparticles has less negative environmental impacts and serves as an exceptional tool for improvement in environmental research.

References

1. Claude, D., Houssemeddine, G., Andriy, B., Jean-Marc, C., Whole cell algal biosensors for urban waters monitoring. *Novatech.*, 7, 1507–1514, 2007.
2. Slifonova, O., Burlage, R., Barkay, T., Bioluminescent sensors for detection of bioavailable Hg(II) in the environment. *Appl. Environ. Microbiol.*, 59, 3083–3090, 1993.
3. Mahmoud, A.E.D. and Fawzy, M., Nanosensors and nanobiosensors for monitoring the environmental pollutants, in: *Waste Recycling Technologies for Nanomaterials Manufacturing*, pp. 229–246, 2021.
4. Saini, R., Hegde, K., Brar, *et al.*, Advances in whole cell-based biosensors in environmental monitoring, in: *Tools, techniques and protocols for monitoring environmental contaminants*, pp. 263–284, Elsevier, Amsterdam, Netherland 2019.
5. Gui, Q., Lawson, T., Shan, S., Yan, L., Liu, Y., The application of whole cell-based biosensors for use in environmental analysis and in medical diagnostics. *Sensors*, 17, 7, 1623, 2017.
6. Bilitewski, U. and Turner, A. (Eds.), *Biosensors in environmental monitoring*, CRC Press, Boca Raton, FL, 2000.
7. Hamimed, S., Mahjoubi, Y., Abdeljelil, N., Gamraoui, A., Othmani, A., Barhoum, A., Chatti, A., Chemical sensors and biosensors for soil analysis: principles, challenges, and emerging applications. *Adv. Sensor Tech.*, 669–698, 2023.
8. Hashemi Goradel, N., Mirzaei, H., Sahebkar, A., Poursadeghiyan, M., Masoudifar, A., Malekshahi, Z.V., Negahdari, B., Biosensors for the detection of environmental and urban pollutions. *J. Cell. Biochem.*, 119, 1, 207–212, 2018.
9. Thakur, A. and Kumar, A., Recent advances on rapid detection and remediation of environmental pollutants utilizing nanomaterials-based (bio) sensors. *Sci. Total Environ.*, 834, 155219, 2022.
10. Aguilar-Pérez, K.M., Heya, M.S., Parra-Saldívar, R., Iqbal, H.M., Nanobiomaterials in-focus as sensing/detection cues for environmental pollutants, in: Case Studies in Chemical and Environmental Engineering, vol. 2, p. 100055, 2020.
11. Rana, M., Yilmaz, T., Cohen, S., Beyhan, S., Argun, A.A., A novel biosensor for ultrasensitive detection of fungal genes. *Biosens. Bioelectron.*, 222, 114986, 2023.
12. Gavrilaş, S., Ursachi, C. Ş., Perţa-Crişan, S., Munteanu, F.D., Recent trends in biosensors for environmental quality monitoring. *Sensors*, 22, 4, 1513, 2022.
13. Sharma, P., Pandey, V., Sharma, M.M.M., Patra, A., Singh, B., Mehta, S., Husen, A., A review on biosensors and nanosensors application in agroecosystems. *Nanoscale Res. Lett.*, 16, 1–24, 2021.

14. Sellami, K., Couvert, A., Nasrallah, N., Maachi, R., Abouseoud, M., Amrane, A., Peroxidase enzymes as green catalysts for bioremediation and biotechnological applications: A review. *Sci. Total Environ.*, 806, 150500, 2022.
15. Jia, X., Wang, K., Li, X., Liu, Z., Liu, Y., Xiao, R., Wang, S., Highly sensitive detection of three protein toxins via SERS-lateral flow immunoassay based on SiO_2@ Au nanoparticles. *Nanomed. Nanotechnol. Biol. Med.*, 41, 102522, 2022.
16. Long, F., Zhu, A., Shi, H., Recent advances in optical biosensors for environmental monitoring and early warning. *Sensors*, 13, 10, 3928–13948, 20132013.
17. Marrazza, G., Piezoelectric biosensors for organophosphate and carbamate pesticides: a review. *Biosens.*, 4, 3, 301–317, 2014.
18. Noll, M.R., Trace elements in terrestrial environments: Biogeochemistry, bioavailability, and risks of metals. *J. Environ. Qual.*, *32*, 374, 2003.
19. Rathnayake, I.V.N., Megharaj, M., Naidu, R., Green fluorescent protein based whole cell bacterial biosensor for the detection of bioavailable heavy metals in soil environment. *Environ. Tech. Innov.*, 23, 101785, 2021.
20. Rodriguez, B.B., Bolbot, J.A., Tothill, I.E., Urease–glutamic dehydrogenase biosensor for screening heavy metals in water and soil samples. *Anal. Bioanal. Chem.*, 380, 284–292, 2004.
21. Bae, J.W., Seo, H.B., Belkin, S., Gu, M.B., An optical detection module-based biosensor using fortified bacterial beads for soil toxicity assessment. *Anal. Bioanal. Chem.*, 412, 3373–3381, 2020.
22. Potadar, R.S., Moholkar, D.N., Pawar, K.D., Nanosensors for the detection of heavy trace metals in soil, in: *Nanosensors for Smart Agriculture*, pp. 329–353, Elsevier, Elsevier, Amsterdam, Netherland 2022.
23. Mathur, S., Singh, D., Ranjan, R., Genetic circuits in microbial biosensors for heavy metal detection in soil and water. *Biochem. Biophysic. Res. Commun.*, 652, 131–137, 2023.
24. Cova, C.M., Rincón, E., Espinosa, E., Serrano, L., Zuliani, A., Paving the way for a green transition in the design of sensors and biosensors for the detection of volatile organic compounds (VOCs). *Biosens.*, 12, 2, 51, 2022.

18

Transcriptome-Guided Characterization of Molecular Resources in Mussels

Snigdha Baliarsingh[1], Mariadoss Selvanayagam[2], Hrudayanath Thatoi[3], Shailesh Saurabh[4], Yong Seok Lee[5,6] and Bharat Bhusan Patnaik[1,5*]

[1]*P.G. Department of Biosciences and Biotechnology, Fakir Mohan University, Nuapadhi, Balasore, Odisha, India*
[2]*Loyola Institute of Frontier Energy (LIFE), Loyola College, Chennai, Tamil Nadu, India*
[3]*Centre for Industrial Biotechnology Research, Siksha 'O' Anusandhan, Deemed to be University, Kalinga Nagar, Bhubaneswar, India*
[4]*ICAR-Central Institute of Freshwater Aquaculture (CIFA), Kausalyaganga, Bhubaneswar, Odisha, India*
[5]*Korea Native Animal Resources Utilization Convergence Research Institute (KNAR), Soonchunhyang University, Asan, Chungnam, South Korea*
[6]*Department of Biology, College of Natural Sciences, Soonchunhyang University, Asan, Chungnam, South Korea*

Abstract

Mussels act as ecosystem engineers by modifying the aquatic habitat, making it more suitable for the sustenance of their species and other organisms. Moreover, the mussel species are targeted as functional genomics resources mainly attributed to their robust immune system. Further, from the conservation standpoint, few mussel species are sensitively placed. Next-generation sequencing, particularly whole-genome and transcriptome-guided assemblies, enables precise diagnosis of molecular aspects of mussels. This aids host–pathogen dynamics and environmental biomonitoring and informs conservation strategies. The transcriptome approach is more appealing as it provides precise measurement of transcript levels and gene expression information. In the last decade, continued evolution of whole-transcriptome and experiment-specific transcriptome studies have characterized groups of transcripts involved in growth, immunity, and reproduction.

**Corresponding author*: drbharatbhusan4@gmail.com

Hrudayanath Thatoi, Sonali Mohapatra, Swagat Kumar Das and Sukanta Kumar Pradhan (eds.) Applied Biotechnology and Bioinformatics: Agriculture, Pharmaceutical Research and Environment, (407–436)

Further, enhanced information on the transcripts involved in the response and adaptation of mussels to external conditions has been obtained. This review enlists the growing paradigm of information related to the mussel molecular resources obtained through the high-throughput transcriptome approach. The study presented here would be crucial as the combination of molecular data with traditional physiological and population studies would be needed for shellfish aquaculture, their management, and the improvement of genetic strains.

***Keywords*:** Mussels, transcriptome, molecular resources, environmental biomonitoring, conservation

18.1 Introduction

Mussels are bivalve mollusks belonging to the families Mytilidae [marine mussels such as the blue mussel (*Mytilus edulis*), New Zealand green-lipped mussel (*Perna canaliculus*), capax horse mussel (*Modiolus capax*), yellow mussel (*Mytilus citrinus*) etc.] and Unionidae (freshwater mussels such as *Cristaria plicata, Lamellidens marginalis*, and many more among the 750 species distributed worldwide). A revised checklist of Unionidae mussels in United States and Canada is available that recognizes 298 species under 55 genera [family Margaritiferidae (one genus, five species) and family Unionidae (54 genera, 293 species)] [1]. Commercial exploitation of both Mytilidae and Unionidae species of mussels has been reported [2, 3]. Moreover, many Unionidae species show high rate of extinction due to habitat degradation, global warming, and other anthropogenic interventions [4]. In Europe and United States, the zebra mussel has gained pest status, disrupting aquatic food chain and driving many native freshwater mussel species on the verge of extinction. Overexploitation in search of food and pearl is also one of the prime reasons for extinction in the wild, making the freshwater mussels one of the most vulnerable groups. Further, as reported, the Unionidae species are one of the most endemic species, with a lesser dispersal rate, making them susceptible to human interventions [5]. For instance, the population of *Cristaria plicata*—"the freshwater pearl bivalve"—has declined considerably, and in Korea, it has been declared as endangered and protected by law [6]. *Lamellidens marginalis*, the freshwater mussel actively cultured for pearl production in India, is also one of the most threatened groups in the wild due to anthropogenic factors [7]. Considering the economic potential on one hand and threatened species status on the other hand, it is known that the mussels (especially freshwater species) need effective management decisions for resources utilization and conservation.

As non-model organisms, mussels (both marine and freshwater species) need intervention at the genetic and molecular level to understand their adaptive strategies in the wild habitat, the economically important "fitness" traits, and their efficiency as a bio-monitoring species. Further, an understanding of the molecules at the sequence level could dictate the evolutionary relationships within families and would be useful in scrutinizing the phylogeny. In this regard, the genomic resources are more pertinent as they could provide high-throughput information on the rich repertoire of transcripts and their functional specializations. Genomic research leaps forward with next-generation sequencing (NGS), amplifying capabilities and prominence. NGS technologies enable faster, cost-effective DNA analysis, revolutionizing scientific exploration and discovery across disciplines. Primarily, the short reads produced by second-generation sequencers led by Illumina and the *de novo* assembly platform called Trinity has complemented to unravel mysteries at the genome level. Understanding adaptive divergence, species decline, taxonomy, and much more information about mussels needs the extraction of high-quality DNA. Further, other longstanding technological challenges include the highly heterozygous, repeat-rich mussel genomes. In this context, the genome assembly of mussels' places perplexing questions to researchers. This is the reason for the under-representation of mussel genomes. Limited freshwater mussel reference genome assemblies are publicly available, such as those for the European endangered pearl mussel, *Margaritifera margaritifera* [8, 9], the Iberian dolphin mussel, *Unio delphinus* [10], and the Painter's mussel, *Unio pictorum* [11]. A summary of statistics of the mussel genome assemblies has been depicted in Table 18.1. While these genome assemblies exhibit a high percentage of heterozygosity, it is clear that the overall completeness of the assemblies is well-supported, underscoring their reliability and robustness in genomic analyses. Further, recent developments in the third-generation PacBio Hi-Fi reads have been able to resolve repetitive regions so prominent in the mussel genomes [11]. Chromosomal-level genome assemblies of marine mussels such as the hard-shelled mussel *M. coruscus*; blue mussels such as *M. chilensis*, *M. edulis*, and *M. galloprovincialis*; California ribbed mussel *M. californianus*; and a few others using the PacBio and Oxford Nanopore Technologies have been successful in providing affordable genomic resources for population and evolutionary studies [12–15]. While whole-genome studies and reference genome representation in mussels have improved due to long-read sequencing and Illumina short-read-based correction strategies, it still is a complicated experimental proposition for so many researchers working on mussel physiology and adaptation. Recently, the chromosome-level genome assembly of mussel

Table 18.1 Representation of mussel genomes in the public domain.

Species (common name)	NGS platform	Reads (n)	Assembler	Contigs (n)	Protein-coding genes	BioProject/SRA accession	Sequencing objectives	Reference
Mytilus chilensis **(Blue mussel)**	PacBio Sequel II System (SMRT sequencing), Hi-C libraries, and HiSeq 4000	53.8 Gb (63 million reads), 2n = 14	Bowtie v1.3.1	13,762 (2.19 Gb): with PacBio system 11,868: with Hi-C system	34,530 (6531 avg. transcript length); 1377 (avg. CDS length; 4.92 average of exons)	SRR20593343 (PacBio), SRR20966976 (Hi-C)	Genome adaptation and physiological plasticity to develop sustainable production of mussels	Gallardo-Escarate *et al.*, 2023
Perna viridis **(Green mussel)**	*De novo* genome assembly using an Illumina HiSeq2500 platform	731,873,830 bp of raw reads	Platanus v1.2.4	24,293; 99.4% of genome completeness	NS	PRJDB10567	Identification of genes involved in byssus formation	Inoue *et al.*, 2021

(*Continued*)

Table 18.1 Representation of mussel genomes in the public domain. (*Continued*)

Species (common name)	NGS platform	Reads (*n*)	Assembler	Contigs (*n*)	Protein-coding genes	BioProject/SRA accession	Sequencing objectives	Reference
Dreissena polymorpha **(Zebra mussel)**	PacBio Sequel II System (SMRT sequencing), Hi-C libraries, and HiSeq2500	1,798,019,516 bp	Canu v1.7	2,863	68,018	PRJNA533175	Rich resource for invasive species research	McCartney *et al.*, 2022
Mytilus coruscus **(Hard-shelled mussel)**	Illumina, ONT, and Hi-C ONT and Illumina HiSeq X	1,235,384,620 (Illumina), 11,108,773 (ONT), 832,911,978 (Hi-C) 11,312,815 (ONT), 288,220,402 (Illumina)	Canu v1.5 Jellyfish	NS 10,484	37,478 42,684	PRJNA578350 PRJEB33342	Comparative genomics at the chromosome level Genome-scale selective breeding and the speciation and local adaptation	Yang *et al.*, 2021 Li *et al.*, 2020

(*Continued*)

Table 18.1 Representation of mussel genomes in the public domain. (*Continued*)

Species (common name)	NGS platform	Reads (*n*)	Assembler	Contigs (*n*)	Protein-coding genes	BioProject/SRA accession	Sequencing objectives	Reference
Unio delphinus **(Iberian dolphin freshwater mussel)**	PacBio Sequel II System (SMRT sequencing) and Illumina HiSeq X	~2.5 Gbp	Hifiasm 0.16.1-r375, Flye v.2.8.3, NextDenovo v2.4.0 and Peregrine-2021 v0.4.3	NS	44,382 (CDS); 32,089	PRJNA917855	Investigating the species biology and evolutionary history	Gomes-dos-Santos *et al.*, 2023b
Mytilus galloprovincialis **(Mediterranean mussel)**	Illumina HiSeq2000	1.6 Gb	Jellyfish	NS	10,891	PRJNA262617	Broaden genomic knowledge within the family	Murgarella *et al.*, 2016

(*Continued*)

Table 18.1 Representation of mussel genomes in the public domain. (*Continued*)

Species (common name)	**NGS platform**	**Reads (*n*)**	**Assembler**	**Contigs (*n*)**	**Protein-coding genes**	**BioProject/SRA accession**	**Sequencing objectives**	**Reference**
Margaritifera **(European freshwater pearl mussel)**	Illumina PE and Illumina long-insert size MP reads processed in Illumina NovaSeq6000 PacBio Sequel II System (SMRT sequencing), Hi-C lib	494 Gb (PE reads) and 76 Gb (MP reads) 103 Gb (PacBio)	Meraculous v2.2.6 Next Denovo v2.4.0	265,718 265,718	35,119 40,165	PRJNA678877 PRJNA925505	Biological and evolutionary features New improved genome assembly	Gomes-dos Santos, 2021a Gomes-dos Santos *et al.*, 2023a
Mytilus californianus **(California ribbed mussel)**	Omni-C and PacBio HiFi	5.1 million reads	CCGP assembly v3.0	176	NS	PRJNA777198	Biological and evolutionary features	Paggeot *et al.*, 2022

Abbreviations: SMRT, Single-molecule real-time sequencing; ONT, Oxford Nanopore Technologies; NS, not specified; Gb, Gigabases.

species has been advocated. This facilitates comparative genomic studies on chromosome rearrangements across different species and an understanding of the evolution of gene families [12, 16].

NGS technologies involving the transcriptome sequence have provided cost-effective and reliable answers to non-model species conservation including the threatened mussel species. It has also provided unprecedented coverage of the coding sequence, thousands of potential marker loci, elucidation of fitness attributes, and gene expression signatures of growth, immunity, and reproduction [17, 18]. Such rapid development of molecular resources with biological and commercial significance has led to extensive use of transcriptome sequencing for targeted functional outcomes. The Threatened Species Initiative (TSI) in Korea actively contributes to unraveling molecular resources of endangered mollusks, including the mussel species, through the utilization of Illumina-based transcriptome sequencing techniques [6, 19]. The development of bioinformatics pipelines for rapid annotation of sequences proves the availability of large datasets from molluscan genomes and transcriptomes [20, 21]. Transcriptome sequencing has revealed varied contours of marine mussel functional specializations and freshwater mussel conservation [22, 25]. Highlighted studies on transcriptome-derived molecular resources have been compiled in this review that could be a good reference for future research in this direction. Further, the transcriptome holds promise for the future with the availability of structural and functional genomics resources.

18.2 Species of Mussels Sequenced at the Transcriptome Level

Non-model organisms such as mussels need more attention at the genotype level to elucidate their fitness phenotypes, thus essential to understand their biology, adaptation, and evolutionary successes. For many species, almost no genetic resources are available, and this is a deterrent to their rescue in wild habitats and/or identification from newer habitats. Most of the freshwater mussels are vulnerable (especially order Unionida), and the lack of molecular resources is hindering the implementation of successful conservation strategies. Genomic tools are now providing comprehensive and accurate information on genetic variation, population structure and dynamics, and adaptative traits that prioritize conservation efforts on mussels [26–28]. Moreover, leveraging high-throughput datasets enables the development of predictive models. These models aid in evaluating the consequences of human-induced interventions, such as land-use alterations, resource depletion, and climate change on vulnerable taxa [29]. Figure 18.1 depicts an overview of transcriptome initiatives involving mussels. Broadly, the applications of transcriptome initiatives involve an understanding of stress-responsive physiology, local environment adaptability, and elucidation of reproduction-related traits, apart from the more sensitive conservation aspect.

Cristaria plicata, a freshwater mussel classified under the order Unionida and family Unionidae, exhibits limited geographic range. Despite this, it holds significant popularity in China for both the pearl industry and medicinal applications [30, 31]. However, the species is red-listed in Korea, primarily due to habitat fragmentation and loss [32]. With the availability of the mitochondrial genome for the species, molecular taxonomy studies have been made possible with the application of DNA barcoding [33]. The visceral mass transcriptome of *C. plicata* revealed candidate genes associated with immunity, stress responses, sex determination, and reproduction [6]. Further, the mantle transcriptome of *C. plicata* was studied for the identification of the biomineralization-promoting genes [34].

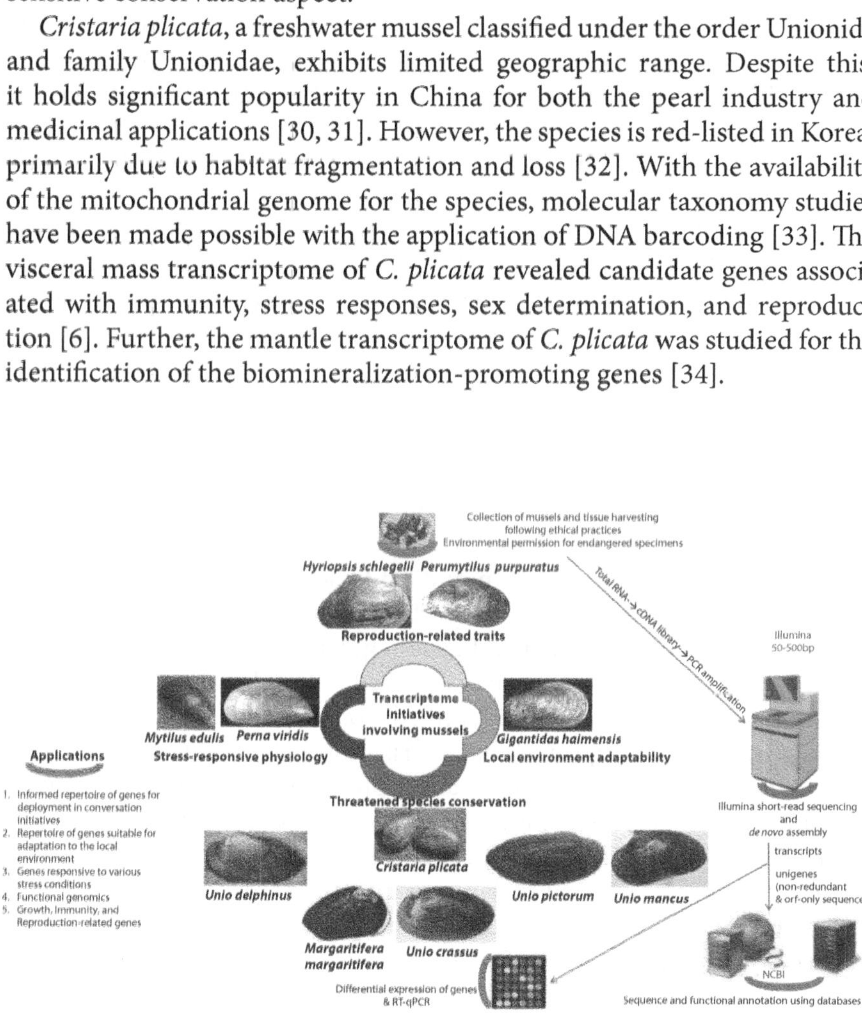

Figure 18.1 An overview of the transcriptome initiatives involving mussels. Illumina short-read sequencing and de novo assembly was utilized to explore the functional transcripts and to obtain differential expression profile of the transcripts under specific adaptation challenges.

The threatened European freshwater mussels are also sensitive to human-mediated threats such as climate change and habitat fragmentation. The gill transcriptomes of five European freshwater mussels such as *Margaritifera margaritifera* (listed as Endangered by IUCN) [35], *Unio delphinus* (Near Threatened by IUCN) [36], *Unio crassus* (Endangered by IUCN) [37], *Unio pictorum*, and *Unio mancus* (Endangered by IUCN) [38] (all under *Unio* genus) have been studied with respect to conservation. Among all these species, only *M. margaritifera* had a draft genome assembly based on two distinct NGS libraries, such as Illumina paired-end reads and the Illumina Mate-pair reads [8]. Hence, the transcriptome-derived information is considered significant for deciphering the adaptability of these species to cope with current or future threats and to protect such group of highly threatened species.

Hyriopsis schlegelii, the Biwa pearl mussel thrives in Chinese cultivation. Originally from lake Biwa, Japan, it faces critical endangerment, documented in the Japanese Red Data Book (Ministry of Environment 2005), highlighting the urgent need for conservation measures to protect this valuable freshwater mussel species. The quality of pearls from the cultured pearl bivalve *H. schlegelii* in China is compromised by the nature of its gonads. Due to the absence of a reference genome, the details of sex determination and species development remain a mystery, with significant gaps in the understanding of these intricate biological processes. The transcriptome of *H. schlegelii* was investigated, aiming to unravel the foundations of hermaphroditism and understand the mechanisms governing gonadal development [39]. Earlier, sex-determination genes in *H. schlegelii* have been identified and gene regulatory mechanisms that control sex determination have been discovered [40]. Likewise, the sex-determination genes have also been identified from the triangle sail mussel, *H. cumingii* using transcriptome characterization [41]. Focus on reproductive-related genes using *de novo* transcriptome assembly was also envisaged in the marine mussel *Perumytilus purpuratus*, largely distributed in the Pacific and Atlantic coasts. In this case, sperm-process-related transcripts were discovered and sperm–egg interactions were studied [42].

Understanding the local adaptation and adaptive strategies of mussels in diverse environments remain inconclusive. Transcriptome assemblies have systematically probed genes associated with antioxidant defense, innate immune systems, and metabolism, providing substantial insights into the adaptive mechanisms of mussels in unfamiliar environments. The transcriptome of cold seep mussel *Gigantidas haimaensis* is notable as it explored the adaptation strategies in the *in situ* and laboratory environments supporting life responses to environmental changes [24]. Further, a combinatorial approach of proteomic and transcriptomic analysis of *G. haimaensis* gonads evaluated the intersex mechanism seen frequently in mollusks [43].

Mytilus edulis, the blue mussel, and *Perna viridis*, the Asian green mussel, are economically significant species examined for their stress-responsive physiology. The blue mussel is a bioindicator for monitoring coastal water pollution. Transcriptome studies have helped in deciphering markers for the physiological tolerance of the species to environmental insults [44–46]. Moreover, the mantle transcriptome of a few species under the *Mytilus* genus has been investigated for the identification of proteins involved in biomineralization and pigmentation [23]. Further, the mantle transcriptome of blue mussel response was investigated during a major spawning event, paying particular attention to sex variations/differences and culture depth [47]. On the other hand, *Perna viridis* indigenous to the Indo-Pacific region has been sequenced at the transcriptome level to highlight tissue-specific response during specific toxicity regimes [48]. Foot-specific transcriptomic responses in *P. viridis* have identified genes involved in byssus formation (a proteinaceous holdfast to tightly attach to underwater substrates) [49]. Overall, the comprehensive transcriptome analysis of both freshwater and marine mussel species has expanded our understanding of the gene repertoire crucial for physiological adaptation. This includes molecular resources associated with metabolism, immune responses, and physiological processes, providing vital insights to conserving the species in disrupted natural habitats.

18.3 Transcriptome Pipeline for Mussel Molecular Resources

The pipeline adopted for *de novo* assembly of mussel transcriptomes can be categorized under pre-NGS, NGS, and post-NGS workflow. The raw read data obtained from the sequencer need to be publicly discoursed under the NCBI sequence read archive (SRA) submission. Before contemplating the transcriptome pipeline, the choice of the sequencer and the assembler, researchers need to prioritize the sequencing objectives (based on the species and the experimental protocols to be followed). As reviewed, most of the marine mussels have been studied at the transcriptome level towards deciphering the robust immune system, adaptation to wild habitats (that is fragmented due to increased human interventions including climate change), and resource mobilization. On the contrary, the freshwater mussels have been studied under the context of conservation and strategies to improve the germplasm. In case the transcriptome objectives include conservation and cataloguing of molecular resources, the collection of such specimens needs the prior approval of environment, biodiversity, or forest officials. In most of such cases, these species are categorized under the threatened species initiatives of the government and are protected by law. The sequencing of the whole transcriptome of such mussel species should also highlight the suggested outcomes of the study and importance in the context of a nation's sustainable progress. For instance, the pearl bivalve *C. plicata* is endangered in Korea, but in China, it has a wider presence due to freshwater pearl farming, use as a model species for aquaculture production, and for medicinal purposes [6]. Table 18.2 shows a methodological summary of mussel transcriptomes studied to unravel molecular resources. Apart from the *C. plicata* visceral mass (includes digestive, nervous, excretory, reproductive, and respiratory systems) transcriptome, all the other work focused on specific tissues such as mantle, foot, muscle, and gonads. Many reports are available that have studied the visceral mass transcriptome of threatened molluscan species to unravel the genomic and genetic resources for deployment in informed conservation management decisions [19]. Mantle transcriptomes in mollusks have been explored to discover biomineralization-related proteins and provide critical information for understanding shell formation [50–52].

Under the pre-NGS pipeline, total RNA isolation, RNA quality control (QC), and cDNA library preparation was prioritized. While Trizol reagent-based extraction was popular, researchers also used other commercialized kits for total RNA extraction. The quality of RNA is the limiting factor for purposeful cDNA library preparation and therefore an RNA integrity number of >7 is the benchmark for proceeding to sequencing. Nanodrop and Bioanalyzer were used commonly for RNA QC while the Illumina-based mRNA-Seq and TrueSeq sample preparation kits were used for the dscDNA library preparation (Table 18.2). In Illumina-based NGS, the pre-NGS workflow after dscDNA library preparation involves enriching with sequencing adapters and subsequent PCR amplification. The majority of mussel transcriptomes, like those of other non-model mollusks, have been produced using the Illumina-based short-read sequencing platform [53–55]. Illumina platforms are the preferred second-generation NGS platforms for building reference transcriptomes of non-model species including mollusks and, with Trinity *de novo* assembly pipeline, perform the best in characterizing the molecular resources and their differential expression [56, 57]. Recently, full-length transcriptomes of a few mussel species have been characterized using a combination of third-generation long-read PacBio

Table 18.2 A methodological summary of Next-Generation Sequencing (NGS) studies to unravel molecular resources from *de novo* assembled mussel transcriptome.

Species→ Workflow ↓	***Cristaria plicata***		***Mytilus edulis***		***Margaritifera margaritifera***	***Unio crassus***	***Unio pictorum***	***Unio mancus***	***Unio delphinus***	***Gigantidas haimaensis***	***Perna viridis***	***Hyriopsis schlegelii***	***Perumytilus purpuratus***
Tissues	**Mantle**	**Visceral mass**	**Mantle**	**Whole muscle tissues**	**Gill**					**Muscle**	**Foot**	**Male, female, and hermaphrodite gonad**	**Male gonad**
Pre-NGS workflow													
Total RNA isolation	RNAsimple Total RNA kit (Tiangen Biotech, China)	Trizol (Invitrogen)	RNeasy Mini Kit (Qiagen)	RNeasy Mini Kit (Qiagen)	NZY Total RNA Isolation kit (NZY Tech)					N.S	Trizol (Invitrogen)	Trizol (Invitrogen)	E.Z.N.A Total RNA Kit II (Omega Biotek)
Purity and integrity of RNA	Qubit 2.0 Fluorometer/ Bioanalyzer RNA 6000 Nano Kit	Nanodrop-2000 and Bioanalyzer 2100	Nanodrop ND-1000	Nanodrop-1000 and Bioanalyzer RNA 6000 nano kit	DS-11 Series Spectrophotometer/Fluorometer					Bioanalyzer 2100 and Quawell Q5000	N.S.	Bioanalyzer 2100	N.S.

(*Continued*)

Table 18.2 A methodological summary of Next-Generation Sequencing (NGS) studies to unravel molecular resources from *de novo* assembled mussel transcriptome. (*Continued*)

Species→ Workflow ↓	*Cristaria plicata*		*Mytilus edulis*		*Margaritifera margaritifera*	*Unio crassus*	*Unio pictorum*	*Unio mancus*	*Unio delphinus*	*Gigantidas haimaensis*	*Perna viridis*	*Hyriopsis schlegelii*	*Perumytilus purpuratus*
Tissues	**Mantle**	**Visceral mass**	**Mantle**	**Whole muscle tissues**	**Gill**					**Muscle**	**Foot**	**Male, female, and hermaphrodite gonad**	**Male gonad**
cDNA library Construction	NEBNext® Ultra™ RNA Library Prep Kit (NEB)	mRNA-seq sample preparation kit (Illumina)	TruSeq stranded mRNA Sample Preparation Kit (Illumina)	TruSeq stranded mRNA Sample Preparation Kit (Illumina)	As specified with Macrogen Inc.					N.S.	N.S.	SMARTer PCR cDNA Synthesis Kit (Clontech)	Illumina Tru-Seq Stranded Total RNA Library Preparation Kit
NGS													
Sequencing platform	Illumina HiSeq 2500	Illumina HiSeq 2500	Illumina HiSeq 4000	Illumina HiSeq2000	Illumina HiSeq 4000					N.S.	Illumina HiSeq2000	PacBio SMRT and Illumina Novaseq	Illumina HiSeq 2500
Post-NGS workflow													
Processing of raw reads	Cutadapt and PrinSeq	Sickle and Fastq_filter software	Trimmomatic software v 0.35	FastQC and Trimmomatic software	FastQC v 0.11.8 Trimmomatic software v 0.38 Rcorrector (v 1.0.3) Centrifuge (v 1.0.3-beta)					N.S.	N.S.	Smrtlink v 5.1 (For PacBio SMRT)	Trimmomatic
***De novo* assembly, redundancy removal/ clustering**	Trinity (assembled unigenes)	Trinity and TGICL (assembled unigenes)	Trinity v 2.0.4	Trinity	Trinity v 2.13.2 (assembled transcripts) Blastn (v 2.11.0) and Univec (Mollusca-specific transcripts) Corset v 1.0.9					Trinity (assembled unigenes)	SOAPdenovo-Trans and TGICL (assembled unigenes)	LoRDEC (v 0.9) for correction of errors and CD-HIT v 4.6.8 (transcripts)	Trinity (assembled unigenes)

(*Continued*)

Table 18.2 A methodological summary of Next-Generation Sequencing (NGS) studies to unravel molecular resources from *de novo* assembled mussel transcriptome. (*Continued*)

Species→ Workflow ↓	*Cristaria plicata*		*Mytilus edulis*		*Margaritifera margaritifera*	*Unio crassus*	*Unio pictorum*	*Unio mancus*	*Unio delphinus*	*Gigantidas haimaensis*	*Perna viridis*	*Hyriopsis schlegelii*	*Perumytilus purpuratus*
Tissues	**Mantle**	**Visceral mass**	**Mantle**	**Whole muscle tissues**	**Gill**					**Muscle**	**Foot**	**Male, female, and hermaphrodite gonad**	**Male gonad**
Completeness of transcriptome	N.R.	N.R.	N.R.	BUSCO using the metazoan lineage mode (v 5.34c21f2)	BUSCO (v 3.0.2) eukaryote and metazoan lineage mode					N.R.	N.R.	N.R.	BUSCO (v 3.0.2)
Other targeted analysis	RSEM (for expression of unigenes) and transform it to FPKM value	N.R.	RSEM v1.2.12 (for expression of unigenes and DESeq2 v1.10.1 and edgeR v3.12.1 and RT-qPCR for validation of RNA-Seq analyses)	DESeq2	N.R.					RSEM (for expression of unigenes) and transform it to FPKM values and DESeq (v1.20.0) and RT-qPCR (for validation of RNA-Seq analysis)	RPKM (for expression level of each unigene) and DESeq2 [log_2 (ratio)≥1 and FDR≤0.05] and RT-qPCR (for validation of RNA-Seq analyses)	RSEM (for expression of unigenes) and transform it to FPKM values and DESeq2 [log2 (ratio≥1)]	N.R.

(*Continued*)

Table 18.2 A methodological summary of Next-Generation Sequencing (NGS) studies to unravel molecular resources from *de novo* assembled mussel transcriptome. (*Continued*)

Species→ Workflow ↓	*Cristaria plicata*		*Mytilus edulis*		*Margaritifera margaritifera*	*Unio crassus*	*Unio pictorum*	*Unio mancus*	*Unio delphinus*	*Gigantidas haimaensis*	*Perna viridis*	*Hyriopsis schlegelii*	*Perumytilus purpuratus*
Tissues	**Mantle**	**Visceral mass**	**Mantle**	**Whole muscle tissues**	**Gill**					**Muscle**	**Foot**	**Male, female, and hermaphrodite gonad**	**Male gonad**
Annotation	NCBInr NCBInt TrEMBL SwissProt KEGG GO COG ORF Predictor	PANM DB BLAST2GO GO KEGG COG InterProScan	Trinotate suite v2.0.2 [Homology search to (BLAST/ SwissProt/ Uniref90), Protein signal peptide, protein domain (HMMER/ PFAM), transmembrane domain (SignalP/ tmHMM) and annotation databases (eggNOG/ GO)]	Blast2GO (v 5.2.5), BLASTx nr, Trinnotate (nt, eggnog, TMHMM, SignalP, Swiss-Prot, RefSeq, and TrEMBL), KEGG Mapping	Transdecoder (v 5.3.0) InterProScan (v 5.44.80) Blast-n/p/x DIAMOND software					NCBInr NCBInt SwissProt KEGG GO COG pfam	NCBInr SwissProt KEGG GO COG	NCBInr NCBInt SwissProt KEGG GO	SwissProt and BLAST2GO (GO)

(*Continued*)

Table 18.2 A methodological summary of Next-Generation Sequencing (NGS) studies to unravel molecular resources from *de novo* assembled mussel transcriptome. (*Continued*)

Species→ Workflow ↓	*Cristaria plicata*		*Mytilus edulis*		*Margaritifera margaritifera*	*Unio crassus*	*Unio pictorum*	*Unio mancus*	*Unio delphinus*	*Gigantidas haimaensis*	*Perna viridis*	*Hyriopsis schlegelii*	*Perumytilus purpuratus*
Tissues	**Mantle**	**Visceral mass**	**Mantle**	**Whole muscle tissues**	Gill					**Muscle**	**Foot**	**Male, female, and hermaphrodite gonad**	**Male gonad**
Other bioinformatics analysis	N.R.	Keyword search of BLASTx annotation against PANM DB	N.R.	N.F..	N.R.							CNCI CPC PLEK (for coding potential of transcripts) AnimalTFDB 2.0 (for transcription factors) and MISA (for SSR analysis)	N.R.
NCBI submission													
Data records	SRA SRP068075	SRA BioProject PRJNA293023	SRA BioProject PRJNA557478	SRA BioProject PRJNA914271	SRA BioProject PRJNA839062					Science databank	SRA BioProject SRP077042	SRA BioProject PRJNA806470	SRA BioProject PRJNA343253
References	Wang *et al.*, 2017	Patnaik *et al.*, 2016	Galladi *et al.*, 2021	Hall *et al.*, 2023	Gomes-dos-Santos *et al.*, 2022					Zhang *et al.*, 2022	Zhang *et al.*, 2017	Zeng *et al.*, 2022	Briones *et al.*, 2018

Abbreviations: NGS, Next-generation sequencing; nt, Non-redundant nucleotide database; nr, Non-redundant protein database; NCBI, National Center for Biotechnology Information; GO, Gene Ontology; KEGG, Kyoto Encyclopedia of Genes and Genomes; COG, Clusters of Orthologous Gene; RSEM, RNA-Seq by Expectation-Maximization; RPKM, Reads Per Kilobase per Million Mapped Reads; BUSCO, Benchmarking Universal Single-Copy Orthologs; FPKM, Fragments Per Kilobase of Transcript Per Million Mapped Reads; ORF, Open Reading Frame; SMRT, Single-Molecule Real-Time; signal, Signal Peptide; CNCI, Coding–Non-Coding Index; CPC, Coding Potential Calculator; SSR, Simple Sequence Repeat; SRA, Sequence Read Archive; N.S., Not Specified; N.R., Not Reported.

(single-molecule SMRT) sequencing and the Illumina platform [39, 58]. The PacBio SMRT sequencing platform provides an ideal strategy for identifying genetic subtypes, as well as novel genes and the isoforms, thus taking into account the alternative splicing effects. Illumina HiSeq2500, 4000, and NovaSeq (paired-end approaches) were the preferred Illumina platforms for sequencing the mussel transcriptomes and were also applied in other mollusk transcriptome pipelines [19, 42, 59].

The post-NGS pipeline for decoding the mussel transcriptomes included the raw reads QC, *de novo* assembly/redundancy removal/clustering, assessing the completeness of transcriptome, annotation, and bioinformatics analysis. After obtaining the raw reads from the sequencer, it is processed to obtain clean reads by removal of adapter sequences and other ambiguous sequences, for which Trimmomatic was the preferred choice. Additional QC checks of the raw reads were performed with Rcorrector and Centrifuge in case of European endangered mussels [25]. Under raw data cleanup, Rcorrector [60] was used to correct the random sequencing errors, and Centrifuge [61] was used to screen the corrected raw reads against the NCBI nt database so that the contaminants can be excluded from the dataset. For the majority of transcriptomes, *de novo* assembly of clean reads was executed using Trinity. The utilization of Trimmomatic and Trinity assembly for trimming and validation of Illumina short reads has been demonstrated for optimized clean reads [62]. The TGICL package and/or CD-HIT have been used to define non-redundancy by clustering the transcripts. Further, prior to clustering, the TransDecoder program has been utilized to screen the ORF-only transcripts, and such an enrichment of ORF-containing sequences will be critical in the final discovery of novel homologs for the species [63]. The TransDecoder program to screen the ORF-only transcripts in the gill transcriptomes of European endangered freshwater mussels has been reported [25]. Further, the BUSCO (Benchmarking Universal Single-copy Orthologs) analysis for benchmarking the completeness of transcriptome is a prerequisite in the post-NGS transcriptome pipeline. The gill transcriptome of freshwater mussels, the whole-muscle tissue transcriptome of *M. edulis*, and the gonadal transcriptome of *P. purpuratus* have used the BUSCO analysis with eukaryote and metazoan lineage mode [25, 42, 43]. The completeness of the transcriptomes was optimum as the BUSCO complete scores (single + duplicated) were above 90% in all the mussel species, suggesting full repertoire of genes [25].

The bioinformatics pipeline utilized for the annotation of transcripts after *de novo* assembly of mussel transcriptome includes a BLAST-based scrutiny against public databases. While BLASTx against NCBI nr protein database was preferred for annotation of transcripts, the protostome database (PANM-DB; panm.sch.ac.kr/PANM.html) is recommended for BLASTx-based annotations of transcripts involving molluscan, arthropod, and nematode species [64]. Most of the non-model molluscan species sequenced at the transcriptome level under the threatened species initiative of Korea (including the freshwater mussel *C. plicata* transcriptome) have utilized PANM-DB as the primary protein database for annotation of transcripts [6, 55]. As the molluscan, arthropod, and nematode sequences in PANM-DB (in multi-FASTA format) have been derived from NCBI nr and regularly validated (ver 5.0 release in 2022), the database can be applied in many transcriptome applications for faster and more efficient annotation than NCBI nr. Further, for the functional specializations of the transcripts, COG, GO, KEGG, and InterProScan databases were utilized preferably using the BLAST2GO suite. As exceptions, the *M. edulis* mussel transcriptome and the European endangered freshwater mussels' transcriptome utilized the Trinotate suite v2.0.2 and DIAMOND software, respectively [25, 47]. As an extension of bioinformatics analysis, a keyword search of BLASTx annotation of PANM-DB was used for retrieving the molecular resources of *C. plicata* related to immunity, growth, and reproduction [6]. Moreover, genetic markers such as the simple sequence repeats (SSRs) have also been derived from a few of the mussel transcriptomes [6, 39]. The differently expressed genes (DEGs) at the transcriptome level provide an approximation of the interplay of genes pertaining to an experimental context. RSEM (RNA-seq by Expectation-Maximization) is a tool to obtain abundance estimates of transcripts. In majority of *de novo* assembled mussel transcriptomes wherein DEGs are targeted, the RSEM tool provided an abundance estimation of transcripts. As shown in Table 18.2, DEGs were targeted to provide insights into hermaphroditism in *Hyriopsis schlegelii* [39], responses to cadmium stimulation in *P. viridis* [65], and adaptive mechanism to environmental stresses in *G. haimaensis* [24]. Furthermore, the DEGs offer an enriched perspective on biological processes and pathways, particularly in relation to immunity, growth, and reproduction. In this regard, the transcriptome pipeline for generating the molecular resources in mussels has been responsible for understanding mussel biology, adaptation characteristics, and applications in the environment.

18.4 Mussel Transcriptome Assembly and Annotation

The *de novo* assembly of the mussel transcriptomes has been successful in providing sequences that are non-redundant and, when annotated to public protein and nucleotide databases, helps to identify homologs implicated in a variety of physiological processes. However, prior to *de novo* assembly and annotation, the raw reads were pre-processed to clean reads as described before in the post-NGS workflow (Table 18.2). In most of the mussel transcriptomes described in Table 18.3, approximately more than 98% of raw reads were processed as clean reads, except for *P. purpuratus* gonad transcriptome wherein approximately 83% of clean reads were processed [42]. The clean reads represent the trimming of unambiguous sequences and Illumina-derived adapter sequences. After *de novo* assembly, the contiguous sequences or contigs (mostly overlapping sequences) were either directly used for DEG analysis and functional annotation or further processed to unigenes by removing redundancy and selecting ORF-only sequences. For instance, in the threatened species initiative of Korea, the contigs were first screened for ORF-only sequences using the TransDecoder suite and later were clustered to unigenes using the TGICL package. The unigenes so obtained were non-redundant and annotated to homologous sequences in the public databases [19]. On the contrary, in the study of threatened European freshwater mussels, TransDecoder-based coding transcripts were directly annotated to public databases using the Blast server and InterProScan [25]. Generally, over 30% of all contigs received an annotation hit in this case. Approximately 22% of all unigenes of mantle and visceral mass transcriptomes of *C. plicata* were annotated, while more than 45% and 50% of *M. edulis* mantle and *G. haimaensis* muscle transcripts were annotated against NCBI [6, 24, 34, 48]. Further, 67.6% of differentially expressed transcripts (DETs) of *M. edulis* whole-muscle tissue were annotated to Blast2GO nr followed by the TrEMBL database [46].

Table 18.4 summarizes the outputs from the other assembled mussel transcriptomes in the last few years with sequencing objectives as variable as adaptation [50, 66], antimicrobial response [67, 68], metamorphosis [69], and innate immunity [70–73]. Significantly, software such as Hisat2, CLC genomics workbench, and CAP3 were utilized in the transcriptome assembly process apart from the Trinity suite. In the case of *M. coruscus* mantle and developmental stages, transcriptome analysis Hisat2 was used to map the clean reads to the genome [66, 69]. Significantly, the pearl sac transcriptome before and after implantation in the case of *H. cumingii* was utilized to understand the immunological response [74].

Table 18.3 Comparison of mussel transcriptome assembly and annotation.

Species (tissue) → Statistics ↓	Cristaria plicata (Mantle) (Visceral mass)		Mytilus edulis (Mantle) (Whole muscle tissues)		Margaritifera margaritifera (Gill)	Unio crassus (Gill)	Unio pictorum (Gill)	Unio mancus (Gill)	Unio delphinus (Gill)	Gigantidas haimaensis (Muscle)	Perna viridis (Foot)	Hyriopsis schlegelii (Gonad)	Perumytilus purpuratus (Gonad)
Raw reads (subreads)	43,313,570	286,152,584	2.3 billion (approx.)	N.S.	131,051,306	132,002,266	104,108,396	100,704,688	112,439,686	N.S.	278,818,353	36.04 Gb (12,478,014)	126,380,260
Clean reads	43,235,090 (99.81% of raw reads)	281,322,837 (98.31% of raw reads)	98.7% of billion reads	2,917,849,670	129,369,332 (98.71% of raw reads)	130,122,324 (98.57% of raw reads)	103,069,704 (99% of raw reads)	99,844,362 (99.14% of raw reads)	111,114,412 (98.82% of raw reads)	238,614,839	274,596,005 (98.48% of raw reads)	N.S.	105,419,693 (83.41% of raw reads)
Contigs	108,856	453,931	518,691	2,819,874	470,852	169,668	68,670	65,620	82,542	187,368	N.S.	359,889	385,288
Unigenes	98,501	374,794	296,118	N.R.	N.R.	N.R.	N.R.	N.R.	N.R.	84,050	105,127	N.R.	314,741
All annotated	21,946 (22.27% of unigenes)	84, 274 (22.48% of unigenes)	133,468 NCBI (45.07% of unigenes)	67.6% of DET to B2G nr	71,046 (15.08% of contigs)	51,937 (30.61% of contigs)	24,194 (35.23% of contigs)	24,775 (37.75% of contigs)	32,688 (39.6% of contigs)	42,169 (50.17% of unigenes)	32,268 (30.69% of unigenes)	157,781	43,756 (13.9% of unigenes)
					NCBI BLASTp/x/n hits								NCBI BLAST SwissProt

Abbreviations: N.S., Not Specified; N.R., Not Reported; DET, Differentially expressed transcripts; B2G, Blast2GO.

Table 18.4 Summary of mussel transcriptomics in the last 5 years (2019–2023)* using Next-Generation Sequencing (NGS) platforms.

Species (tissue)	NGS platform	Reads (*n*)	Assembler	Contigs (*n*)	Others	BioProject SRA accession	Sequencing objectives	Reference
***Mytilus coruscus* (mantle)**	Illumina NovaSeq6000	c-72.84 Gb	Hisat2 to map the clean reads to genome of *M. coruscus* and Trinity	86,924	NR	PRJNA918971	Adaptation in estuarine environment	Liao *et al.*, 2023
***Mytilus galloprovincialis* (hemocyte)**	Illumina NovaSeq6000	27.9 million reads	CLC Genomics Workbench v.20	101,154	NR	PRJNA939420	Antiviral response	Saco *et al.*, 2023
***Mytilus coruscus* (developmental stage-specific)**	Illumina NovaSeq6000	??	Hisat2 v2.0.4	37,478	NR	SRP300586	Metamorphosis	Wang *et al.*, 2023
***Solenaia oleivora* (gill)**	Illumina HiSeqX10	r-319,773,008 c-313,830,592	*De novo* assembly	NS	u-212,905	PRJNA755745	Antioxidant and immune functions	Zhang *et al.*, 2021

(*Continued*)

Table 18.4 Summary of mussel transcriptomics in the last 5 years (2019–2023)* using Next-Generation Sequencing (NGS) platforms. (*Continued*)

Species (tissue)	NGS platform	Reads (*n*)	Assembler	Contigs (*n*)	Others	BioProject SRA accession	Sequencing objectives	Reference
***Hyriopsis cumingii* (hepatopancreas)**	Illumina HiSeq	c-36,800,120–40,439,110	Trinity	NS	u-245,638	PRJNA727624	Immune response to bacterial infection	Yang *et al.*, 2021
***Hyriopsis cumingii* (pearl sac)**	Illumina HiSeq 2500	c->45 million reads	CAP3	339,311	u-293,863	GSE123069	Immunological response after implantation	Huang *et al.*, 2019

18.5 Conclusions and Future Perspectives

The *de novo* transcriptome assembly in non-model species has heralded unprecedented coverage of transcripts, most of which provide critical functional resources for understanding species biology, adaptability of the species in the wild and modified habitat, and their evolution. Lately, *de novo* assembly of mussel transcriptome has provided novel insights to "fitness genes" that can be used to assess the population dynamics of the species, their occurrence from newer habitats, and adaptation traits for informed conservation planning. While this is true for a few of the threatened species of freshwater mussels, the marine mussels have been exploited to understand the robust immune system and applications to environmental risk assessment and biomarker discovery. Further, the discovery of genetic resources is most pertinent to pearl mussels in order to establish robust breeds with superior pearl quality, disease resistance, and growth traits.

Acknowledgments

This study was supported by the Fakir Mohan University Seed Research Grant Letter No. FMU-R&D-261-2023/9822; Dated 31.10.2023. We also acknowledge the National Research Foundation of Korea (NRF-2021R1A6A1A03039503/ NRF-2017R1D1A3B06034971) and Soonchunhyang University Research Fund for support.

References

1. Williams, J.D., Bogan, A.E., Butler, R.S. *et al.*, A revised list of the freshwater mussels (Mollusca: Bivalvia: Unionida) of the United States and Canada. *Freshwater Mollusk Biol. Conserv.*, 20, 2, 33–58, 2017.
2. Guillou, E., Cyr, C., Laplante, J.F. *et al.*, Commercial performance of blue mussel (*Mytilus edulis*, L.) stocks at a microgeographic scale. *J. Mar. Sci. Eng.*, 8, 6, 382, 2020.
3. Saurabh, S., Pradhan, S., Suman, S., Recent Trends in Freshwater Pearl Farming in India, in: *Update on Malacology*, S. Ray and S. Mukherjee (Eds.), Intech Open, London, UK, 2021.
4. Bogan, A.E., Global Diversity of Freshwater Mussels (Mollusca, Bivalvia) in Freshwater. *Hydrobiologia*, 595, 139–147, 2008.

5. Sano, I., Saito, T., Miyazaki, J.I. *et al.*, Evolutionary history and diversity of Unionoid mussels (Mollusca: Bivalvia) in the Japanese Archipelago. *Plankton Benthos Res.*, 15, 2, 97–111, 2020.
6. Patnaik, B.B., Wang, T.H., Kang, S.W. *et al.*, Sequencing, *de novo* assembly, and annotation of the transcriptome of the endangered freshwater pearl bivalve, *Cristaria plicata*, provides novel insights into functional genes and marker discovery. *PloS One*, 11, 2, e0148622, 2016.
7. Pavan-Kumar, A., Varshney, S., Suman, S. *et al.*, Complete mitochondrial genome of freshwater pearl mussel *Lamellidens marginalis* (Lamarck, 1819) and its phylogenetic relation within Unionidae family. *Mol. Biol. Rep.*, 49, 10, 9593–9603, 2022.
8. Gomes-dos-Santos, A., Lopes-Lima, M., Machado, A.M. *et al.*, The Crown Pearl: a draft genome assembly of the European freshwater pearl mussel *Margaritifera margaritifera* (Linnaeus, 1758). *DNA Res.*, 28, 2, dsab002, 2021.
9. Gomes-dos-Santos, A., Lopes-Lima, M., Machado, A.M. *et al.*, The Crown Pearl V2: an improved genome assembly of the European freshwater pearl mussel *Margaritifera margaritifera* (Linnaeus, 1758). *GigaByte*, 1–14, 2023a.
10. Gomes-dos-Santos, A., Lopes-Lima, M., Machado, M.A. *et al.*, PacBio Hi-Fi genome assembly of the Iberian dolphin freshwater mussel *Unio delphinus* Spengler. *Sci. Data*, 10, 1, 340, 17932023b.
11. Gomes-dos-Santos, A., Froufe, E., Machado, A.M. *et al.*, A PacBio Hi-Fi genome assembly of the painter's mussel *Unio pictorum* (Linnaeus, 1758). *Genome Biol. Evol.*, 15, 7, evad116, 2023c.
12. Yang, J.L., Feng, D.D., Liu, J. *et al.*, Chromosome-level genome assembly of the hard-shelled mussel *Mytilus coruscus*, a widely distributed species from the temperate areas of East Asia. *GigaScience*, 10, 4, giab024, 2021.
13. Simon, A., Three new genome assemblies of blue mussel lineages: North and South European *Mytilus edulis* and Mediterranean *Mytilus galloprovincialis*. *BioRxiv*, 2022–09, 2022.
14. Paggeot, L.X., DeBiasse, M.B., Escalona, M., *et al*, Reference genome for the California ribbed mussel, *Mytilus californianus*, an ecosystem engineer. *J. Hered.*, 113, 6, 681–688, 2022.
15. Gallardo-Escárate, C., Valenzuela-Muñoz, V., Nuñez-Acuña, G. *et al.*, Chromosome-Level Genome Assembly of the Blue Mussel *Mytilus chilensis* reveals Molecular Signatures Facing the Marine Environment. *Genes*, 14, 4, 876, 2023.
16. Ferreira, J.G.R.N., Americo, J.A., do Amaral, D.L.A.S., *et al.*, A chromosome-level assembly supports genome-wide investigation of the DMRT gene family in the golden mussel (*Limnoperna fortunei*). *GigaScience*, 12, giad072, 2023.
17. Eldem, V., Zararsiz, G., Taşçi, T. *et al.*, Transcriptome analysis for non-model organism: Current Status and Best Practices, in: *Applications of RNA-Seq and Omics Strategies-From Microorganisms to Human Health*, F.A. Marchi, P.D.R. Cirillo, E.C. Mateo (Eds.), pp. 1–19, Intech Open, London, UK, 2017.

18. Chalifa-Caspi, V., RNA-Seq in non-model organisms. *Methods Mol. Biol.*, 2243, 143-167, 2021.
19. Patnaik, H.H., Sang, M.K., Park, J.E. *et al.*, A review of the endangered mollusks transcriptome under the threatened species initiative of Korea. *Genes Genomics*, 45, 8, 969–987, 2023.
20. Caurcel, C., Laetsch, D.R., Challis, R. *et al.*, MolluscDB: a genome and transcriptome database for molluscs. *Phil. Trans. R. Soc. B.*, 376, 1825, 20200157, 2021.
21. Jeong, J.Y., Park, J.E., Song, D.K. *et al.*, PANM DB ver 5.0: an update of the PANM database for invertebrate NGS data analysis. *Kor. J. Malacol.*, 38, 125–128, 2022.
22. Uliano-Silva, M., Americo, J.A., Brindeiro, R. *et al.*, Gene discovery through transcriptome sequencing for the invasive mussel *Limnoperna fortunei. PloS One*, 9, 7, e102973, 2014.
23. Malachowicz, M. and Wenne, R., Mantle transcriptome sequencing of *Mytilus* spp. and identification of putative biomineralization genes. *Peer J.*, 6, e6245, 2019.
24. Zhang, H., Yao, G., He, M., Transcriptome analysis of gene expression profiling from the deep sea *in situ* to the laboratory for the cold seep mussel *Gigantidas haimaensis. BMC Genom.*, 23, 1, 1–3, 2022.
25. Gomes-dos-Santos, A., Machado, A.M., Castro, L.F. *et al.*, The gill transcriptome of threatened European freshwater mussels. *Sci. Data*, 9, 1, 494, 2022.
26. Meek, M.H. and Larson, W.A., The future is now: Amplicon sequencing and sequence capture usher in the conservation genomics era. *Mol. Ecol. Res.*, 19, 4, 795–803, 2019.
27. Hohenlohe, P.A., Funk, W.C., Rajora, O.P., Population genomics for wildlife conservation and management. *Mol. Ecol.*, 30, 1, 62–82, 2021.
28. Formenti, G., Theissinger, K., Fernandes, C. *et al.*, The era of reference genomes in conservation genomics. *Trends Ecol. Evol.*, 37, 3, 197–202, 2022.
29. McCartney, M.A., Auch, B., Kono, T. *et al.*, The genome of the zebra mussel, *Dreissena polymorpha*: a resource for comparative genomics, invasion genetics, and Biocontrol. *G3 (Bethesda)*, 12, 2, jkab423, 2022.
30. Dong, Z.G. and Li, J.L., Biodiversity and conservation of freshwater mollusks. *Acta Hydrobiol. Sin.*, 28, 4, 440–444, 2004.
31. Kondo, T., Monograph of Unionoida in Japan (Mollusca: Bivalvia), in: *Malacological Society of Japan*, vol. 3, pp. 23–25, 2008.
32. Kim, S.B., *Korean Red List of Threatened Species*, Ed. 2nd, National Institute of Biological Resources, Incheon, Korea, 2014.
33. Lee, J.H., Choi, E.H., Kim, S.K. *et al.*, Mitochondrial genome of the cockscomb pearl mussel *Cristaria plicata* (Bivalvia, Unionoida, Unionidae). *Mitochondrial DNA*, 23, 1, 39–41, 2012.
34. Wang, X., Liu, Z., Wu, W., Transcriptome analysis of the freshwater pearl mussel (*Cristaria plicata*) mantle unravels genes involved in the formation of shell and pearl. *Mol. Genet. Genom.*, 292, 343–352, 2017.

35. Moorkens, E., Cordeiro, J., Seddon, M.B. *et al.*, Margaritifera margaritifera (Freshwater Pearl Mussel), in: *The IUCN Red List of Threatened Species*, 2017, https://www.iucnredlist.org/species/12799/128686456.
36. Araujo, R., Unio delphinus, in: *IUCN Red List of Threatened Species*, vol. 3, p. 2011, Version 2015, https://www.iucnredlist.org/species/195510/8975648.
37. Lopes-Lima, M., Kebapçı, U., Van Damme, D., Unio crassus, in: *The IUCN Red List of Threatened Species*, p. e.T22736A42465628, 2014.
38. Lopes-Lima, M. and Seddon, M.B., Unio mancus, in: *IUCN Red List of Threatened Species*, p. e.T22737A42466471, 2014.
39. Zeng, Q., Hu, B., Blanco, A.H., Full-length transcriptome sequences provide insight into hermaphroditism of freshwater pearl mussel *Hyriopsis schlegelii*. *Front. Genet.*, 13, 868742, 2022.
40. Shi, J., Hong, Y., Sheng, J., *De novo* transcriptome sequencing to identify the sex-determination genes in *Hyriopsis schlegelii*. *Biosci. Biotechnol. Biochem.*, 798, 1257–1265, 2015.
41. Zhang, A.J., Liu, S.L., Zhu, J.Y. *et al.*, Transcriptome analysis of the freshwater pearl mussel, *Hyriopsis cumingii* (Lea) Using Illumina paired-end sequencing to identify genes and markers. *Iranian J. Fisheries Sci.*, 15, 1, 479–496, 2016.
42. Briones, C., Nuñez, J.J., Pérez, M. *et al.*, *De novo* male gonad transcriptome draft for the marine mussel *Perumytilus purpuratus* with a focus on its reproductive-related proteins. *J. Genom.*, 6, 127, 2018.
43. Shi, Y., Yao, G., Zhang, H. *et al.*, Proteome and Transcriptome Analysis of Gonads Reveals Intersex in *Gigantidas haimaensis*. *BMC Genom.*, 23, 1, 174, 2022.
44. Martino, P.A., Carlon, D.B., Kingston, S.E. *et al.*, Blue mussel (genus *Mytilus*) transcriptome response to simulated climate change in the Gulf of Maine. *J. Shellfish Res.*, 38, 3, 587–602, 2019.
45. Tang, X.W., Yu, R., Fan, M.H. *et al.*, Physiological and transcriptome analysis of *Mytilus coruscus* in response to *Prorocentrum lima* and microplastics. *Front. Mar. Sci.*, 9, 1087667, 2022.
46. Hall, S., Méthé, D., Stewart-Clark, S., Clark, F. *et al.*, Size and site specific transcriptomic responses of blue mussel (*Mytilus edulis*) to acute hypoxia. *Mar. Genom.*, 71, 101060, 2023.
47. Gallardi, D., Xue, X., Mercier, E. *et al.*, RNA-Seq analysis of the mantle transcriptome from *Mytilus edulis* during a seasonal spawning event in deep and shallow water culture sites on the northeast coast of Newfoundland, Canada. *Mar. Genom.*, 60, 100865, 2021.
48. Leung, P.T., Ip, J.C., Mak, S.S. *et al.*, *De novo* transcriptome analysis of *Perna viridis* highlights tissue-specific patterns for environmental studies. *BMC Genom.*, 15, 1, 1–14, 2014.
49. Inoue, K., Yoshioka, Y., Tanaka, H. *et al.*, Genomics and Transcriptomics of the green mussel explain the durability of its byssus. *Sci. Rep.*, 11, 1, 5992, 2021.

50. Shi, Y., Xu, M., Huang, J. *et al.*, Transcriptome analysis of mantle tissues reveals potential biomineralization-related genes in *Tectus pyramis* Born. *Comp. Biochem. Physiol. Part D: Genom. Proteom.*, 29, 131–144, 2019.
51. Song, N., Li, J., Li, B. *et al.*, Transcriptome analysis of the bivalve *Placuna placenta* mantle reveals potential biomineralization-related genes. *Sci. Rep.*, 12, 1, 4743, 2022.
52. Carducci, F., Biscotti, M.A., Mosca, A. *et al.*, The Mantle Transcriptome of *Chamelea gallina* (Mollusca: Bivalvia) and Shell Biomineralization. *Animals*, 12, 9, 1196, 2022.
53. Kang, S.W., Patnaik, B.B., Hwang, H.J. *et al.*, Sequencing and *de novo* assembly of visceral mass transcriptome of the critically endangered land snail *Satsuma myomphala*: annotation and SSR discovery. *Comp. Biochem. Physiol. Part D: Genom. Proteom.*, 21, 77–89, 2017.
54. Kang, S.W., Patnaik, B.B., Park, S.Y. *et al.*, Transcriptome analysis of the threatened snail *Ellobium chinense* reveals candidate genes for adaptation and identifies SSRs for conservation genetics. *Genes Genom.*, 40, 333–347, 2018.
55. Patnaik, B.B., Chung, J.M., Hwang, H.J. *et al.*, Transcriptome analysis of air-breathing land slug, *Incilaria fruhstorferi* reveals functional insights into growth, immunity, and reproduction. *BMC Genom.*, 20, 1–26, 2019.
56. Feldmeyer, B., Wheat, C.W., Krezdorn, N. *et al.*, Short read Illumina data for the *de novo* assembly of a non-model snail species transcriptome (*Radix balthica*, Basommatophora, Pulmonata), and a comparison of assembler performance. *BMC Genom.*, 12, 1–12, 2011.
57. Lu, X., Li, J., Yang, J. *et al.*, *De novo* transcriptome of the desert beetle *Microdera punctipennis* (Coleoptera: Tenebrionidae) using Illumina RNA-Seq technology. *Mol. Biol. Rep.*, 41, 7293–7303, 2014.
58. Liao, X., Liu, Y., Han, T. *et al.*, Full-length transcriptome sequencing reveals tissue-specific gene expression profile of mangrove clam *Geloina erosa*. *Front. Physiol.*, 13, 851957, 2022.
59. Ip, J.C., Mu, H., Chen, Q. *et al.*, AmpuBase: a transcriptome database for eight species of apple snails (Gastropoda: Ampullariidae). *BMC Genom.*, 19, 1, 1–9, 2018.
60. Song, L. and Florea, L., Rcorrector: efficient and accurate error correction for Illumina RNA-Seq reads. *GigaScience*, 4, 1, s13742–015, 2015.
61. Kim, D., Song, L., Breitwieser, F.P. *et al.*, Centrifuge: rapid and sensitive classification of metagenomic sequences. *Genome Res.*, 26, 12, 1721–1729, 2016.
62. Sewe, S.O., Silva, G., Sicat, P. *et al.*, Trimming and validation of Illumina short reads using Trimmomatic, Trinity assembly, and assessment of RNA-Seq data. *Methods Mol. Biol.*, 2443, 211–232, 2022.
63. Mun, S., Kim, Y.J., Markkandan, K. *et al.*, The whole-genome and transcriptome of the Manila clam (*Ruditapes philippinarum*). *Genome Biol. Evol.*, 9, 6, 1487–1498, 2017.

64. Jeong, J.Y., Park, J.E., Song, D.K. *et al.*, PANM DB ver 5.0: an update of the PANM database for invertebrate NGS data analysis. *Kor. J. Malacol.*, 38, 125–128, 2022.
65. Zhang, X., Ruan, Z., You, X. *et al.*, *De novo* assembly and comparative transcriptome analysis of the foot from Chinese green mussel (*Perna viridis*) in response to cadmium stimulation. *PloS One*, 12, 5, e0176677, 2017.
66. Liao, Z., Liu, F., Wang, Y. *et al.*, Transcriptomic response of *Mytilus coruscus* mantle to acute sea water acidification and shell damage. *Front. Physiol.*, 14, 1–23, 2023.
67. Saco, A., Rey-Campos, M., Novoa, B. *et al.*, Transcriptomic response of mussel gills after a *Vibrio splendidus* infection demonstrates their role in the immune response. *Front. Immunol.*, 11, 615580, 2020.
68. Saco, A., Rey-Campos, M., Novoa, B. *et al.*, Mussel antiviral transcriptome response and elimination of *viral haemorrhagic septicemia* virus (VHSV). *Fish Shellfish Immunol.*, 136, 108735, 2023.
69. Wang, Y.Q., Liu, Q., Zhou, Y. *et al.*, Stage-specific transcriptomes of the mussel *Mytilus coruscus* reveals the developmental program for the planktonic to benthic transition. *Genes*, 14, 2, 287, 2023.
70. Venier, P., Varotto, L., Rosani, U. *et al.*, Insights into the innate immunity of the Mediterranean mussel *Mytilus galloprovincialis. BMC Genom.*, 12, 69, 2011.
71. Rey-Campos, M., Moreira, R., Valenzuela-Munoz, V. *et al.*, High individual variability in the transcriptomic response of Mediterranean mussels to *Vibrio* reveals the involvement of myticins in tissue injury. *Sci. Rep.*, 9, 3569, 2019.
72. Zhang, T., Yao, J., Xu, D. *et al.*, Gill physiological and transcriptomic response of the threatened freshwater mussel *Solenaia oleivora* to salinity shift. *Comp. Biochem. Physiol. Part D: Genom. Proteom.*, 40, 100913, 2021.
73. Yang, J.L., Feng, D.D., Liu, J. *et al.*, Chromosome-level genome assembly of the hard-shelled mussel *Mytilus coruscus*, a widely distributed species from the temperate areas of East Asia. *GigaScience*, 10, 4, giab024, 2021.
74. Huang, D., Shen, J., Li, J. *et al.*, Integrated transcriptome analysis of immunological responses in the pearl sac of the triangle sail mussel (Hyriopsis cumingii) after mantle. *Fresh Shellfish Immunol.*, 90, 385–394, 2019

Index

Printed and bound by CPI Group (UK) Ltd, Croydon, CR0 4YY

17/09/2026

14964275-0008